The World's 20 Greatest Unsolved Problems

John R. Vacca

PRENTICE HALL
Professional Technical Reference
Upper Saddle River, NJ 07458
www.phptr.com

A CIP catalog record for this book can be obtained from the Library of Congress.

Editorial/production supervision: *MetroVoice Publishing Services*
Cover designer: *Anthony Gemmellaro*
Publisher: *Jeffrey Pepper*
Editorial assistant: *Linda Ramagnano*
Marketing manager: *Robin O'Brien*

Publishing as Prentice Hall Professional Technical Reference
Upper Saddle River, New Jersey 07458

Prentice Hall books are widely used by corporations and government agencies for training, marketing, and resale.

Prentice Hall PTR offers excellent discounts on this book when ordered in quantity for bulk purchases or special sales. For more information, please contact:

U.S. Corporate and Government Sales
1-800-382-3419
corpsales@pearsontechgroup.com

For sales outside of the U.S., please contact:

International Sales
1-317-581-3793
international@pearsontechgroup.com

Printed in the United States of America
3 4 5 6 7 8 9 10 11—CRW—0807060504
Third printing, August 2004

ISBN 0-13-142643-5
Pearson Education LTD.
Pearson Education Australia PTY, Limited
Pearson Education Singapore, Pte. Ltd.
Pearson Education North Asia Ltd.
Pearson Education Canada, Ltd.
Pearson Educación de Mexico, S.A. de C.V.
Pearson Education—Japan
Pearson Education Malaysia, Pte. Ltd.

To Max Ary (Omniplex) and Scott Huggins (Spitz, Inc.),
without whose support, encouragement, and contributions,
this book would not have been possible.

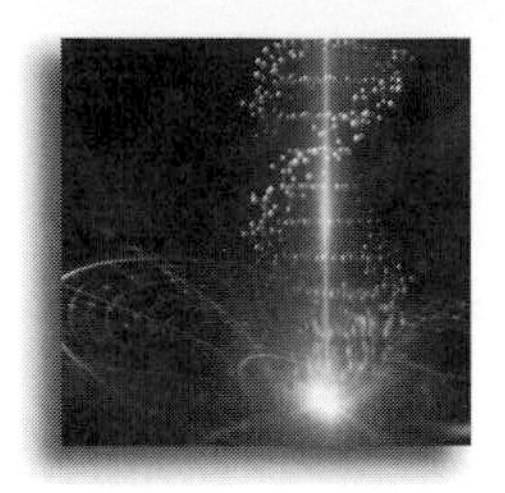

Contents

Foreword

The sense of mystery and excitement associated with the exploration of the fundamental concepts of science is fueled by the feeling that the more we learn about something, the more we realize what we do not know. The response this analytic process generates often serves to deepen the desire for understanding and promotes philosophical reflection. This book captures that exquisite feeling by carefully and systematically presenting many of the most intriguing and complex questions in science that humanity has yet to resolve. It truly provides "food for thought."

At times it may seem that the basic principles of physics, chemistry, biology, astronomy and beyond, are well known to us and all that remains is the act of filling in the details. However, as elegantly presented in the following work, many of the important issues in these areas of research have only begun to be addressed. Although this situation means that we have yet to solve an array of key problems in disciplines such as medicine, energy production, and the prediction of natural disasters (such as earthquakes), it also indicates that for those individuals interested in a career in science and research, there is still much to do.

Besides identifying an outstanding collection of the greatest unsolved problems in science, the author has also utilized a powerful tool for assisting the reader in developing potential solutions to these problems and in generating approaches by which these hypotheses may be ultimately tested. This tool involves the inclusion of comments and opinions by scientists currently working in the various fields. Because the perspectives of others who have carefully considered any given problem can often synergize with one's own thoughts on the matter, this format is sure to spark the imagination. Thus, I invite you to read, dream, and enjoy.

—Paul J. Bertics, Professor
Department of Biomolecular Chemistry
University of Wisconsin

Preface

Science has extended life, conquered disease, and offered new sexual and commercial freedoms through its rituals of discovery, but many unsolved problems remain. Although science has pushed aside many demons and demigods, and revealed a cosmos more intricate and awesome than anything produced by pure imagination, there are new troubles in the peculiar form of paradise that it has created. This has precipitated new questions about whether it has the popular support to meet the future challenges of food and water, urban sprawl, disease, pollution, security, energy, and education.

Even while the public hungers for new gadgets and drugs, it seems increasingly intolerant of grand, technical fixes. In areas like genetic engineering, germ warfare, global warming, nuclear power, and the proliferation of nuclear arms, the public has also come to fear the potential consequences of unfettered science and technology.

Due to tension between science and the public, new barriers have been thrown up to research involving deadly pathogens, stem cells, and human cloning.

With the environmental movement of the 1960s, some of the doubts about science began. Also, traditional beliefs have been disturbed by science, which has caused an even deeper unease. Stunned by the increasing vigor of fundamentalist religion worldwide on the big issues of everyday life, some scientists wonder if old certainties have rushed into a sort of vacuum left by the inconclusiveness of science.

Recent opinion surveys have also gauged disaffection with science. Recently, a Harris poll found that the percentage of Americans who saw scientists as having very great prestige had declined twelve percentage points in the last quarter-century, down to 55 from 67 percent. While half of Americans believe in ghosts and a third believe in astrology, another recent Harris poll found that most of them also believe in miracles. These results are hardly an endorsement of scientific rationality in the United States.

Research priorities have become increasingly politicized in this atmosphere of ambivalence. It seems warranted to ask a question that runs counter to centuries of Western thought: Does science really matter? Do people care about it anymore? As the world marches into a century born amid fundamentalist strife in oil-producing nations and a divisive political climate in the United States and abroad, more sophisticated scientific credos like Darwin's theory of evolution are being challenged at every turn.

Do People Really Care about Science Anymore?

So, for a long time, science has mattered a lot. In the past century, advances in food, public health, and medicine helped raise life expectancy in the United States from roughly 50 to 82 years. Now exceeding seven billion, the world population between 1950 and 2000 has more than tripled. Biology played a large role in discovering the structure of DNA, making test-tube babies, and curing diseases. And, offering the hope of new treatments for cancer and other diseases, the decoding of the human genome is leading scientists toward a detailed understanding of how the body really works.

Breakthroughs in physics produced discoveries in digital electronics and in the subatomic world. American ingenuity won the space race, put men on the moon, probed distant planets, and lofted hundreds of satellites, including the Hubble Space Telescope. However, major problems arose quite quickly: for example, deactivation of the Hubble Space Telescope; acid rain; environmental toxins; the Bhopal chemical disaster in India; nuclear waste; global warming; the ozone hole; fears over genetically modified food; the fiery destruction of two space shuttles; not to mention the curse of junk e-mail (spam). Such troubles are only the tip of the iceberg, but nonetheless, they have helped feed social disenchantment with science.

The physical sciences began to lose luster and funding when the Cold War ended. After spending $3 billion, Congress killed physicists' pre-eminent endeavor, the Superconducting Super Collider, an enormous particle accelerator.

According to the National Science Foundation (NSF), industry spending on research soared to three times that of the federal government, about $290 billion in 2003. One result of all this, is that Americans see less news about the fundamental building blocks and great shadowy vistas of the universe, and more about drugs, cell phones, advanced toys, innovative cars, and engineered foods.

The main exceptions to the downward trend in the federal science budget are for health and weapons. In 2003, spending on military research hit $69 billion, higher in fixed dollars than during the Cold War.

In the meantime, other countries are spending more on research. This is taking some of the glory that America once monopolized. According to CHI Research, Japan, Taiwan, and South Korea now account for more than a quarter of all American industrial patents. Countries that make up the European

Union are working together on what will be the world's most powerful atom smasher. The British recently flew the first probe in a quarter century to look for evidence of life on Mars. (However, the United States recently landed two Mars rovers, named Spirit and Opportunity, whose mission is to look for water and signs of life in the Martian soil and rocks.)

Threats and Challenges

Cancer and the AIDS epidemic still darken many lives, despite the explosion in the life sciences. Furthermore, the flowering of biotechnology has fed worries about genetically modified foods and organisms, as well as the pending reinvention of what it means to be human. Many people worry that the growing power of genetics will sully the sanctity of human life.

Recently, the Bush administration's Council on Bioethics issued a report warning that biotechnology in pursuit of human perfection could lead to unintended and destructive ends. Science experts worry about terrorists using advances in biology for intentional harm, perhaps on vast new scales.

The physical sciences seem to be adrift. Without the space race and the Cold War, and perhaps facing intrinsic limits, as well as declining budgets, the physical sciences are in a state of diminishing returns.

Scientists still top the list of 23 high-status professions, ahead of doctors, teachers, lawyers, and athletes, despite the decline in prestige recorded in the recent Harris poll. Contradictions are perennially identified by the NSF through polls. According to NSF's latest numbers, 92 percent of adult Americans say they are very or moderately interested in science discoveries. Despite those poll results, only half of survey respondents knew that the Earth takes a year to revolve around the sun.

Drawing the Battle Lines of Evolution

About two-thirds of Americans believe that alternatives to Darwin's theory of evolution should be taught in public schools, alongside the bedrock concept of biology itself. Many scientists are jarred by this two-thirds number. Today, the organized opposition to the mainstream theory of evolution has become vastly more sophisticated and influential than it was 28 years ago. The leading foes of Darwin espouse a theory called *intelligent design*. This theory holds that purely random natural processes could never have produced humans. These opponents are led by a relatively small group of people with various academic and professional credentials, including some with advanced degrees in science and even university professorships.

Backers of intelligent design say they are simply pointing out the shortcomings in Darwin's theory. Scientists have publicly rallied in response. In 2003, they staved off an effort by the Texas State Board of Education to have intelligent design taught alongside evolution.

Actually, through the development of technologies and medicines, science has sold itself from the start as something more than a utilitarian exercise. Einstein's theories (which often used religious and philosophical language to explain discoveries), seemed to tell humanity something fundamental about the fabric of existence.

Eye on the Future

Industry looks to short-term goals and has proven highly adept at using science to take care of itself and consumers. A far more uncertain issue is whether the federal government can successfully address issues of human welfare that lie well beyond the industrial horizon—years, decades, and even centuries ahead. Well, I wouldn't hold your breath on that one.

As oil becomes increasingly scarce, an urgent goal here is to develop new sources of energy, which will become vitally important. Another is to better understand the nuances of climate change, for instance, how the sun and ocean affect the atmosphere. Such work is in its infancy. Another is to develop ways of countering the spread of germ weapons and nuclear arms.

In areas like waste, water use, congestion, highways, hazard mitigation, and pollution control, the world will also need a new science of cities to help coordinate planning. The number of urban dwellers is expected to grow from three billion now to six billion by 2036.

There are also some worries by scientists about a significant shift in the demographics of American graduate schools in science and engineering. By 2000, according to the latest figures from the NSF, the number of foreign students in full-time engineering programs had soared so high that it exceeded, for the first time, the steeply declining number of American students.

Whether the complex challenges of today generate a new era of scientific greatness, may depend on how a deeply conflicted public answers the question of whether science still matters. Some experts warn that if support for science falters and if the American public loses interest in it, such apathy may foster an age in which scientific elites ignore the public will and global imperatives for their own narrow interests, producing something like a dictatorship of the lab coats.

Who This Book is For

This book is primarily targeted toward the domestic and international scientific community. This book is also valuable for those involved in teaching science around the world, as well as, the average technical reader or science novice. Basically, the book is targeted for all types of people and organizations around the world that are involved in scientific research. Whether it's physicist Hideo Mabuchi's discoveries about the interface between quantum mechanics and everyday life, or conservation biologist Gretchen Daily's rigorous assessments of ecosystems and economics, this new generation of scientists are probing the frontiers of knowledge, and have already shown the promise (or the work) that makes senior scientists applaud in awe.

What's So Special about This Book?

Much of what we know about the world has been learned during the past few centuries, but some mysteries still remain. Apparently, Nature has not revealed to us all her secrets, but scientists are hard at work trying to decipher them.

The purpose of this book is to show the moderately technical reader that in an age where scientific discovery is everyday news, there still remains, in fact, many scientific problems that have never been solved. Actually, there are so many unsolved problems in all areas of science that the task of creating a complete list of them would be another unsolved problem.

Many of these unsolved problems are well documented in this book. The book also stresses that many problems in the natural sciences are also unsolved and that we are far from understanding the nature around us.

Throughout this book, you will learn how to penetrate the concepts and quantitative methods of the physics of fluctuations into biological sciences and medicine. This book also brings together the views (through extensive interviews) of many scientists (some controversial) from physics, biophysics, biomedical engineering, biology, and medicine. But, this list also includes:

- archaeology
- astronomy
- astrophysics

- biology
- botany
- chemistry
- cosmology
- energy
- environmental science
- geology
- genetics
- gravity
- human origins
- mathematics
- medicine
- neuroscience
- paleontology
- physics
- space science
- zoology

This book also includes a wide range of advanced scientific theories, algorithms, and techniques that can be used to access an entire world of knowledge. Those responsible for teaching science at all levels require an in-depth knowledge of these unsolved problems to allow them and their students to explore vast new frontiers, from a galaxy 12 billion light-years away, to the smallest genetic switch inside a human cell.

In this book, you will gain extensive knowledge of the problems surrounding the most challenging scientific projects in history. You will be shown ongoing research that is attempting to solve a range of scientific problems, including the origin of life itself. In other words, you will learn about the latest advanced scientific techniques in use today.

The book presents the results of extensive interviews that were conducted with more than 60 of the world's greatest scientists (including Stephen Hawking) about how they will go about solving the problems in their respective fields. These scientists (40 of whom are Nobel laureates) were chosen because they are already glowing stars in their respective fields. Each of them and their successors are expected only to shine brighter as they move through the next 40 years and light the paths to scientific enlightenment.

Finally, this book leaves little doubt that it is nothing less than a manual/guide of the greatest unsolved scientific problems for the 21st century. No question, it will benefit all scientists and others interested in the constant changes of science and technology. One of the main goals of science and this book is to reduce ignorance and superstition of mankind!

Organization of the Book

The book is organized into seven parts composed of 20 chapters. It provides a step-by-step discussion of 20 of the greatest unsolved problems in science. Given this wealth of unsolved problems in science, some 60 scientists were interviewed to predict what unanswered questions and/or unsolved problems will dominate their fields over the next 40 years. Their replies expressed not only excitement at the pace of discovery, but also a broad concern about the use of new technologies and scientific information. Almost 100 researchers were surveyed to find these scientists, asking for nominations of the persons who are the best-of-the-best in their respective fields around the globe, and who have demonstrated a once-in-a-generation insight. The response left me very optimistic. The talent pool is vast. I could have easily included 600 scientists without compromising the quality of the group.

Part I: Astronomy and Cosmology

This part of the book first covers the unsolved problem of "Astronomy: The Mystery of Dark Matter." It discusses what scientists think dark matter really is. If it's invisible, then how does one sense its existence? A discussion ensues about how galaxies would simply disintegrate, unless there was dark matter surrounding them to sort of keep them together. Scientists also provide answers to some of the speculations surrounding the composition of dark matter, as well as why dark matter is important.

The unsolved problem of "Cosmology: The Creation of the Universe" is covered next. Here, scientists provide answers about how much they know about the beginning of the universe; what is the ultimate quest in cosmology; is anything known about the universe before one-trillionth of a second; and what needs to be known to understand creation?

Finally, Part I concludes with a discussion of "Theoretical Cosmology and Particle Physics: The Cosmological Constant Problem." A discussion of quintessence follows.

Part II: Physics and Astrophysics

Part II begins by giving you an overview of the unsolved problem of "Gravity: The Construction of a Consistent Quantum Theory of Gravity." Here, scientists provide answers to the following questions:

- What are some of the unusual features of quantum mechanics?
- What is the difficulty?
- How would our understanding of gravity be affected?
- Why is this problem important?

A discussion of how general relativity produces gravity is also presented.

Next, this part covers the unsolved problem of "Particle Physics: The Mechanism That Makes Fundamental Mass." Once again, here, scientists provide answers to the following questions:

- What is the standard model?
- What is the problem?
- Does "electroweak breaking" affect the macroscopic world?
- When are scientists likely to solve this problem?

In addition, an update is presented on the possible discovery of the "god particle."

Part II also discusses "Particle Physics and Astrophysics: The Solar Neutrino Problem." This problem is partially solved but not fully understood. Scientists provide answers as to why there is strong evidence that neutrinos have mass and that electron neutrinos emitted in the core of the sun transform into other neutrinos via oscillations on their way to the Earth. But, what are neutrinos? According to scientists, there are three types, or "flavors," of neutrinos, one of which is associated with the electron, one of which is associated with the muon, and one of which is associated with the tau lepton. When neutrinos are initially produced, they spin in a very specific way. Also, an update on the Solar Neutrino Problem (SNO) measurement of neutrino oscillations and masses is presented.

This part covers the "Astrophysics: The Source of Gamma-Ray Bursts" unsolved problem. Here, scientists answer the following questions:

- What is a gamma-ray burst?
- Why do scientists study gamma-ray bursts?

- What do we know, and what don't we know?
- How will the mystery be solved, or has it been?

A discussion of changes in the gamma-ray burst problem is also presented.

Next, Part II presents the unsolved problem of "Theoretical High-Energy Physics: The Unification of the Basic Forces." Again, scientists answer the following questions:

- What does unification mean?
- What are some examples of unification?
- Which forces have not been unified?
- Are other forces such as friction unified?
- What is the best candidate for unification?
- What is a grand unification theory (GUT)?

Finally, Part II covers the unsolved problem of "Solid State Physics: The Mechanism Behind High-Temperature Superconductors."

Part III: Biology and Paleontology

Part III opens up by showing you how the unsolved problem in "Biology: How the Basic Processes of Life are Carried Out by DNA and Proteins" is being addressed. The following questions are answered by scientists:

- What is DNA?
- What is the challenge?
- How important is it to understand DNA?
- How will this knowledge affect our lives?

Next, this part discusses the unsolved problem of "Biology: Protein Folding." It covers how scientists plan to unravel the mystery of protein folding by looking at the early studies first, and then noticing that certain diseases are characterized by extensive protein deposits in certain tissues. With regards to further studies on this unsolved problem, scientists are working in relative obscurity. These protein biochemists have discovered how a completely unfolded protein, with hundreds of millions of potential folded states to choose from, consistently found the correct one—and did so within seconds to minutes. Temperature sensitivity is discussed next, as well as Familial Amyloidotic Polyneuropathy (FAP), Alzheimer's disease, Mad Cow disease, and other species. Nevertheless, despite the examples of FAP, Alzheimer's disease, and Mad Cow disease, in which the problem derives from accumulation of toxic, insoluble gunk, many human diseases arise from protein misfolding, which leaves too little of the normal protein to do its job properly.

Finally, Part III concludes with a discussion of the unsolved problem of "Paleontology: How Present-Day Microbiological Information Can Be Used to Reconstruct 'The Ancient Tree of Life.'"

Part IV: Neuroscience

Part IV opens up with a discussion of the unsolved problem of "Free Will." In this part, scientists answer the questions: How is it that humans have the freedom to decide and act? What plays a role in free will?

Finally, this part of the book concludes with the unsolved problem of "Consciousness." Here, scientists explain the unsolved problems of:

- Sleep and dreaming, especially lucid dreaming
- Voluntary control of internal states (biofeedback, meditation, etc.)
- Enhancement of creativity and learning

- Artificial intelligence applications in neuroscience (artificial neural networks, human/computer interfaces, etc.)
- Relationships between brain states and electromagnetic fields
- Psychoneuroimmunology (effects of psychological states upon the immune system)
- Extraordinary human abilities (communicative and energetic anomalies associated with altered states of consciousness)

Part V: Geology

This chapter covers the unsolved problem of "The Dynamics of the Inner Earth." Revealing Earth's deepest secrets is the primary theme of the discussion here. In other words, in work that promises to advance understanding about the origin and dynamics of Earth's iron-rich inner core and the generation of the planet's magnetic field, scientists have found that the elastic properties of iron are quite different at extremely high temperatures than at low temperatures.

Finally, Part V concludes with a discussion of the unsolved problem of "Earthquake Predicting." A scientific analysis is conducted of the seismic risk issue: earthquake prediction. The goal of earthquake prediction is to give warning of potentially damaging earthquakes early enough to allow appropriate response to the disaster, enabling people to minimize loss of life and damage.

Part VI: Chemistry

This part of the book discusses the unsolved problem of "How Microscopic Atomic Forces Produce Various Macroscopic Behaviors." The areas of physical chemistry, microscopic systems, and macroscopic systems are also discussed.

Finally, Part VI concludes with coverage of the unsolved problem of "The Fabrication and Manipulation of Carbon-Based Structures (Fullerenes)." Nanostructures' fabrication from carbon nanocones and the design and functionality of a "nanoplotter" are also examined.

Part VII: Energy

This last part of the book discusses the unsolved problem of "Electrical Energy: Free Energy—The Quantum Mechanical Vacuum." The final secret of free energy and the free electrical energy work of Nikola Tesla are discussed at length.

The final chapter of this book is composed of the unsolved problems of "Nuclear Fusion and Waste." Nuclear fusion energy—the harvesting of energy from the same reaction that powers the sun—is discussed at length. The dream of harvesting energy from the same reaction that powers our sun has been around since 1920, when Arthur Eddington suggested that the energy of the sun and stars was a product of the fusion of hydrogen atoms into helium. The nuclear fusion reaction involves the binding together of hydrogen atoms, creating helium. Current fusion research was big news in 1989 when it was reported that scientists had achieved fusion at room temperatures with simple equipment. A fusion power plant—a full-scale fusion reactor capable of generating 1,000 MW (1 MW = 1 million watts) of electricity, comparable to a conventional nuclear power plant—would be a very large and complex machine. The advantages/disadvantages of fusion are also discussed.

Finally, this chapter also discusses how to permanently store nuclear waste or eliminate it altogether. This has been one of the hardest challenges of the last 50 years in America. The storage of nuclear waste in other countries is also discussed.

Conventions

This book has several conventions to guide you through the text and help you find important facts, notes, cautions, and warnings:

Sidebars: We use sidebars to highlight related information, give an example, discuss an item in greater detail, or help you make sense of the swirl of terms, acronyms, and abbreviations so abundant to the subject. The sidebars are meant to supplement each chapter's topic. If you're in a hurry on a cover-to-cover read, skip the sidebars. If you're quickly flipping through the book looking for juicy information, read only the sidebars.

Notes: A note highlights a special point of interest about a topic.

Caution: A caution tells you to watch your step to avoid any problems.

Warning: A warning alerts you to the fact that a problem is imminent or will probably occur.

—John R. Vacca

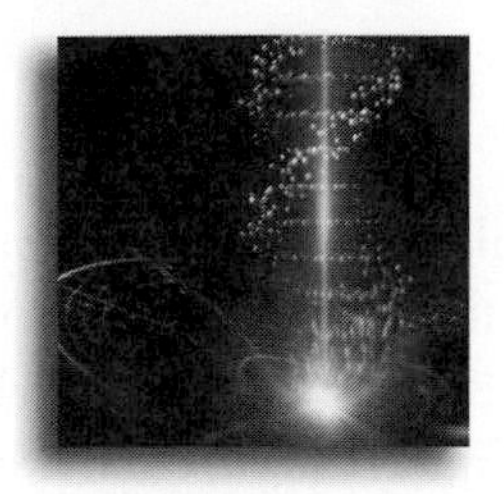

Acknowledgments

There are many people whose efforts on this book have contributed to its successful completion. I owe each a debt of gratitude and want to take this opportunity to offer my sincere thanks.

A very special thanks to my editor and publisher Jeff Pepper, without whose continued interest and support would not have made this book possible. And, editorial assistant Linda Ramagnano, who provided staunch support and encouragement when it was most needed. Special thanks to my technical science editors, Dr. William McNary and Dr. Ernest Abers, who ensured the technical accuracy of the book, was indispensable. Thanks to my production editor, Anne Garcia; project manager, Scott Suckling; and copyeditor, Sharon Jehlen whose fine editorial work has been invaluable. Thanks also to my marketing managers, Robin O'Brien and Dan DePasquale, whose efforts on this book have been greatly appreciated. And, a special thanks to Paul Bertics of the University of Wisconsin, who wrote the foreword for this book. Finally, thanks to all of the other people at Prentice Hall whose many talents and skills are essential to a finished book.

Thanks to my wife, Bee Vacca, for her love, her help, and her understanding of my long work hours.

I wish to thank all of the scientists: Stephen Hawking, Alan Guth, Steven Weinberg, Stephen Wolfram, Nima Arkani-Hamed, Scott Gaudi, Richard Muller, Lee Smolin, Andrea Ghez, Young-Kee Kim, Nikola Pavletch, Leroy Hood, Mary-Claire King, Joshua Lederberg, Peter Schultz, Craig Ventor, Julie K. Bartley, Paul Bowen, Suk-Joo Choh, Susan J. Smith, Peter A. Koehler, Dorinda R. Ostermann, Antonio Damasio, Miroslav Backonja, Paul J. Bertics, Edwin R. Chapman, Dr. Gerd Steinle-Neumann, Lars Stixrude, Ronald Cohen, Oguz Gülseren, Gerd Steinle-Neumann, Ruth Ludwin, Andrew Cooksy, Gerald T. Babcock, Warren F. Beck, Dr. Robert I. Cukier, Marcos Dantus, James L. Dye, James Geiger, James F. Harrison, Katharine C. Hunt, Paul M. Hunt, George E. Leroi, Paul F. Mantica, John L. McCracken, William C. McHarris, David J. Morrissey, Piotr Piecuch, Dr. Lynmarie A. Posey, Alexander Tulinsky, David Weliky, Benjamin L. Lawson, Olga A. Shenderova, Donald W. Brenner, Tom E. Bearden, Joe, Newman, Tom Napier, Harold Aspden, Dr. Scott Kniffin, J. S. Chang, D. Chettle, Wm. J. Garland, D. P.Jackson, Marilyn Lightstone, W.F.S. Poehlman, William Prestwich, Mamdouh Shoukri, G.J. Evans, R.E. Jervis, Masahiro Kawaji, W.S. Andrews, Brent J. Lewis, Daniel Rozon, Robin A. Chaplin, Dr. Esam Hussein, Derek H. Lister, Frank R. Steward, Stephen J. Bushby, Ben Rouben, George Schmidt, Gerald Smith, Steven Howe, Francis Thio, T. Kammash, Jon Nadler, and John M. Horack, who answered questions and contributed quotes that were necessary for the completion of this book.

Finally, a very, very special thanks to Web designer, Karin Ellison, without whose support and hard work in contacting all of the preceding scientists to obtain their contributed quotes, would not have made this book possible. And to Tom Pelly, graduate assistant to Stephen Hawking, for his encouragement and help in obtaining a contributed quote for this project from Dr. Hawking.

Astronomy and Cosmology

Astronomy: The Mystery of Dark Matter

"In the beginning... God said, 'Let there be light'; and there was light... and God separated the light from darkness."
—Genesis 1:1–4

During the past 30 years, astronomers have discovered that most of the material in the observable universe is invisible. In other words, it's impossible to see the material with a telescope since it does not interact with light. Such material is called "dark matter."

Dark matter can best be described as areas in space (between stars and galaxies) composed of material that appears as "voids" or "nothingness," but these voids can be measured due to their gravitational pull and interaction with the surrounding stars and galaxies. If you, as an observer, were to enter this void in a spacecraft that could withstand the gravitational forces on its hull, you would probably see nothing but pitch-black darkness when looking out a portal or window.

Astronomers stumbled upon this profound cosmic mystery while carefully measuring the speed of rotation of galaxies. They estimated that about 80% of the mass of our galaxy (or all observable galaxies for that matter) consists of this mysterious material.

Initially, astronomers detected dark matter through its gravitational effects. They could estimate what the rotation speed should be by calculating the mass of all visible stars and gas, thereby determining the gravity of the galaxy. Much to their surprise, the measurements showed that most galaxies are rotating faster than they should. In other words, astronomers observed that dark matter causes the stars in the outer regions of a galaxy to orbit faster than expected. Not a little faster. Much faster! More than twice as fast. This meant that, according to the theory of gravity, these galaxies should be flying apart. Yet clearly, they are not.

So, what can the answer be? Is it possible that most galaxies are surrounded by some "dark" form of matter that cannot be observed by radio, infrared, optical, ultraviolet, X-ray, or gamma-ray telescopes? Could Einstein's theory of gravity, which has proved to be correct in all cases so far, be somehow wrong?

"Einstein's theory of general relativity has never been tested on cosmological scales," says physicist Dr. Richard Muller (University of California at Berkeley), "so I would be very surprised if it turned out to be correct, at least without modifications. We know it works in the lab, and for objects the size of stars, but there is no reason to think that we can extrapolate this theory to the size of galaxies or galactic clusters. So we should not be surprised if we are surprised!" [2]

According to Astronomer Dr. Scott Gaudi (Harvard-Smithsonian Center for Astrophysics), "There are really only two logical possibilities to explain the discrepancy between the rotation speed of galaxies (or the velocity dispersion of clusters), and the amount of mass we observe directly. If one accepts Einstein's theory of general relativity, then there must exist some form of matter that contains mass in Galaxies and clusters that 'holds them together.' The other possibility is that general relativity is wrong or incomplete, and we are incorrectly calculating the action of gravity for these systems. Although not conventional, this possibility is not completely ridiculous, because our theory of gravity has generally not been tested in the regimes that are important in Galaxies and clusters. This is the basis of the most widely considered alternative to dark matter, namely Milgrom's alternative theory of gravity called MOdified Newtonian Dynamics (MOND). MOND has had some successes, but has not been developed into a full theory, and therefore is difficult to either test or refute." [3]

One of the mysteries of dark matter is why stars are traveling so much faster when they are near dark matter. Furthermore, an even greater mystery is why dark matter seems to be present in even greater concentrations in clusters of galaxies.

X-ray telescopes have discovered vast clouds of multimillion-degree gas in clusters of galaxies. These hot gas clouds increase the mass of the cluster, but not enough to solve the mystery. In fact, they provide an independent measurement of dark matter. The measurement shows that there must be at least four times as much dark matter as all the stars and gas astronomers observe, or the hot gas would escape the cluster.

According to theoretical physicists, astrophysicists, and cosmologists ("theoretical" being the operable word here, since there was no one to actually observe it), dark matter helped material to clump together through gravity to produce galaxies and galaxy clusters during the creation and evolution of the universe. Based on today's knowledge of the observable universe, astronomers are rather frustrated that they do not know what makes up 96% of its (universe) contents. The frustration that astronomers feel is what has become known as the "dark matter mystery"—the greatest unsolved problem in astronomy!

What Really Is Dark Matter?

So, what really is dark matter? If astronomers could fully answer that question, the writing of this chapter would not be necessary. Nevertheless, during the last couple of years, some progress has been made to solve this enduring riddle of cosmology—the distribution of the universe's dark matter (see sidebar, "Mapping Dark Matter and X-Raying").

NOTE

Dark matter is any form of matter which does not emit, absorb, or otherwise interact with electromagnetic radiation (radiation consisting of electric and magnetic waves that travel at the speed of light: like light, radio waves, gamma rays, and X-rays). By "matter," I mean some form of particles that can cluster under the force of gravity; this is to be distinguished from "dark energy," which is smoothly distributed throughout space. The existence of dark matter has been inferred from its gravitational effects on the dynamics of luminous tracers in galaxies and galaxy clusters, and on the bending of light rays (gravitational lensing) when viewing distant objects billions of light years away. See the Glossary for detailed definitions of dark matter and dark energy.

Dark Matter Candidates

Scientists are considering a number of possible dark matter candidates that include Massive Astrophysical Compact Halo Objects (MACHOs), Weakly Interacting Massive Particles (WIMPs), and hydrogen gas objects. A "pro and con" analysis is given for each of the candidates.

Massive Astrophysical Compact Halo Objects

Examples of MACHOs include brown dwarf stars, white dwarf stars, neutron stars, and black holes.

Brown Dwarf Stars

Brown dwarf stars have a mass that is less than eight percent of the mass of the sun. Basically, that's too low to produce the nuclear reactions that make stars shine.

Recently, astronomers have found some objects that are either brown dwarf stars or very large planets around other stars. Observations of the brightening and then dimming of distant stars (thought to be due to the gravitational lens effect of a foreground star) may also provide further evidence for a large population of brown dwarfs in our galaxy. However, there is as yet no evidence that brown dwarfs are anywhere near as abundant as they would have to be to account for the dark matter in our galaxy.

White Dwarf Stars

White dwarfs are the final condensed states of small to medium sized stars. Our sun will eventually become one.

White dwarfs are known to exist and to be plentiful. Maybe they could be plentiful enough to explain the dark matter if young galaxies produced white dwarfs that cool more rapidly than present theory predicts.

No good alternative to the present theory exists. Also, the production of large numbers of white dwarfs implies the production of a large amount of helium, which is not observed.

Neutron Stars or Black Holes

Neutron stars or black holes are the final condensed states of large and very large stars. Our sun will never become one of these.

Neutron stars can be dark, especially black holes, which are totally dark, except for a negligible amount of so-called Hawking radiation. Hawking radiation occurs because empty space, or the vacuum, is not really empty. It is actually a sea of virtual particles of every type that pop into and out of existence for a very brief time. This is possible because according to the uncertainty principle of quantum mechanics, energy may always be borrowed from the vacuum, but it must be repaid quickly. The greater the amount borrowed, the quicker it must be repaid.

These objects are expected to be much scarcer than white dwarfs. Also, the processes that produce these objects release a lot of energy and heavy elements; there is no evidence of such a release.

Weakly Interacting Massive Particles

WIMPs examples include exotic subatomic particles such as axions, massive neutrinos, neutralinos, and photinos. One of the key efforts that has brought cosmologists and particle physicists together is the quest to find the missing universe. The leading dark matter candidates are neutrinos or two other kinds of particles: neutralinos and axions (predicted by some physics theories but never detected). All three of these particles are thought to be electrically neutral. In other words, they are stable enough to have survived from the earliest moments after the Big Bang, but unable to absorb or reflect light.

Theoretically, WIMPs could have been produced in the Big Bang origin of the universe in the right amounts and with the right properties to explain the dark matter. On the other hand, no one has ever observed even one of these particles, let alone enough of them to explain the dark matter.

Hydrogen Gas

Dark matter in galaxies may not be so exotic or even very dark. According to astronomers, most or all of it may be ordinary molecular hydrogen (H_2) gas, which, unlike atomic hydrogen (H), is invisible except at certain infrared wavelengths.

Seventy to seventy-five percent of the visible matter in the universe is in the form of hydrogen, the simplest element. It may be possible that the dark matter is numerous small clouds of hydrogen gas. Nonetheless, it is very difficult to hide hydrogen gas from the probing, sensitive eyes of radio, infrared, optical, ultraviolet, and X-ray telescopes.

Mapping and X-Raying Dark Matter

With data from the Australian/British 2dF Galaxy Redshift Survey, astronomers have come a few steps closer to cracking the mystery behind the elusive dark matter that dominates the universe. Researchers have found that dark matter is distributed just like galaxies on a large scale.

Mapping Dark Matter

Attempting to map the distribution of dark matter (see Figure 1.1) [1] with a telescope is a bit like locating the poles of street lights at night. The street lights are visible, but not the poles. Recently, a team of astronomers has pinned down where the dark matter is located, using new statistical analysis techniques and distant microwave observations.

Figure 1.1 M104 (Sombrero Galaxy). To understand the mass distribution and the dynamics of a galaxy, astronomers need to know the size of the galaxy, what the distribution of mass is in a galaxy, and what the kinematics of stars are in the outer halo as well as in and near the plane. With that knowledge in hand, astronomers will be able to eventually determine the nature of the dark matter and the formation history of a galaxy.

One study by astronomers at Cambridge University contrasted fluctuations in the 2dF galaxy distribution with that of the Cosmic Microwave Background (CMB). This is the radiation left over from the Big Bang. In other words, according to astronomers, due to the force of gravity (see Chapter 4), small density fluctuations produced shortly after the Big Bang evolved into present-day galaxies.

A second study by astronomers at Rutgers University has discovered that dark matter and visible galaxies generally stick together. They form sheets, filaments, and vast voids.

Astronomers have continued to debate whether galaxies trace out the dark matter. The placement of strong constraints on theories of how and where galaxies form is contingent on the affirmation of this result.

This discovery by astronomers has led them to find about seven times as much dark matter as ordinary matter. Still, this is only about one quarter of the matter needed to bring the expansion of the universe to a halt.

X-Raying Dark Matter

As previously discussed, using NASA's Chandra X-ray observatory (see Figure 1.2), [1] a team of astronomers from the Institute of Astronomy at Cambridge University in England has detected huge clouds of hot gas roiling in the hearts of several massive galaxy clusters as shown in Figure 1.3. [1] Lying at distances ranging from 1.5 billion to 4 billion light-years, the Cambridge team selected a sample of five of the largest known galaxy clusters. This selection was based upon prior data and images obtained from the Canada-France-Hawaii telescope atop Mauna Kea and the Hubble Space Telescope. Resulting in what amounts to high-temperature portraits, or maps, of these gigantic objects, the astronomers aimed Chandra at the clusters to soak up their X-ray emissions. In other words, according to astronomers, to prevent the hot gas from escaping these clusters, the temperature maps are used to determine the mass needed.

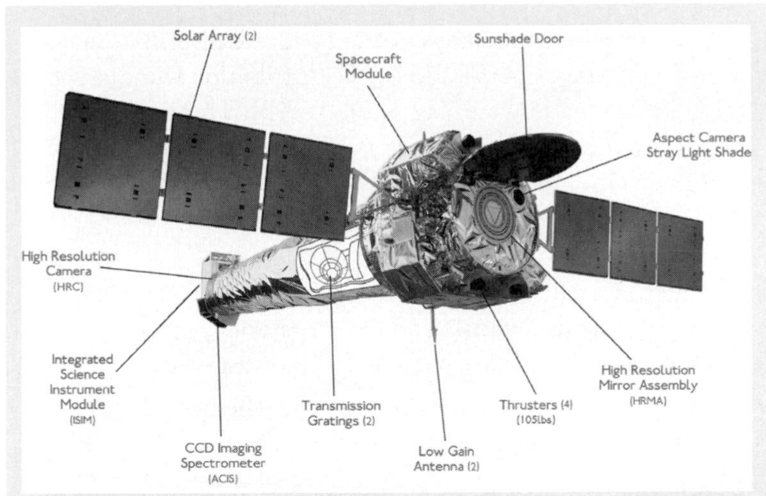

Figure 1.2 Labeled Chandra spacecraft.

Despite the lack of visible mass to gravitationally bind the gas there, the gas remains trapped within these clusters. This has given the scientists ample reason to think much of it is ensnared by surrounding, yet invisible, material.

According to astronomers, the stars in the galaxies and hot gas together contribute only about 13% of the mass of the total cluster. Dark matter must make up the rest.

Furthermore, most of the universe's matter is part and parcel of this dark, elusive stuff, according to astronomers. Pinpointing its presence has proven extremely tricky, especially since it reveals itself only indirectly by gravitational influences on visible matter.

Figure 1.3 This montage shows two sets of Chandra X-Ray Observatory images (left) and Hubble Space Telescope images (right) of the giant galaxy clusters Abell 2390 and MS2137.3-2353 in the Cambridge survey. The clusters are located 2.5 and 3.1 billion light years from Earth, respectively. The X-ray emission comes from the multimillion-degree gas that fills the clusters. Chandra provides detailed temperature maps for this gas and allows astronomers to precisely determine the masses of the clusters. Most of the mass is in the form of dark matter. The Hubble Space Telescope optical images show the distribution of galaxies in the central regions of the clusters. The gravity of the dark matter associated with the brightest galaxies in the clusters, and the cluster as a whole bends light from distant background galaxies to produce the giant arcs seen in the images. The Hubble Space Telescope data place independent constraints on the masses of the clusters that confirm the Chandra results.

Regardless of the progress that has been made, all of the ordinary matter astronomers can find only accounts for about four percent of the observable universe. Astronomers know this by calculating how much mass would be needed to hold galaxies together and, at the same time, cause them to move about the way they do when they gather in large clusters. Looking at how gravity bends the light from distant objects is another way to weigh the unseen matter. The results from these measurements tells astronomers that most of the universe is invisible.

So, if dark matter is invisible, how does one sense its existence? As previously explained, astronomers have detected dark matter indirectly through its gravitational pull. With that in mind, this part of the chapter presents a theoretical and factual question-and-answer discussion composed of the thoughts and ideas of some very prominent theoretical physicists, astrophysicists, and cosmologists.

The Invisible Universe: How Important Is It?

When you really think about it, the concept of dark matter altogether is pretty absurd. Why? Because scientists in this field indicate that 96% of most galaxies (including our own) in the observable universe are made out of dark matter.

So, with that in mind, the first question is, obviously, how do you know for sure that dark matter is out there? In other words, what experimental hard evidence and/or data do you have that proves the existence of this rather incredible theory?

"Accepting Einstein's theory of relativity," states Dr. Scott Gaudi, "there is really no way of escaping the existence of dark matter. There are numerous lines of evidence supporting its existence, including Galaxy rotation curves, cluster velocity dispersions, the Cosmic Microwave background, cosmological structure formation, etc. All of these are experimental hard evidence for the existence of some sort of dark matter, but of course all of these are indirect. There is not direct evidence for dark matter (i.e., there is no direct detection of dark matter)." [3]

According to Dr. Richard Muller, "The main reason we know about stars is because they shine. We know about the planets because they are close. Anything else is hidden. The idea of dark matter isn't absurd. What is strange is that we assume that there is nothing other than stars. If the Universe were full of bricks, we wouldn't see them. They would be 'dark matter'. The only reason that people think that the dark matter isn't bricks is because of theory. Ordinary matter would affect our calculations of the early moments of the big bang. But that could all be wrong." [2]

So, would galaxies simply disintegrate without dark matter surrounding them to keep them together? In other words, would the outer part of a galaxy fly off into intergalactic space? Would the core of a galaxy remain as a unit? Also, would galaxies be much smaller?

"If one were to somehow magically remove all the dark matter from Galaxies instantaneously," says Dr. Gaudi, "then the outer part of the Galaxy would indeed fly apart. The center of most (but not all) Galaxies would remain mostly intact, as they are rotationally supported by visible matter. Indeed, such remnants of this magical removal of dark matter would generally be much smaller." [3]

What are some of the speculations as to the composition of dark matter? Since neutrinos have mass, are they a component of dark matter? Are black holes and undiscovered, exotic elementary particles other possibilities?

NOTE

Neutrinos are electrically neutral particles with negligible mass. They are produced in many nuclear reactions such as in beta decay (nuclear decay by emission of an electron or a positron).

"For many years," continues Dr. Gaudi, "the two leading contenders for dark matter were baryonic and non-baryonic dark matter. Baryonic dark matter would be made up of particles that make up ordinary stuff, i.e., protons, neutrons, etc., whereas non-baryonic dark matter would be made up of exotic particles, such as WIMPs. Now, the general consensus is that baryons cannot make up the majority of dark matter because our 'tally' of all the baryons in the universe, acquired through other means, falls short of the amount needed. Therefore, non-baryonic dark matters seems the most likely. Black holes can make up the dark matter, if they are primordial, however they cannot have been formed from stars, because they would then be essentially baryonic." [3]

Why is dark matter important? Did dark matter play a crucial role in galaxy formation during the evolution of the cosmos? Will dark matter also determine the ultimate fate of the universe?

According to Dr. Gaudi, "Dark matter is important because it provides the vast majority of the dynamical mass in the universe, and the most important force that governs the formation and evolution of the universe is gravity. Therefore, the way the universe looks today is mainly a function of the amount and properties of dark matter. Dark matter (and dark energy) determine the ultimate fate of the universe; they determine whether the universe will expand forever, or eventually stop expanding and recontract." [3]

Surely, all the photons emitted since the beginning of time, and not absorbed by the one atom per cubic meter, can easily account for 95% of undescribed manifest gravity and are gravitationally equivalent to 19 atoms per cubic meter.

"Photons absolutely cannot account for the dark matter," replies Dr. Gaudi. "Although they were, at one point in the history of the universe, dynamically important, the importance of the energy density of photons falls off more quickly than that of matter, and therefore photons soon became much less dynamically important than matter, and are essentially negligible today. Furthermore, if photons were a large constituent of the energy density of the universe, galaxies and clusters could not have formed as we see them today, because photons behave differently than non-relativistic matter." [3]

So, is it possible that dark matter is not dark at all? "It need not be any darker than planets," says Dr. Muller. [2]

Could parallel universes be the answer to the dark matter mystery? According to Dr. Muller, "parallel universe theory is not serious theory. It is what physicists play with when they are being idle. In the history of science, such wild speculations have never led to productive science. Theory unconstrained by observation has always been sterile, and I see no reason why that won't continue." [2]

Could the supermassive black holes discovered at the center of nearly all visible galaxies possibly make up for the lack of mass observed in the universe? And since so little is known about the interior regions of a black hole (where the majority of physics and mathematics may not even be applicable), could the computations of the mass of these large bodies be miscalculated?

According to Dr. Gaudi, "Supermassive black holes in the center of galaxies constitute a very small fraction of the mass needed to explain dark matter. Further, they are located in the center of galaxies, and dark matter is located in the outskirts of galaxies. One might postulate that the dark matter in the halos of galaxies is composed of supermassive black holes, but this is ruled out by the thinness of spiral disks and the longevity of stellar clusters. The mass of supermassive black holes is computed from their dynamic effects far from the regions inside the event horizon where the physics is uncertain. Therefore, in order for these mass determinations to be in error, our understanding of gravity in normal regimes must be in error." [3]

Astronomer and Physicist, Dr. Andrea Ghez (IGPP UCLA), concurs with Dr. Gaudi that "the supermassive black holes at the center of galaxies contribute a very small fraction of the dark mass inferred to exist within our universe. They are therefore not major players in the universe's dark matter problem. In spite of the uncertainties of details of the state of matter in the interior regions of a black hole, the mass, as inferred from the motions of nearby stars and gas, should not be effected." [4]

"We know that the dark matter can't be hidden at the center of the galaxies, since the rotation curves show it is distributed," replies Dr. Muller. "Also, the physics of the interior of black holes is not mysterious. If we accept the theory of black holes, then being inside one would be difficult to notice. In fact, we may be inside one right now—called the Universe." [2]

Why does the production of neutron stars and stellar black holes imply the release of a lot of energy and heavy elements? And, why does this make them unlikely dark matter candidates?

"Neutron stars and black holes are believed to be formed from the death of massive stars," explains Dr. Gaudi. "Massive stars produce an enormous amount of energy in the form of light, and also produce a large amount of heavy elements as they evolve. We have reasonable good measurements of both the amount of light that was produced in the past, and the amount of heavy elements that have been produced. In order for neutron stars or black holes to make up all of dark matter, the amount of light and heavy elements would exceed the observed amounts." [3]

Can WIMPs solve the dark matter problem? "WIMPs, or something like them," replies Dr. Gaudi, "seem likely to be the best candidates for dark matter." [3]

Regarding the dark matter Mystery, did the equation for the mass of the galaxy include planetary zones for the stars?

According to Dr. Muller, "Planets contribute very little to the mass of the solar system." [2] Dr. Gaudi concurs, "Planets do not constitute a significant fraction of dark matter." [3]

Would it be possible for some of the missing mass to be in the form of extra-solar planets? "You mean planets without stars?" asks Dr. Muller. "It is possible, and ruled out only by theoretical calculations of the big bang. Such calculations are not trustworthy." [2]

Dr. Gaudi explains it this way: "It is known that the missing mass cannot be composed of extrasolar planets. First, planets are composed of baryonic material, and baryons do not comprise enough mass density in the universe to account for dark matter. This is known from other lines of evidence, such as measurements of the inhomogeneities in the cosmic microwave background. Furthermore, gravitational lensing experiments aimed at detecting baryonic dark matter in the halo of our galaxy have ruled out planetary-mass objects as a significant constituent of the dark matter in our Galaxy." [3]

Galaxies rotate as a solid body and are actually a "dipping elliptical trajectory" of the stars. So why are we looking for the missing matter when the "falling" of stars through galactic gas is more reasonable?

Finally, according to Dr. Gaudi, "Velocity anisotropies cannot explain dark matter. Furthermore, there are measurements of the velocity dispersion of stars in the Galactic disk that show they have primarily circular motions (i.e., they are not on elliptical trajectories)." [3]

Nevertheless, it's still tempting to say that the universe must be full of dark clouds of dust or dead stars and be done with it. But still, in view of the above comments, there are persuasive arguments that this is not the case. First of all, almost every attempt to find missing clouds and stars has failed, even though there are ways to spot even the darkest forms of matter. Secondly, and more convincing, cosmologists can make very precise calculations of the nuclear reactions that occurred right after the Big Bang. The expected results can then be compared with the actual composition of the universe. Those calculations show that the total amount of ordinary matter, composed of familiar protons and neutrons, is much less than the total mass of the universe. Whatever the rest is, it isn't like the stuff of which we're made.

Finally, whether or not the universe dies in fire or ice, or the universe dies with a fiery implosion (to burst inward) or the big chill—it ultimately depends on how much dark matter there is out there. Unfortunately, scientists still don't have a clue as to what dark matter is really all about.

Conclusion

As you can see from the preceding discussions and comments from the scientists that were interviewed, the nature of dark matter has been and is still being hotly debated. But a growing consensus among astronomers is that it is formed by a curious elementary particle that abets gravity with its mass, but only weakly engages normal (so-called baryonic) matter. Ironically, scientists have dubbed these particles WIMPs, which belies their true strength on the universal stage.

Galaxy clusters offer a good model of the universe as a whole because they encompass huge chunks of real estate. Given astronomers' findings of scant normal matter in galaxy clusters, it has been determined that we live in a low-density universe, which will only continue to expand.

Such a conclusion is consistent with recent studies of the large-scale distribution of galaxies, the cosmic background, and distant supernovae. The findings may also contribute to a more exacting measurement of the universe's total mass density.

Nevertheless, astronomers have been forced to the improbable conclusion that the universe is filled with an exotic form of matter invisible to human eyes and to advanced particle detectors, even after years of theory and observation. As previously explained, scientists working with the Chandra X-ray Observatory have rendered this notion more tangible by making precise (albeit indirect) measurements of so-called dark matter hovering around a galaxy cluster four billion light-years away.

In the clouds of hot X-ray-emitting gas that fill large clusters of galaxies is one of the most fruitful places to hunt down dark matter. The energy of the emission depends on the gravitational forces acting on the gas, which in turn depend on the cluster's total mass. This is the way astronomers have used Chandra to weigh a galaxy cluster in the constellation Draco, as shown in Figure 1.4.[1] Stars, gas, and dust can account for about 20% of the measured mass; the rest must be dark matter, according to astronomers.

Chandra's high-resolution optics have allowed astronomers to make a detailed map of all gravitating material within the cluster. The map offers the best look yet at how dark particles clump together and shows that dark matter is more common in the center of the cluster. Future Chandra observations may clarify whether more than gravity binds the particles of dark matter. As you have seen from the previous comments made by the interviewed scientists, there are so many different explanations for what dark matter could be. Right now they know so little.

Finally, the question must be asked again: Will dark matter ultimately determine the fate of the universe? The universe is currently expanding (that is, the very fabric of space is stretching), thereby causing distant galaxies to move away from one another. If a lot of dark matter is present, then its gravitational

Figure 1.4 The cluster of galaxies (EMSS 1358+6245), about 4 billion light-years away in the constellation Draco, is shown in this Chandra image. When combined with Chandra's X-ray spectrum, this image allowed scientists to determine that the mass of dark matter in the cluster is about four times that of normal matter. The relative percentage of dark matter increases toward the center of the cluster. Measuring the exact amount of the increase enabled astronomers to set limits on the rate at which the dark matter particles collide with each other in the cluster. This information is extremely important to scientists in their quest to understand the nature of dark matter, which is thought to be the most common form of matter in the universe.

pull will eventually cause the expansion to cease. Space would then collapse upon itself, drawing all the galaxies together in a tremendous implosion. The situation would be similar to the Big Bang, but in reverse. The universe would heat up and become full of light and radiation. If not too much dark matter is present, then the universe will continue its expansion forever. Galaxies would become separated by enormous distances, and the universe would cool to frigid temperatures. Eventually, all stars would burn out, and the universe would become cold and black. Thus, the world will either end in fire or ice. The amount of dark matter in the universe determines which of these two possibilities will occur. "The truth is still out there!"

References

[1] "Astronomy Picture of the Day," [http://apod.gsfc.nasa.gov/apod/archivepix.html], NASA Headquarters, 300 E Street SW, Washington DC 20024-3210/ Jet Propulsion Laboratory, 4800 Oak Grove Drive, Pasadena, California 91109/NASA/Goddard Space Flight Center, Greenbelt, MD 20771, USA, 2003.

[2] Dr. Richard Muller, Physicist, Physics Department, University of California, Berkeley, CA 94720.

[3] Dr. Scott Gaudi, Astronomer, Harvard-Smithsonian Center for Astrophysics, 60 Garden Street, Cambridge, MA 02138.

[4] Dr. Andrea Ghez, Physicist and Astronomer, Department of Physics and Astronomy/IGPP, UCLA.

Cosmology: The Creation of the Universe

"I am a creationist; I refuse to believe that I could have evolved from humans."
—Anonymous

Little time is wasted in banishing pervasive darkness in the Old Testament. On the first day of creation, "Let there be light" does the trick once and for all. But, according to astronomers, the universe didn't begin to shine again in a full panoply of stars and galaxies until hundreds of millions of years later. Basically, the universe went dark soon after its explosive birth. Now, astronomers are pondering the exact source of the universe's glow and are glimpsing into creation's second dawn. A stupendously violent first generation of monster stars is one leading possibility here.

Outshining everything that followed, the fireball of the Big Bang was plenty bright, which astronomers believe went off 15 billion years ago as shown in Figure 2.1.[1] However, within about 400,000 years, the newborn universe would have left a dull murk made mostly of hydrogen gas, and eventually would have cooled and faded to black. The universe that astronomers see today was built by the mighty forces that would have writhed in that murky gloom.

The scattered gas was then pulled into filaments and collapsing globs by gravity from ordinary matter, as well as other strange and invisible material. Next, the filaments and collapsing globs

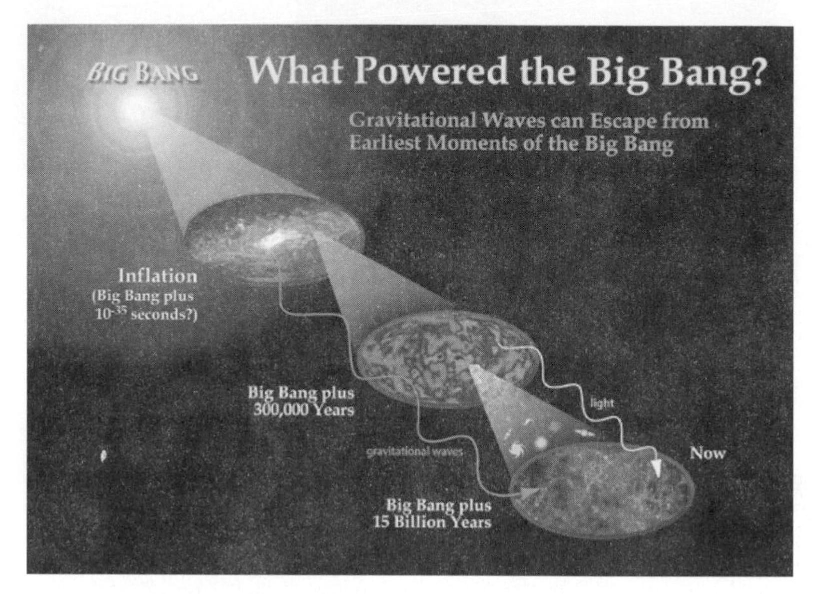

Figure 2.1 Quantum fluctuations during the Big Bang are imprinted in gravitational waves, the cosmic microwave background, and in the structure of today's universe. Studying the Big Bang means detecting those imprints.

eventually burst into flame in stars and other points of light, as they grew denser. Then, the thin ionized gas filling the universe today was gradually converted into the surrounding hydrogen by ultraviolet rays from stars and other exotic celestial objects of that time. (See sidebar, "Lighthouses of the Universe".)

Lighthouses of the Universe

Thanks to an automated telescope in New Mexico that has been surveying deep space since 1998 (in a project called the Sloan Digital Sky Survey), the picture shown in Figure 2.1 [1] got a big boost in late 2001. Astronomers concluded they had detected faint signals from the most distant quasars ever seen, after sifting through the data. Powered by black holes millions of times more massive than the sun, these brilliant beacons in early galaxies are so far away that their light left them when the universe was no more than a billion years old (less than a tenth its present age).

When a 28-astronomer Sloan Digital Sky Survey team used the giant Keck Telescope in Hawaii to analyze the quasars' light, it found shadows that hinted the team was seeing the very end of the cosmic dark ages. The quasars would have burned brightly at ultraviolet wavelengths, but those wavelengths are missing. The finding implies that the quasars were shining during the era of reionization that ended the dark ages, when foglike banks of neutral hydrogen gas still lingered and blocked some of the light. A second team also believes it has seen signs of remnant wisps of hydrogen fog in one of the same quasars.

The Sloan Digital Sky Survey team now knows why reionization happened. They need to find out what did it. What made that first light? Both teams think quasars were too rare and generally formed too late to have been the main actors. That leaves stars—probably huge and rambunctious stars, with masses 100 times or more that of the sun. Such stars could have been born early in the dark ages, perhaps just 50 to 100 million years after the Big Bang. They would have burned expanding bubbles of reionized gas until the entire universe was cleared over hundreds of millions of years.

Several projects are afoot to look for faint infrared or radio signals from these first lights. A billion-dollar Next Generation Space Telescope (now called the James Webb Space Telescope [see Figure 2.2 [1]], nearly 10 times as powerful as today's Hubble Space Telescope [see Figure 2.3 [1]], is on the drawing boards, and astronomers are sketching plans for colossal radio arrays (one more than 200 miles wide) and optical and infrared telescopes with mirrors the size of a football field.

This first-generation star theory is further reinforced by observations by astronomers during the last decade. Sky maps of the radiation relic of the Big Bang (first by NASA's Cosmic Background Explorer [COBE] satellite and more recently by other experiments, including Antarctic balloon flights and NASA's Wilkinson Microwave Anisotropy Probe [WMAP—see

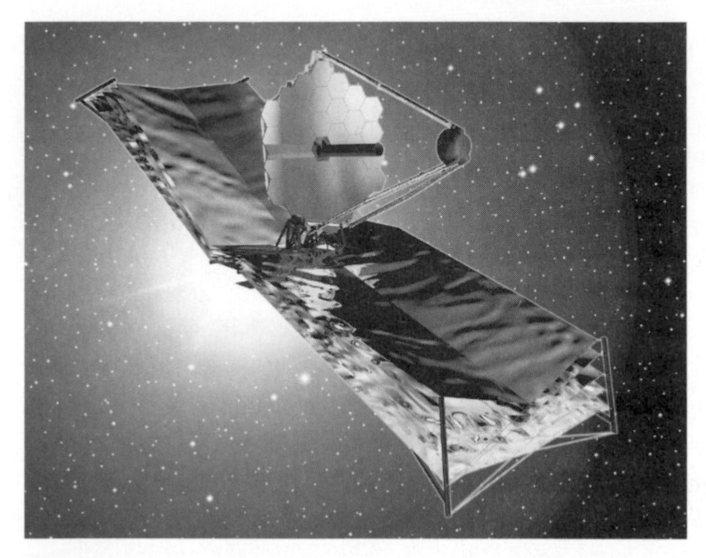

Figure 2.2 James Webb Space Telescope.

Figure 2.3 Hubble Space Telescope.

Figure 2.4 [1]]) have displayed the wrinkles imprinted on the universe in its first moments, as shown in Figure 2.5.[1] Gravity has pulled these wrinkles into the lumpy universe of galaxies and planets astronomers see today. Yet still unanswered are these questions: Why was the Universe so smooth before? And what made the tiny but all-important wrinkles in the first place?

"Indeed," remarks astronomer Dr. Gaudi, "these are two very important questions; whose answers are driving the very forefront of research in physics and astronomy. We are beginning to obtain empirical data that may address the answers to these questions."[5]

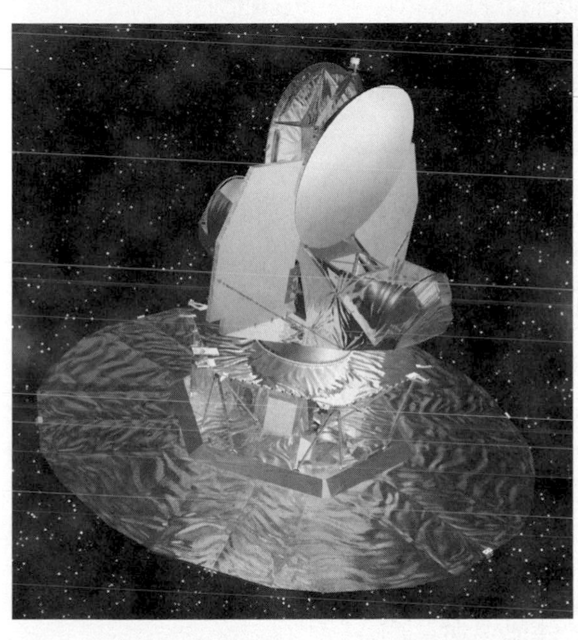

Figure 2.4 Launched on June 30, 2001, WMAP maintains a distant orbit about the second Lagrange Point, or "L2," four times farther than the moon and a million miles from Earth. The composite/aluminum spacecraft is 150 inches (3.8 meters) high by 198 inches (5 meters) wide. WMAP weighs 1,850 pounds (840 kilograms) and is supplied with 419 watts of power. The original WMAP mission lifetime was 27 months: three months of transit to L2 and 24 months of observing time. WMAP is still functioning, and its mission has been extended to improve the already great results.

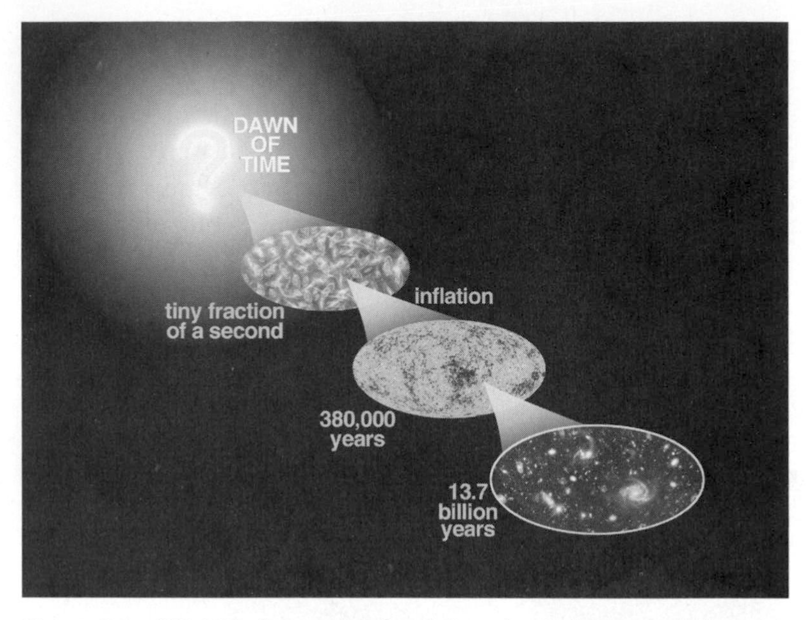

Figure 2.5 WMAP observes the first light to break free in the infant universe, the afterglow of the Big Bang. This light emerged 380,000 years after the Big Bang. Patterns imprinted on this light reflect the conditions set in motion a tiny fraction of a second after the Big Bang. In turn, the patterns are the seeds of the development of the structures of galaxies astronomers now see billions of years after the Big Bang.

Einstein's theories led to the Big Bang model, but they are silent on these questions as well as the simplest: What powered the Big Bang? Modern theoretical ideas that try to answer these questions predict that the wrinkles COBE discovered arose from two kinds of primordial particles: of the energy field that powered the Big Bang, and gravitons, fundamental particles of space and time.

"The origin of the universe is one of the greatest unsolved puzzles in cosmology and physics," remarks Dr. Gaudi. "It is indeed true that General Relativity does not address what happened before and during the very first few fractions of a second of the universe. It is likely that the laws of gravity as we know them break down during this time, and we will need to understand quantum gravity before we can begin to understand what happens during the very early universe." [5]

According to physicist Dr. Richard Muller, "we don't know" what powered the Big Bang. "One can plausibly argue that the kinetic energy of the Big Bang is cancelled by the negative gravitational potential energy. If that is the case, then the Big Bang did not require any energy. I don't think it's correct that Einstein's equations led to the Big Bang. That was discovered by Hubble. Gamov's theory really didn't require relativity theory." [2]

So, with that in mind, how much do scientists really know about the beginning of the universe? How did the universe really begin? Let's take a look!

How Did the Universe Begin?

The unimaginably hot and dense universe that existed at the Big Bang must have been a place where distinctions between gravity, strong force, particles, and antiparticles had no meaning. Of course, this could only be true if all four forces of nature are really a single force that takes on different complexions at temperatures below several million degrees. Einstein's theories of matter and space–time (which depend upon more familiar benchmarks) cannot explain what caused the hot primordial pinpoint of the universe to inflate into the universe astronomers see today. They don't even know why the universe is full of matter. Energy in the early universe should have produced an equal mix of matter and antimatter according to current physics ideas—which would later annihilate each other. Some very helpful and mysterious mechanism tipped the scales in favor of matter, leaving enough to produce galaxies full of stars.

The good news is that a few clues were fortunately left behind by the primordial universe. One is the afterglow of the

Big Bang: the cosmic microwave background radiation. For several decades now, wherever astronomers looked at the edges of the universe, that weak radiation measured the same. Astronomers believed such uniformity meant that the Big Bang commenced with an inflation of space–time that unfolded faster than the speed of light.

However, more recent careful observations show that the cosmic background radiation is not perfectly uniform. There are minuscule variations from one small patch of space to another that are randomly distributed. Could random quantum fluctuations in the density of the early universe have left this fingerprint? Very possibly, according to many other cosmologists. They now believe the lumps of the universe (vast stretches of void punctuated by galaxies and galactic clusters) are probably vastly magnified versions of quantum fluctuations of the original, subatomic-size universe.

So, what really is the ultimate quest in cosmology? In other words, the big questions are: How was our Universe born? And what happened when the universe first emerged just after the very instant of creation? Also, is anything known about the universe before one-trillionth of a second? At very early times, it is suspected that the fabric of space underwent a tremendous stretching that caused the radiation and material in the Universe to disperse. This idea is called inflation.

Inflation

One important question remains more than two decades after cosmic inflation transformed cosmology: Is the universe a bigger place because of it? To help answer that question, an unknown young Stanford University physicist named Alan Guth hit upon a grand idea so powerful that he called it *inflation.* [3] The idea of inflation seemed to be just what the doctor ordered to mend gaping holes in the Big Bang theory at the time—1979. The standard Big Bang model in 1979 adequately explained the presence of the CMB left over from the primordial blast and the origin of light elements (see Figure 2.6).[1] But the standard model fell short on answering a handful of important questions: Why is the universe so uniform on the largest length scales? What caused fluctuations in the matter density of the universe that ultimately led to the formation of stuff like galaxies, planets, and people? And why does the universe contain so much stuff?

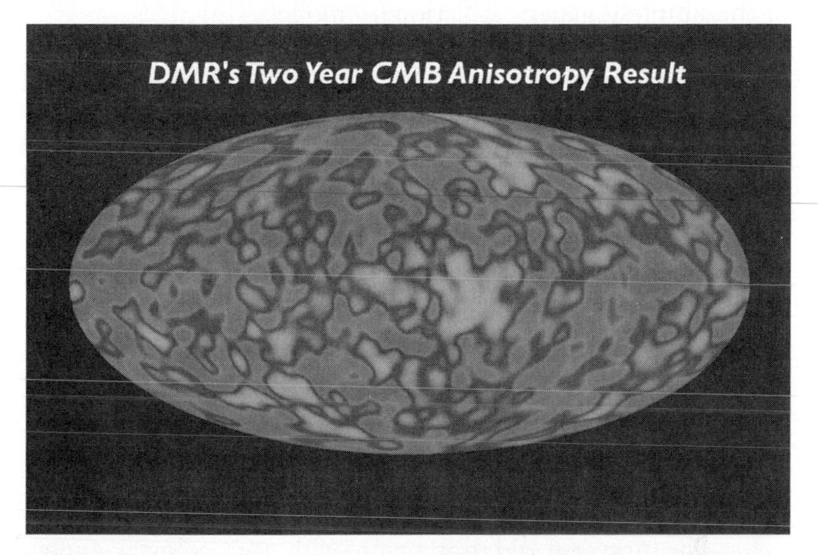

Figure 2.6 Following subtraction of the dipole anisotropy and components of the detected emission arising from dust (thermal emission), hot gas (free-free emission), and charged particles interacting with magnetic fields (synchrotron emission) in the Milky Way Galaxy, the CMB anisotropy can be seen. CMB anisotropy (tiny fluctuations in the sky brightness at a level of a part in one hundred thousand) was first detected by the COBE DMR instrument. The CMB radiation is a remnant of the Big Bang, and the fluctuations are the imprint of density contrast in the early universe. This image represents the anisotropy detected in data collected during the first two years of DMR operation. Ultimately the DMR was operated for four years.

"When my colleagues and I were trying to calculate the spectrum of density perturbations from inflation in 1982," explains renowned physicist Alan Guth (MIT), "I never believed for a moment that it would be measured in my lifetime. Perhaps the few lowest moments would be measured, but certainly not enough to determine a spectrum. But I was wrong. The fluctuations in the CMB have now been measured to exquisite detail, and even better measurements are in the offing. So far everything looks consistent with the predictions of the simplest, generic inflationary models." [3]

It is now worth investigating how well inflation has stood the test of time, especially after more than two decades after Guth's moment of insight. [3] Inflation is doing well on the one hand—widely considered an integral component of the new Big Bang cosmology. However, inflation still suffers from vagueness on the other hand. So, does the basic idea of inflation still look attractive? And do cosmologists now know the real details of inflation or the mechanism that drives it, or is that still a mystery?

"The inflationary mechanism produces an entire universe starting from essentially nothing," says Dr. Guth, "so one would naturally want to ask where the energy for this universe comes from. The answer is that it comes from the gravitational field. The universe did not begin with this colossal energy stored in the gravitational field, but rather the gravitational field can supply the energy because its energy can become negative without bound. As more and more positive energy materializes in the form of an ever-growing region filled with a high-energy scalar field, more and more negative energy materializes in the form of an expanding region filled with a gravitational field. The total energy remains constant at some very small value, and could in fact be exactly zero. There is nothing known that places any limit on the amount of inflation that can occur while the total energy remains exactly zero. Note that while inflation was originally developed in the context of grand unified theories, the only real requirements on the particle physics are the existence of a false vacuum state, and the possibility of creating the net baryon number of the universe after inflation."

On the other hand, "Inflation is an attractive theory because it has no competitors," states Dr. Richard Muller. "We know no other way to explain the uniformity of the black body radiation. But it has not yet made a prediction that is verified, so it is premature to say that it has passed even one test, let alone the test of time." [2]

Today, as astronomers must decide whether inflation will prevail, require modification, or be replaced, its trial period is quickly coming to a close. As you'll see, the possibilities are nearly infinite.

Doubling Up the Acceleration

Inflation's roots are pretty straightforward as cosmological concepts go. Matter at very high energies (like those expected during the formation of the universe) can assume a variety of unusual forms (including one state that turns gravity upside-down), according to a basic prediction from particle physics (see Chapters 5 and 6 for detailed discussions on particle physics). Massive particles repel rather than attract each other in this case. Furthermore, matter in such a topsy-turvy state possesses another strange property: Even as the volume of space it inhabits expands like mad, its density remains constant.

So, such a space-filling, antigravity substance can be described by a scalar field in the parlance of particle physics. Scalar fields are merely a way of representing an array of numerical values at various points in space, even though they may sound exotic. For example, you've got an air pressure scalar field if you can quantify a particular parameter at multiple locations, such as air pressure at every spot in a room. Furthermore, scalar fields (as far as particle physicists are concerned) are subject to quantum fluctuations. Also, a region of space can expand rapidly if the fluctuations in the antigravity scalar field become large enough. Like a rising loaf of bread on steroids, if even the tiniest patch of space is filled with an antigravity scalar field for a brief moment, it will swell uniformly and exponentially.

An antigravity scalar field dominated the universe when it was just 10^{-35} seconds old, according to Guth's inflationary model. Then smaller than the size of a proton, fluctuations in this field caused the budding universe to expand faster than the speed of light. The universe then doubled in size 100 times over in the course of roughly a trillionth of a trillionth of a trillionth of a second. At this point, inflation ground to a halt as the antigravity material became unstable almost as quickly as it began, eventually releasing energy transformed during the inflation-induced expansion into radiation and matter. Finally, at the pace astronomers observe today, the universe continued to expand at a much more moderate rate.

Tapioca and Disorder

Inflation has never been immune from criticism, despite its initial success. Initially, Guth mistakenly pictured the phase transition between the antigravity era and the radiation-dominated era that followed as something akin to water boiling. In this scenario, the post-inflation universe was filled with "bubbles" that never had a chance to collide and merge owing to the rapid expansion. According to Guth, the formation of these bubbles would have made things so inhomogeneous that it would look nothing like the universe astronomers see today. [3]

In order to overcome this problem, a "new inflation" theory was devised in late 1981 by Andrei Linde (a physicist at Stanford) and independently by Paul Steinhardt and Andreas Albrecht (Princeton and the University of California, Davis, respectively). Guth's theory of boiling water was replaced by a milder, slower transition, which was more like the congealing of tapioca. [3] Each bubble was given the chance to expand to fantastic dimensions (far bigger than the observable universe), thus slowing down the pace of inflation. If this view is right, are we living inside a bubble so large that astronomers couldn't possibly see the edge?

According to Dr. Guth, "Inflation does not end everywhere at once, but instead inflation ends in localized patches, in a succession that continues *ad infinitum*. Each patch is essentially a whole universe—at least its residents will consider it a whole universe—and so inflation can be said to produce not just one universe, but an infinite number of universes. These universes are sometimes called bubble universes, but I prefer to use the phrase "pocket universe," to avoid the implication that they are approximately round." [3]

By adjusting the behavior of the antigravity scalar field (which physicists also describe in terms of its potential energy), Paul Steinhardt and his colleagues developed another prescription for stalling inflation. It's like a ball on top of a hill, when the field is at its maximum potential energy. In the same way that a ball will naturally roll downhill, its potential energy becomes kinetic energy as the system moves toward equilibrium. So inflation could be ultimately halted, if kinetic energy, in turn, is converted to radiation and then some of that radiation later gets converted into matter that fills the universe, as a bonus.

Nevertheless, after realizing that the universe need not have been spawned from a hot, dense phase as the Big Bang model predicts, Andrei Linde later rejected new inflation. In 1983, he proposed a modification called "chaotic inflation."

NOTE

Incidentally, Steinhardt and Guth, and others, think new inflation may still be viable.

Chaotic Inflation

Out of sheer randomness, the basic premise of chaotic inflation is that the scalar field can have a variety of energy values. In some places, the scalar field happens to possess more potential energy, while at other places it is sitting at its "ground state," near the potential energy minimum. The region with excess potential energy will inflate and become exponentially large, whereas the region of space near the minimum stays small and doesn't inflate. Thus, the concept (despite its name) has nothing to do with "chaos theory," which describes unpredictable systems that grow increasingly disordered over time. So, is the opposite really taking place here? Do you start with chaos and end with perfection?

"We think inflation is almost always eternal," says Dr. Guth. "I think the inevitability of eternal inflation in the context of new inflation is really unassailable—I do not see how it could possibly be avoided, assuming that the rolling of the scalar field off the top of the hill is slow enough to allow inflation to be successful. The argument in the case of chaotic inflation is less rigorous, but I still feel confident that it is essentially correct. For eternal inflation to set in, all one needs is that the probability for the field to increase in a given Hubble-sized volume during a Hubble time interval is larger than 1/20. Thus, once inflation happens, it produces not just one universe, but an infinite number of universes." [3]

Extended Inflation

In 1989, a new way to end inflation by proposing changes in the behavior of gravity at high energies to slow the universe's expansion was introduced by Steinhardt and his graduate student, Daile La. It was called "extended inflation." Since then, a host of other inflationary models have been proposed, including "hyper-extended inflation," "open inflation," and "two-round inflation." All of these models attempt to explain three things:

- How does inflation start?
- What drives inflation?
- How do you end inflation?

According to Dr. Guth, "Once a region of false vacuum materializes, the physics of the subsequent evolution is rather straightforward. The gravitational repulsion caused by the negative pressure will drive the region into a period of exponential expansion. The exponential expansion dilutes away any particles that are present at the start of inflation, and also smooths out the metric. The expanding region approaches a smooth de Sitter space, independent of the details of how it began. Eventually, however, the inflation field at any given location will roll off the hill, ending inflation. When it does, the energy density that has been locked in the inflation field is released. Because of the coupling of the inflation to other fields, that energy becomes thermalized to produce a hot soup of particles, which is exactly what had always been taken as the starting point of the standard Big Bang theory before inflation was introduced. From here on the scenario joins the standard Big Bang description. The role of inflation is to establish dynamically the initial conditions that otherwise would have to be postulated." [3]

It might seem surprising that models missing key details (such as the energy scale at which inflation proceeds) can perform so well. (See sidebar, "High-Altitude Balloon Experiments.") Can astronomers make accurate predictions, even though they don't fully understand the mechanism behind inflation?

The mechanisms that lead to eternal inflation in both new and chaotic models are fully understood by Dr. Guth. He states: "Although the infinity of pocket universes produced by eternal inflation are unobservable, it is argued that eternal inflation has real consequences in terms of the way that predictions are extracted from theoretical models. Although inflation is generically eternal into the future, it is not eternal into the past: it can be proven under reasonable assumptions that the inflating region must be incomplete in past directions, so some physics other than inflation is needed to describe the past boundary of the inflating region." [3]

Furthermore, how do astronomers go about separating the most plausible models from the more dubious ones? Can they rule out most inflationary theories simply by studying the microwave background in greater detail? Also, can those tests help narrow down the field? Will astronomers find a unique solution from any experiments that are currently planned?

According to Dr. Guth, one of the "consequences of eternal inflation is the possibility that it offers to rescue the predictive power of theoretical physics. Here I have in mind the status of string theory, or the theory known as M-theory, into which string theory has evolved. The theory itself has an elegant uniqueness, but nonetheless it appears that the vacuum is far from unique. Since predictions will ultimately depend on the properties of the vacuum, the predictive power of string/M-theory may be limited. Eternal inflation, however, provides a possible mechanism to remedy this problem. Even if many types of vacua are equally stable, it may turn out that there is one unique metastable state that leads to a maximal rate of inflation. If so, then this metastable state will dominate the eternally inflating region, even if its expansion rate is only infinitesimally larger than the other possibilities. One would still need to follow the decay of this metastable state as inflation ends. It may very well branch into a number of final low-energy vacua, but the number that are significantly populated could hopefully be much smaller than the total number of vacua. All of this is pure speculation at this point, because no one knows how to calculate these things. Nonetheless, it is possible that eternal inflation might help to constrain the vacuum state of the real universe, perhaps significantly enhancing the predictive power of M-theory." [3]

Gravity Waves Surfing

Meanwhile, astronomers' best chance for studying the conditions that triggered inflation is the gravity waves produced in the universe's first moment. Because gravity waves would span the entire universe and it would take an instrument at least that big to detect them, there's no prospect of seeing the most energetic gravity waves directly. Instead, the strategy is to look for the subtle impressions these waves would make on the CMB. (See sidebar, "High-Altitude Balloon Experiments.") The recently launched Microwave Anisotropy Probe (MAP) might have a prayer of seeing that faint imprint, and the higher-resolution Planck spacecraft (scheduled for launch in March 2007) will have an even better shot (see Figure 2.7). [1]

High-Altitude Balloon Experiments

Inflation has received a welcome boost from several research collaborations observing the ubiquitous CMB—(remnant radiation from the creation of the universe) in the last few years. Just like inflation, as it turns out, small temperature variations in the CMB speak volumes about the geometry of the universe and its origin. What's more, if the theory didn't mesh with CMB observations, inflation would suffer a fatal blow.

Depending on the overall density of matter, the geometry of the universe can take on one of three states:

1. If the density is equal to a benchmark called the critical density, the geometry is flat, like a sheet of paper.
2. If the density is above or below critical, the geometry is positively curved (like the surface of a sphere) or negatively curved (like the surface of a saddle), respectively.

3. Multi-wavelength observations indicate that the universe is very likely flat.

Inflation predicts that the geometry of the universe should be flat, because it takes a small, homogenous region of the universe and expands the region exponentially until it becomes flat. Recently, other research strongly suggests a flat universe: the high-altitude balloon experiments conducted by BOOMERanG and MAXIMA, which show the fluctuations that give the CMB its characteristic mottled uniformity cluster around one degree in angular size.

Figure 2.7 The European Space Agency's Planck spacecraft will continue the work of COBE and MAP in studying the structure of the cosmic microwave background as an indicator of the large-scale structure and history of the universe. Planck is scheduled for launch in 2007, sharing a rocket with the Herschel Space Observatory. Its baseline mission calls for two complete scans of the sky during 15 months of observations, but it is designed to be operational for up to five years.

Equivalent to showing when (and at what temperature) the expansion began is any future detection of the imprint left by primordial gravity waves that could reveal the energy scale of inflation. Thus, the physics governing inflation tells researchers about how the shape of the gravity curve relates directly to the energy of the antigravity scalar field. So, are these gravity waves the only pieces of evidence remaining from the early, early universe? Until somebody comes up with another idea, they are astronomers' only hope of measuring anything that tells them about this epoch.

Will astronomers have to wait a decade to learn anything new? Also, will MAP tell astronomers exactly how flat space is? Furthermore, will the error bars shrink a lot in the next two years? And is it not a foregone conclusion that inflation will still agree?

So, what still needs to be known to understand creation? Are there alternate inflation theories? Let's take a look.

Alternative Ways of Understanding Creation

What else can explain the riddles the theory of creation was designed to address, if not inflation? In other words, how can the same problems be resolved in an observationally distinguishable and conceptually different way? Recently, Paul Steinhardt and three collaborators have offered a new "ekpyrotic" scenario to solve the creation problem. Their model, inspired by string theory, postulates that the expanding universe was created from the collision of two three-dimensional worlds moving along a hidden extra dimension. (See sidebar, "My Cup Runneth Over with Universes.") The model's predictions are straightforward, although the idea sounds wacky: At long wavelengths, the gravity wave signature in the microwave background should drop off sharply. By contrast, inflation predicts a "scale-invariant" signature with roughly equal amplitude waves on all scales.

My Cup Runneth Over with Universes

Alexander Vilenkin of Tufts University in 1983 made the astonishing suggestion that almost all inflation models are "eternal"—it never ends, once the process starts. Ultimately spawning an infinite number of "pocket universes," inflation is like a chain reaction—stopping in one region of space only to start in another. So, to prove this theory, what would it take to create a universe under laboratory conditions if it were possible?

Creating Universes

You need about thirty pounds of stuff to create a universe in the laboratory. What kind of stuff? Any kind: silver will do as well as coal, or carrots, or cement blocks. The important thing, and the most difficult to achieve, is that you crush this material to the super-hot and super-dense conditions thought to exist within the inferno of the Big Bang. So let the vises groan; let the fires of the laboratory beat down.

Suddenly a transformation: The thirty pounds of stuff winks out of view, amidst the brilliant light of this fireball. In its place appears a miniature black hole and within the black hole is something even more extraordinary: an expanding universe. In an enormous rush, space in this infant universe is incandescently hot and expands and fills with still more matter. This expansion cools the vapors, which eventually coalesce into stars, galaxies, and worlds upon worlds. Oceans sparkle and volcanoes belch. Across broad beaches unimaginable beings move in the light of their sun, while overhead yawn vast expanding gulfs of intergalactic space.

As for the original thirty pounds of matter, it served its function as a seed—a seed from which an entirely new universe sprung. So, you can do it again if you have another thirty pounds of stuff lying around.

A Nutshell Vantage Point to a New Universe

The last thirty years have witnessed an extraordinary shift in scientists' ideas about the universe as a whole. The idea that our cosmos might be but one among many universes or multi-universes has recently been gaining ground among astronomers, while it had been thought that ours was the only universe. And if that is not mind-boggling enough, imagine that each one of these universes has an infinite number of parallel universes (see Chapter 4 for a detailed discussion of parallel universes).

If astronomers are right, we are descendants of a seed, and a small one at that. But are there other potential seeds lying around? Indeed there are, because all matter is composed of the same elementary particles, so anything at all is capable of doing the trick. The question naturally arises whether it might be possible to produce a universe in a laboratory: What sort of device would astronomers need to assume the godlike role of creator?

Some sort of "vise" is needed on the one hand. Characteristic of conditions during the Big Bang, this would consist of an immense array of battering rams capable of bearing down upon the seed material and crushing it to the enormous temperatures and densities. For example, a seed will catastrophically implode into the required configuration, if it is placed at the very center of an array of hydrogen bombs situated on a tiny moon orbiting Neptune or Uranus and timing the bombs to explode simultaneously.

So what would happen if by an extraordinary effort astronomers succeeded in doing all of this? Would the seed then evolve into a second universe, complete with its own condensation energy, its own inflation, and the subsequent creation of matter?

The actual situation is a good deal more complex, as it turns out. There is a crucial difference between the conditions in the Big Bang and any such simulation. Whereas during the Big Bang this seed was part of a cosmos that underwent the very same compression and heat, a simulation involves only the seed itself.

What happens in a simulation depends on one's vantage point. For example, imagine a hypothetical observer located on the surface of the seed—let's say a stowaway roach. As the blast wave bears down, this insect sees the array of exploding bombs suddenly vanish from sight—and not just the array, but the moon on which the array stands, the space surrounding the moon, and, indeed, the solar system and the entire universe. The roach suddenly finds itself immersed in a universe in the throes of its own Big Bang. This universe is undergoing the very sequence of events that took place billions of years ago within our own universe. The miraculous bug is a lone witness to creation for a second, but also a dead one.

However, the perspective would be quite different to those of you orbiting at a safe distance from the simulation. You would see the seed vanish from sight. It would collapse in on itself and form a black hole. This is an object whose gravitational attraction is so intense that not even light can escape. You would also not be able to reach within that cosmos to affect events in any way or see the seed flower into a second cosmos. Furthermore, the seed flower would have expanded in directions that do not even exist in our universe.

So, hidden from its creators, a child universe is born amid fires and cataclysmic pressures, and what is possible in the universe as a whole is also possible as an experiment. At this very moment, could natural processes elsewhere in the cosmos be generating child universes? Do new universes stream outwards from ours as bits of fluff blown from a dandelion in spring? And two more questions: Is our universe the first? Or are we the progeny of some inconceivable parent?

The Chaotic Universe

It's a known fact that conditions in nearby lands differ radically from those in far-off lands. Travel far enough, and you encounter strange geographies, flora and fauna, and climates. This may be true of the universe with a vengeance, according to ideas bandied about by astrophysicists.

Astrophysicists might encounter places where the very laws of physics are different, if they were capable of building a spacecraft that could reach not just the planets or stars, but far beyond the most distant galaxies that astronomers' telescopes can detect. Were astrophysicists able to travel off in one direction, they might come upon regions in which the usual elementary particles are absent, their places taken by other particles with natures they can only guess at. Voyaging in another direction, astrophysicists might find a space that, except for their presence, is utterly devoid of matter. In yet another direction, the four fundamental forces of nature might be stronger than in the astrophysicists' part of the universe or perhaps weaker. Elsewhere, space might possess four, five, or even ten spatial dimensions.

Some cosmologists propose that the universe during the Big Bang was actually chaotic in nature, and that the condensation energy differed from place to place. Here it was large, there larger, but over there it was zero. Regions of the first sort would undergo normal inflation. Those of the second sort would undergo a far greater degree of inflation and would be quite different from our universe. Indeed, it is possible that some regions inflate forever. And finally, regions of the third sort would undergo no inflation at all. By now, they may have completed their evolution and collapsed into annihilation in a second Big Bang. Meanwhile, the gigantic inflation of our own region would have carried all these separate "mini-universes" far beyond astronomers' view.

This has a crucial bearing on the attempt to unify the forces of nature. Physicists describe all the multitudinous aspects of the world in terms of just four fundamental forces. One is the so-called strong force, which is the source of nuclear energy, binds the atomic nucleus together, and powers the shining of the stars. A second is the so-called weak force, which controls the decay of subatomic particles. Third and far more familiar than the first two is gravity, the universal force of attraction of matter towards other matter. The final force is electromagnetism, which includes light, electricity, and magnetism.

It is hard to see how all of those could be aspects of electromagnetism: the electric flow of a current in a wire, the magnetic attraction that twists a compass needle towards the north, and the glimmer of a distant light. This is why the grand unified theory (GUT) of particle physics is so important in uniting the strong, weak, and electromagnetic forces into one.

NOTE

In 1974 a theory known as the grand unified theory of particle physics was proposed. In this theory all phenomena pertaining to particle collisions are treated as separate manifestations of the same underlying unity. According to the grand unified theory, in such circumstances an utterly new and unfamiliar form of energy appears: the condensation energy. This energy is energy released by a fundamental change in the state of matter. For example, changing water from a liquid state to a vaporous state by boiling requires energy, but the reverse transformation (from vapor to liquid) releases energy. This transformation is known as a phase transition. When the universe expanded and cooled during the Big Bang, it underwent a phase transition similar to the one that water vapor undergoes when it cools and condenses into liquid. The condensation energy released during the phase transition drove the inflation.

Grand unified theory asks why neutrons, protons, and electrons (as opposed to some other subatomic particles) are the ones out of which matter is constructed. It asks why the four fundamental forces have their particular strengths (for example, the electric attraction between electrons), and why they exist in the first place. Certain versions even ask why space has three dimensions.

The answers turn out to depend on the history of the cosmos, and in particular, on the manner in which inflation ended. But because in a chaotic cosmology the cosmos splits into an endless sequence of mini- or multi-universes, each with a different history, the answers are different in different segments. Different laws of physics are contained in all of these segments. The same may also be true of a universe created in the laboratory.

Radical-Looking Universes

The world we live in is an extraordinarily diverse place. Think of the differences between your backyard and the jungles of central Africa. Compare the tropical forests of Central America with arid Death Valley. Extend your vision further to the permanent night of the Mariana Trench or the harsh blaze of sunlight on Mercury. Consider downtown Los Angeles at rush hour; consider the intergalactic voids. The same underlying laws of physics give rise to these radical differences.

NOTE

If you accept the cosmological principle, then all places in the universe are alike. There are no privileged locations, and physical laws operate the same everywhere.

They are different hands dealt in the same card game. But even greater must be the differences between our universe and the alternate universes envisaged by modern physics. When you change the rules of the game, these differences occur.

To describe the physical principles upon which these new worlds operate is one thing, but quite another is to seek to understand what the principles imply. This is a question that scientists have barely begun to address.

What would a world be like whose fundamental forces were different? Consider electricity as an example. This force is not merely confined to things such as light bulbs and motors. It plays a far greater role in the overall nature of things than one would guess at first glance, and is far more ubiquitous. Through its operation, electrical charges repel and attract one another. Also, electrically charged particles (the electrons and protons) are the building blocks of the chemical elements.

Thus, a decisive influence on the elements is the force of electricity. It determines the nature of that matter's chemical reactions, and it determines what kind of matter exists. A cosmos in which electricity is more powerful than in ours is one in which the heavier elements of the periodic table simply do not exist—uranium, for example, and iron, and even the carbon, nitrogen, and oxygen so crucial to life. On the other hand, a cosmos with weak electricity is one containing these elements. But, though capable of the most astonishing chemical reactions, it also contains other elements that are far heavier than the heaviest known to us. For example, in the never-ending metabolism of the cell, chemistry is at work within our bodies. A world with altered chemistry is one in which the nature of enzymes, proteins (see Chapter 10), DNA (see Chapter 11), and the mechanism of heredity (see Chapter 12) are called into question.

Properties of the smallest units of the physical world, the elementary particles, have an uncanny habit of percolating upwards to affect large-scale features. Utilizing the proposed alternative laws of physics, neutrons and protons would not be stable under certain scenarios. Rather, their place would be taken by what, to physicists, are strange and unfamiliar particles. However, the constituents of atomic nuclei of which atoms are constructed, are neutrons and protons, so no atoms could exist in such a universe. Everything from stones to clouds to our bodies is made of atoms in our physical universe. So what is a universe composed of if it is not composed of atoms?

"We are made of atoms only because we are relatively cool," replies D. Muller. "The early universe would have been made of quarks, gluons, electrons, neutrinos, photons, and other exotic particles such as Z and W bosons." [2]

The very number of dimensions of space might be different in different universes, if the more speculative ideas underlying grand unification are correct. The possibility of a fourth dimension has exercised a perpetual fascination and led to the endless conundrums beloved of science-fiction writers. Few of these writers, however, have dared to venture further afield into the possibility of a fifth dimension—or more. And, in any event, they have been asking the wrong question. The question they should ask is what it would be like to be a higher-dimensional creature, not what it would be like for three-dimensional creatures to visit such a world.

A Glimpse of a Universe of Universes

Because the required physical conditions lie far beyond anything that astrophysicists are capable of achieving today, in practical terms, the artificial creation of a universe is not feasible. No furnace and battering ram, no array of hydrogen bombs that astrophysicists could assemble would be remotely capable of mimicking the Big Bang.

Recent studies by astrophysicists have indicated that the required conditions might turn out to be even more extreme. In fact, they might turn out to be infinite—a literally infinite degree of compression and temperature. If true, this would be an absolute obstacle, not just a formidable one. Thus, the possibility of creating a universe for all time would be laid to rest, but whether it is true is right now an open question.

All in all, the grand unification theory has essentially no experimental successes to its credit, although it is spectacularly successful in theoretical terms. To reach the energies required to test the theory, the accelerators used by astrophysicists are not powerful enough (and may never be). It is always dangerous to put too much faith in such things. Nature has a way of being more subtle than the subtlest among us can guess. Similar caveats surround the notion of a universe composed of an endless succession of mini- or multi-universes.

At the very cutting edge of research is where all of these ideas lie. They are not proven concepts. They are speculations, dreams, and hopes, and as science progresses, it would be foolish to claim that they will retain their present form. Nevertheless, astrophysicists stand at a great juncture in the evolution of their ideas about the cosmos. Modern theories of elementary particle physics have given astrophysicists an extraordinary vision: a glimpse of a universe of universes, of a never-ending sequence of parallel worlds. And, more than that, they have posed a major challenge to their powers of imagination.

With regards to the preceding discussion, Vilenkin has continued to explore the implications of this idea. What may seem far-fetched is the idea of a process that pops out universes faster than popcorn kernels. Rather than one created from a unique cosmic event, it seems far more plausible to physicists like Guth that our universe was the result of mass reproduction.

Furthermore, it still leaves open the nagging question of what spawned the material that triggered inflation in the first place, even if astronomers accept the notion that inflation continues without end once started. In other words, inflation does not explain how the universe began, nor does it set out to do so. But, according to Guth, it does provide a physical mechanism for producing a vast region of space–time filled with matter and for taking a subatomic speck. So, in essence, inflation reduces the question of the universe's origin to the question of how astronomers get that subatomic speck. At least this minor question is less daunting to contemplate, although it will also not be easy to answer.

Nevertheless, Steinhardt still likes inflation. According to Steinhardt, it's a wonderful theory that may prove to be right. But as a scientist, he still has to ask: Is there another way to do it? A successful theory can lead to complacency, whereas science can be pushed forward by competing theories that make very different predictions. So, is there another way to prove inflation?

"It's important to remember that, according to the scientific method, no theory can be proven, only disproven," states Dr. Gaudi. "All one can say is that a theory supports all available observational data; as soon as a reliable observation is made that disagrees with the theory, the theory should be discarded (or at least altered). Therefore, there is formally no way of 'proving' inflation. However, corroborating evidence can be gathered that supports the theory, and does not contradict it." [5]

According to Dr. Guth, another "consequence of eternal inflation is that the probability of the onset of inflation becomes totally irrelevant, provided that the probability is not identically zero. Various authors in the past have argued that one type of inflation is more plausible than another, because the initial conditions that it requires appear more likely to have occurred. In the context of eternal inflation, however, such arguments have no significance. Any nonzero probability of onset will produce an infinite space–time volume. If one wants to compare two types of inflation, the expectation is that the one with the faster exponential time constant will always win. A corollary to this argument is that new inflation is not dead. While the initial conditions necessary for new inflation cannot be justified on the basis of thermal equilibrium, as proposed in the original papers by Steinhardt, in the context of eternal inflation it is sufficient to conclude that the probability for the required initial conditions is nonzero. Since the resulting scenario does not depend on the words that are used to justify the initial state, the standard treatment of new inflation remains valid." [3]

So, are alternative models like the ekpyrotic scenario good, so long as they're reasonable? Probably so, since new ideas have to be tested and science has to go on.

"Yes," replies Dr. Gaudi, "alternative theories are always good, provided they are really viable. They force scientists to look critically at the favored theory, sharpen the predictions of the theory, and devise new tests to distinguish between the theories." [5]

Dr. Guth concurs: "The cyclic-ekpyrotic model is touted by its authors (Steinhardt and Turok) as a rival to inflation, but in fact it incorporates inflation and uses it to explain why the universe is so large, homogeneous, isotropic, and flat. [3] [4]

Finally, if inflation is eventually replaced by something completely different, will it have served a valuable purpose? By highlighting what needs to be explained and by providing a framework for discussing properties of the early universe, has it set the stage for what must follow?

"Of course it will have served a valuable purpose," explains Dr. Gaudi. "Whatever the new theory entails, it will have been motivated by the successes and failures of the inflation theory." [5]

Finally, explains Dr. Guth, "Essentially all inflationary models are eternal. In my opinion this makes inflation very robust: if it starts anywhere, at any time in all of eternity, it produces an infinite number of pocket universes. Eternal inflation has the very attractive feature, from my point of view, that it offers the possibility of allowing unique (or possibly only constrained) predictions even if the underlying string theory does not have a unique vacuum." In addition, Dr. Guth also explains "the past of eternally inflating models, concluding that under mild assumptions the inflating region must have a past boundary, and that new physics (other than inflation) is needed to describe what happens at this boundary." He also "describes, however, that our picture of eternal inflation is not complete. In particular, we still do not understand how to define probabilities in an eternally inflating space–time. The bottom line, however, is that observations in the past few years have vastly improved our knowledge of the early universe, and that these new observations have been generally consistent with the simplest inflationary models." [3] [4]

Conclusion

Most cosmologists now agree that the universe expanded from a primordial explosion they call the Big Bang. Physicists talk, after all, about the first three minutes. But, if this makes sense, it makes sense as well to talk of times after the first three minutes. And if time has an origin, and a uniform measure, then we are again within the bounds of Newton's universal clock, marking time throughout the cosmos. It is everywhere approximately 15 billion years after the Big Bang, and it is that time now.

But a universe proceeding from nothing to nowhere by means of an enthusiastic expansion is only one possibility. There are others. Some interpretations of the field equation are realized in a static, but unstable universe, one that simply hangs around for all eternity if it manages to hang around at all. Then again, the universe might be rotating in a void, turning serenely like a gigantic pinwheel. In a universe of this sort, each observer sees things as if he or she were at the center of the spinning, with the galaxies (indeed, the whole universe) rotating about him or her. And finally, this strange assumption, as astrophysicists have demonstrated, satisfies the field equation of general relativity exactly.

References

[1] "COBE Slide Set," [http://lambda.gsfc.nasa.gov/product/cobe/slide_captions.cfm], NASA Headquarters, 300 E Street SW, Washington DC 20024-3210/ Jet Propulsion Laboratory, 4800 Oak Grove Drive, Pasadena, California 91109/NASA/Goddard Space Flight Center, Greenbelt, MD 20771, USA, 2003.

[2] Dr. Richard Muller, Physicist, Physics Department, University of California, Berkeley, CA 94720.

[3] Alan Guth, physicist, (MIT, Cambridge, MA) "Inflation," published in Carnegie Observatories Astrophysics Series, Vol. 2, Measuring and Modeling the Universe ed. W. L. Freedman (Cambridge: Cambridge Univ. Press, 2003), quoted with permission of the author.

[4] Alan Guth, physicist, (MIT, Cambridge, MA) *The Inflationary Universe: The Quest for a New Theory of Cosmic Origins*, published by Perseus Books, 1997 [quoted with permission of the author].

[5] Dr. Scott Gaudi, Astronomer, Harvard-Smithsonian Center for Astrophysics, 60 Garden Street, Cambridge, MA 02138.

3

Theoretical Cosmology and Particle Physics: The Cosmological Constant Problem

"One of the advantages of being disorderly is that one is constantly making exciting discoveries."
—A.A. Milne (1882–1956)

The Universe is a dynamic, evolving place—the cosmic equivalent of the web of biological and physical interactions that shape our own planet. For most of the 20th century, cosmologists held a fairly simple picture of the evolution of the universe: It began with the Big Bang, in which the universe was very hot and extremely dense. From that inception, the universe expanded, meaning that space stretched, thereby causing the temperature to drop and matter and energy to spread out. Eventually, the structures theoretical cosmologists observe today (everything from atoms to galaxy clusters) emerged. The theoretical foundation for the development of the universe, known as the Friedmann–Robertson–Walker cosmological model or the Standard Model of cosmology, underwent an important modification in the late 20th century, and due to recent precise astronomical measurements, it has experienced dramatic changes.

The Matter and Energy Spread

The structure and evolution of the observable universe have revolutionized theoretical cosmologists' understanding of the web of cycles of matter and energy within it. To understand the structure and evolution of the universe, cosmologists use tools from throughout the electromagnetic spectrum to explore diverse astrophysical venues. The Chandra X-ray Observatory has been notable in this regard, opening cosmologists' eyes to the richness of the X-ray universe, as the Hubble Space Telescope has done for the optical part of the spectrum and the Space Infrared Telescope Facility (SIRTF) will soon do for the infrared.

The universe is governed by life cycles of matter and energy (see Figure 3.1) [1], an intricate series of physical processes in which the chemical elements are formed and destroyed, and passed back and forth between stars and diffuse clouds. It is illuminated with the soft glow of nascent and quiescent stars, fierce irradiation from the most massive stars, and intense flashes of powerful photons and other high-energy particles from collapsed objects. Even as the universe relentlessly expands, gravity pulls pockets of its dark matter and other constituents together. Then, the energy of their collapse and the

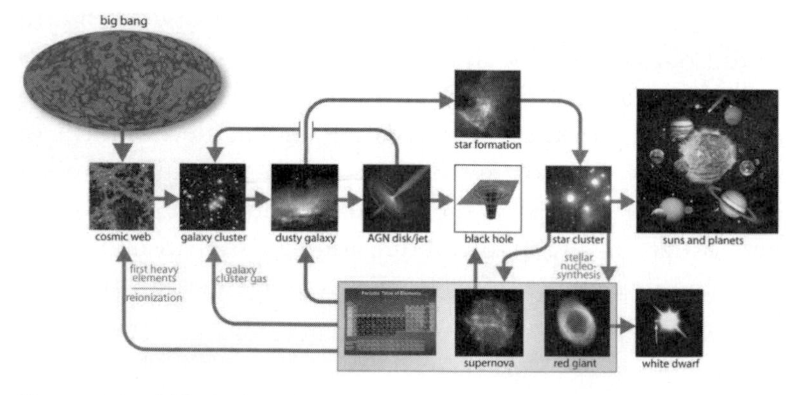

Figure 3.1 Life Cycles of Matter and Energy.

resulting nucleosynthesis, later work to fling them apart once again.

The aim of today's theoretical cosmologists is to understand these cycles and how they created the conditions for our own existence. To understand how matter and energy are exchanged between stars and the interstellar medium, cosmologists must study winds, jets, and explosive events. Their task includes uncovering the processes that led to the formation of galaxies and their dark matter halos. Finally, cosmologists seek to understand the behavior of matter in extreme environments: crushing neutron stars and the sources of gamma-ray bursts and the highest-energy cosmic rays.

The missions of cosmologists today are addressing some of the mysterious issues that surround the cycles of matter and energy. But, to unravel the interlinked cycles, space-based telescopes with additional capabilities are needed:

- To decipher the flows of gas and energy in the first galaxies: a cryogenic, large aperture infrared observatory.
- To uncover how supernovae and other stellar explosions work to create the elements: an advanced Compton telescope and a hard-X-ray spectroscopic imager.
- To map the "invisible" universe of dark matter and gas expelled during the birth of galaxies: a large-aperture telescope for imaging and spectroscopy of optical and ultraviolet light.
- To measure the motions of the hottest and coldest gas around black holes: a radio interferometer in space.
- To see the birth of the first black holes and their effect on the formation of galaxies, and to probe

the behavior of matter in extreme environments: a very large-aperture arc-second X-ray imaging telescope.

- To determine the nature and origin of the most energetic particles in the universe today: a mission to track them through their collisions with Earth.

The cycles that theoretical cosmologists seek to understand are driven by stars and galaxies. Before describing how cosmologists plan to proceed, let's briefly review what you have learned so far about stars and galaxies from Chapters 1 and 2.

Stars: Engines of Change in an Evolving Universe

For a star, mass is destiny—the low-mass stars slowly fuse hydrogen into helium, while massive stars burn fiercely for a brief cosmic moment. Stars about one-half the sun's mass or less have a lifetime that is at least as long as the present age of the universe. The oldest of these stars show theoretical cosmologists that our galaxy once lacked the heavy elements out of which planets and people are made. Stars of later generations, like the sun, inherit a legacy of atoms created by short-lived massive stars when the universe was young. (See sidebar, "Star Light—Star Bright.")

Star Light—Star Bright

The first generation of stars arrived on the scene like a raging wildfire thirsty for kindling and gorged themselves on the primordial stuffing of the universe. Long before the giant elliptical and spiral galaxies that dominate the universe today were more than a glimmer in the cosmic eye, starbursting protogalaxies dominated space with gale-force stellar winds setting off a chain reaction of star formation, and supernovae popping like machine-gun fire in every direction. In the absence of muting effects caused by cosmic dust (the eventual offspring of this premiere generation), only two modes of brightness existed: inky darkness and blinding brilliance.

Lighting Up

Astronomers are closing in on one of the most enigmatic periods of cosmic history—the era when stars first lit up the darkness of space. A recent analysis of a Hubble Space Telescope image hinted that massive stars were born en masse in a giant conflagration a few hundred million years after the Big Bang. Observations confirm the early fireworks from data obtained from the 2MASS ground-based map of the infrared sky. Astrophysicists at NASA's Goddard Space Flight Center in Greenbelt, Maryland, process 2MASS data to filter out nearby stars and galaxies. The remaining glow was two to three times brighter than expected (produced by galaxies that formed when the universe was less than a billion years old). According to astrophysicists, this means galaxies were forming stars at a gigantic rate—more than 30 times faster than in our galaxy today.

At the University of California at San Diego, astrophysicists have created a supercomputer simulation to peer even deeper back in time. Their model indicates that stars started forming 100 million years after the Big Bang, much earlier than previously estimated. Condensed from hydrogen and helium gas (pristine stuff from the Big Bang), those pioneer stars were 100 times more massive than our sun. Before exploding as supernovae, such huge stars would have lived only a few million years. They also synthesized the first heavy elements (the carbon, nitrogen, calcium, and other materials that eventually found their way into planets and life) during that brief time. Cosmologists had little information about when and how the earliest stars formed until now.

So, according to the latest glimpse of the infant universe gathered by astronomers using the NASA/ESA Hubble Space Telescope, quite a controversy has been caused by this new view of the beginning. Nevertheless, astrophysicists have shed some light on an epoch that hadn't received much attention.

Dogma of Controversy

Crucial life-sustaining processes of star formation must have begun sometime between the Big Bang and present epoch, according to astronomers. They seem pretty sure (after using Hubble to peer deep into the past and generate the Hubble Deep Field North in 1995 and its southern counterpart in 1998) that when the universe was about half its present age (roughly seven and one-half billion years ago), that it underwent a massive "baby boom."

However, other astronomers think the reverse. According to new research on galaxies residing in both Hubble Deep Fields, astronomers have found evidence that the "baby boom" took place much more quickly, just a few hundred million years, after the Big Bang. Of course, all of these key assumptions still remain to be proven.

According to astronomers, a concept that's referred to as "missing light" forms the cornerstone for the preceding arguments. Astronomers have now ascertained that rather than representing the cumulative light from an entire galaxy, the light observed in deep-field galaxies is generated only by the biggest and brightest stars and clusters. This is accomplished by carefully examining the color of starlight from galaxies in the deep fields (the youngest most distant galaxies known).

In other words, even with the most powerful telescopes, a significant fraction of each galaxy's light is not represented because it is too dim to be gathered. Much in the same way tourists on safari can spot only the largest members in a distant herd of elephants, astronomers may only be capable of seeing the "tip of the iceberg" when it comes to light from the most distant galaxies. To bolster confidence in their results, astronomers plan to gather more data with Hubble's new Advanced Camera for Surveys. In addition, as a control for their observations, astronomers plan to use supernovae, which don't dim with distance the same way that galaxies do. If all of these theories are proved true, astronomers will have to tweak their own individual theories to account for a much earlier ignition of the star formation process that continues today—albeit now at a much slower rate.

Massive stars create new elements (oxygen, calcium, iron) and return them to space through stellar winds. At the end of these stars' lives, fierce fires forge elements heavier than iron and expel them in the huge explosions called supernovae. The accumulated products of these events become the material for new stars that form in the densest interstellar regions, and which also serve as cradles for organic molecules related to life (see Figure 3.2). [1] Lower mass stars evolve more sedately. As they run out of hydrogen fuel, they slowly expand to become large, cool, "red giant" stars. These stars exude strong "stellar winds" that are the major source for interstellar carbon, oxygen, and nitrogen. Our Earth and our bodies are formed from the chemically processed ejecta of all these stars (see Figure 3.3). [1]

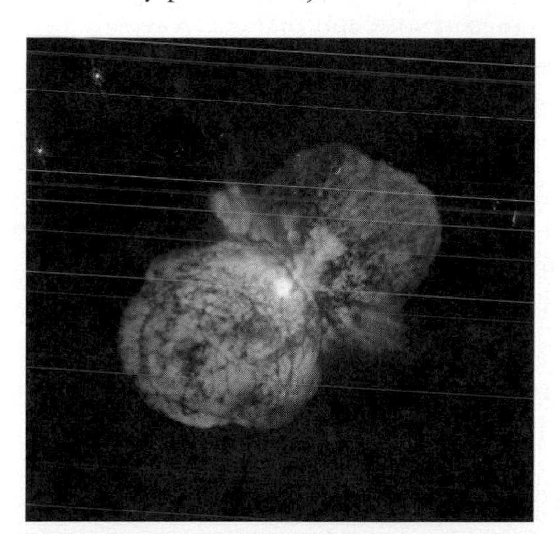

Figure 3.2 A huge, billowing pair of gas and dust clouds is captured in this Hubble telescope picture of the super-massive star Eta Carinae. Eta Carinae suffered a giant outburst about 150 years ago and now returns processed material to the interstellar medium.

Figure 3.3 This HST snapshot of galaxy NGC3079 reveals dramatic activity in its core. Gaseous filaments at the top of a hot bubble of gas are being expelled into intergalactic space. Eventually, some of this gas will rain down on the disk to form new generations of enriched stars.

Galaxies: Bringing It All Together

These stars congregate by the billions, in billions of galaxies, which come in a wide range of sizes and shapes. To explain this rich variety, theoretical cosmologists will trace their evolution from their origins in the early universe to the intricate systems they find today.

According to theoretical cosmologists, when the universe was a much younger and more violent place, super-massive black holes were gorging themselves in a natal feeding frenzy as galaxies formed around them. (See sidebar, "Rip Roaring Black Hole.") The signposts of this process are the quasars and active galactic nuclei. Even relatively quiet galaxies like our own have massive black holes lurking at their centers. What role did black holes play in the evolution of galaxies? According to renowned physicist Dr. Richard Muller, "We don't know what role black holes played in the evolution of galaxies. [2]"

Actually, according to Dr. Scott Gaudi, "I think it remains to be completely understood what role quasars have played in the evolution of galaxies and the universe. However, one area that they clearly did play an important role in is the production of a large amount of ultraviolet radiation that served to partially ionize the neutral hydrogen that fills the universe. The ionization suppress star and galaxy formation for a period of time during the early universe." [3]

Rip Roaring Black Hole

Recently, according to astronomers, two space observatories have provided the first strong evidence of a supermassive black hole stretching, tearing apart and partially gobbling up a star flung into reach of its enormous gravity. The event, though never confirmed, had long been predicted by theory.

Astronomers were drawn to the event by the observation of a powerful X-ray blast. The location of the blast was near the center of a galaxy with a distance of about 700 million light-years from Earth. According to astronomers, the blast was produced by gases from the star (heated to multimillion-degree temperatures) as they fell toward the black hole near the heart of galaxy RX J1242-11.

After veering off course, following a close encounter with another star, astronomers believe a star about the size of our sun neared the black hole. The star then was stretched to the point of breaking by the tremendous gravity of the black hole (estimated to have a mass 100 million times that of our sun).

But, with much more violent results, the effect is the same that the tug of the moon has on the Earth's oceans. Flinging the rest out into space, the black hole consumed an estimated 1 percent of the doomed star.

In order to capture the event, astronomers used NASA's Chandra and the European Space Agency's XMM-Newton X-ray observatories. Normally, in a typical galaxy, similar events are estimated to occur just once every 10,000 years.

In the past, astronomers have seen other similar X-ray blasts before. However, they were never able to pinpoint them at the center of a galaxy, where black holes lurk. Revealed by the new observations, was the characteristic X-ray signature that normally surrounds a black hole.

It is a daunting challenge to try to understand events that happened billions of years ago in faraway places, but theoretical cosmologists can do this in at least three ways:

- They can measure the ages of stars in nearby galaxies to reveal their history of stellar births;
- They can study nearby galaxies still under construction today;
- And they can use powerful telescopes as time machines to see the past directly: As they peer farther out into space, they see back in time.

The Difficulties of Theoretical Cosmology

Modern theoretical cosmology is based on Einstein's fundamental theory of gravity: the theory of general relativity. It supersedes Newton's mechanical picture of gravitation with a fundamental, esthetically pleasing idea: The force of gravity is caused by the curvature of space–time.

Space is dynamic in general relativity. This means that it can stretch, bend, and twist. Thus, a deformation of space–time can be caused by a massive body such as Earth. When a second object, such as an orange, is introduced, it moves in a natural way along geodesics in the curved space–time. The object (the orange) appears to accelerate, because the motion of the object in a curved space–time is not necessarily along a line nor necessarily uniform. It changes its speed or its direction, or both. Newton's version of gravity is almost exactly reproduced by the response of space–time to a massive body (such as the Earth) and the natural motion of an object (such as an orange) in the curved space–time of Einstein. Differences between the two theories are quite small, but the differences increase if one or both of the gravitationally interacting bodies is moving at speeds approaching that of light, or if the curvature of space is great, as is the case near a pulsar or black hole.

It turns out that space should be either stretching or contracting, when general relativity is applied to the universe as a whole. In 1929, the astronomer Edwin Hubble observed that distant galaxies are all moving away from Earth and concluded that the universe is expanding. Galaxies are pulled away from one another as space stretches, and the farther two galaxies are apart, the faster they move away from each other. A balloon with dots on it is the standard analogy. As the bottom stretches when it is inflated, the dots are also pulled away from one another.

Many observations have confirmed that the Universe is expanding, since Hubble's initial findings. The Hubble expansion rate is known as the fractional speed of stretchings.

In one respect, the analogy with a balloon is defective. In order to make it inflate, an outside influence (a human being with lungs) must blow air into it. There are no external influences for the Standard Model of cosmology. So, what causes the expansion of the universe? "Once it starts (the Big Bang), it will be carried by momentum," says Richard Muller. "That is not mysterious," continues Muller. "What we don't know is what (if anything) started it in the first place." [2]

The matter and radiation within it is one answer as to what causes the expansion of the Universe. This matter/radiation is taken to be uniformly distributed in the Friedmann–Robertson–Walker model. At sufficiently large distances, observations of our universe indicate that this is a good approximation. Locally, the universe is lumpy—stars and galaxies being the lumps. But, if viewed from sufficiently far away, such lumps form a rather even distribution. Furthermore, known as the cosmic microwave background radiation (radiation left over from the early universe), it is smooth to one part in 100,000!

New Observational Developments

Measurements of a type Ia supernovae began in 1995 and 1996 that would suggest an astonishing result. Supernovae are stars that explode. They can do so in various ways, and the label "type Ia" indicates a certain kind of supernova. (See sidebar, "Type Ia Supernovae.") Astronomers can see these very bright objects at faraway distances and *use them to measure how fast space is stretching.*

Type Ia Supernovae

Stars explode as supernovae in basically two ways: When a supermassive star runs out of nuclear fuel in its core, it collapses, rebounds, and explodes. The result is a type II supernova. Many stars appear in binary systems. If a white dwarf (a compact dense star) draws in material from its companion, the extra weight can cause the star to collapse. The nuclear chemistry of the star is upset and it explodes. This is a type I supernova. A type Ia supernova is distinguished from other type I supernovae by certain features in its spectrum.

Very faraway supernovae can be seen from telescopes on Earth, since supernovae are very bright, often as bright as an entire galaxy of stars. One can determine the intrinsic brightness of a type Ia supernova from the shape of the light curve. The distance to the supernova can be established by comparing the intrinsic brightness to the apparent (or Earth-based observed) brightness. In short, type Ia supernovae are standard candles that can be used to measure distance.

Easily deduced from its spectrum is the speed of a type Ia supernovae. So, in order to measure the expansion rate of the universe, the speed and distance measurements are used. Finally, astronomers concluded from observations in the late 1990s of distant type Ia supernovae that the expansion rate of the universe was increasing and/or was accelerating!

A sufficient number of type Ia supernovae had been observed by 1998 to conclude that the expansion of the universe was accelerating. (See sidebar, "Type Ia Supernovae.") Since cosmologists thought that the expansion rate should be slowing, this came as a complete surprise.

By introducing a cosmological constant into Einstein's general theory of relativity, the expansion of the universe can be made to accelerate. Also, by introducing a uniform background mass/energy density that has the unusual property of providing a negative pressure that compels the universe to expand increasingly fast, the cosmological constant modifies the Standard Model of cosmology.

The Cosmological Constant Problem

The cosmological constant produces a rather unusual effect on cosmology. According to cosmologists, it's not really considered part of the Friedmann–Robertson–Walker model, but it can be included. It causes a gravitational repulsion that drives matter apart at increasing rates by yielding a "negative pressure." The data from type Ia supernova observations suggest the existence of a cosmological constant.

Type Ia Supernovae and the Cosmological Constant

Because of a fine-tuning problem, theorists have been reluctant to introduce a cosmological constant in the theory. It turns out that if the cosmological constant is non-zero, then it should have a natural value that would cause a very rapid expansion and very dramatic cooling of the universe. The current measured temperature of the universe is 2.725 kelvin. Although extremely cold by human standards (just a few degrees above absolute zero), if the cosmological constant is present, it must be about 10,122 times smaller than the natural value to agree with this temperature measurement. How could the cosmological constant (CC) be set to such a small value with such precision?

"The cosmological constant problem continues to be the most vexing question in theoretical physics," states physicist Dr. Nima Arkani-Hamed (Harvard University Department of Physics), "and whatever its solution, it is bound to change the way we think about the universe in a fundamental way. The basic conundrum is simple: why is the energy of the vacuum so much smaller than a back of the envelope estimate would suggest? [4]"

"There have been a number of different attempts to deal with the problem: (I) The CC is small for a mysterious 'deep' reason related to quantum gravity. Most theorists subscribed to this point of view for a long time. The idea is that there is perhaps some deep symmetry principle that fixes the value of the cosmological constant to zero. The experimental observation of a non-zero though tiny CC, together with the continued absence of any evidence for such a deep principle, makes this explanation seem much less likely. Other attempts to find a deep reason for the smallness of the CC take the tiny size of the cosmological constant and the associated large deSitter universe as an input, and try to explain the largeness of the other scales in particle physics relative to the CC by invoking 'holography' and new non-local interactions in this space. This sort of approach seems promising, and it would be radical and wonderful if something like this was true. But again as of yet, there is very little theoretical evidence for anything like this taking place." [4]

"(II) The CC is actually not small, but gravity is modified at large scales in such a way that the large vacuum energy does not lead to unacceptable large space–time curvature. This is what the CC problem seems to be crying out for—a modification of gravity at large scales that to a large extent 'degravitates' the vacuum energy. This loose motivation has spurred many attempts at modifying gravity at large distances, with many ideas exploiting the possibility that we can live on a membrane in a higher dimensional space, trapping gravity along our 4 dimensions over some scales. The idea would be that the large vacuum energy of our world—which would be reflected in the large tension of our membrane, would mostly curve the higher dimensional space, and not our own four dimensions." [4]

"It turns out to be difficult to modify gravity at large distances, however, in a theoretically consistent and experimentally viable way. Moreover, it is not yet clear that large-*distance* modifications are enough to do the job, because the universe must have enough *time* to know that the CC should be cancelled, and this time is typically so long that it would not have been cancelled over the lifetime of the universe. Other tentative suggestions invoke an acausal modification of gravity, where what is cancelled is the deep future value of the CC. It is not clear that this idea can be made consistent with the apparent locality and causality of our world in a way compatible with quantum mechanics." [4]

"(III) Anthropic explanations—we live where we can live. Given the status of the above attempts, Weinberg's anthropic explanation of the CC looks great. The idea is simple: suppose that the CC actually had different values in different parts of the universe. Over vast stretches of this 'megaverse', the CC would have its natural value, but in these same regions the universe would be expanding so quickly that no structure could ever begin to form. Just the minimal requirement of structure formation limits the size of the CC to be only about

100 times larger than what is observed. So, why don't we observe a large CC? Because such a large CC wouldn't have let any structure evolve and the universe would be empty of inhabitants to ask the question. This is very reasonable in itself—after all, in our own universe, most of space is empty. We live on a rock that occupies a tiny volume in the larger space. Is this a huge fine-tuning and a pressing problem of theoretical physics? Of course not! It is clear that we will find ourselves in a place with structure. This picture further predicts that a non-zero CC should be seen, since there is no reason for us to live in an atypical part of the universe, as long as there is structure. It has further been recently realized in string theory that there are many possible vacuum states with differing values of the cosmological constant, and this "landscape" offers a natural home for anthropic reasoning." [4]

"Many physicists are repulsed by anthropic arguments, as they appear to limit in principle what we can predict about the fundamental properties of the universe. This seems wrong to me—the history of physics is filled with questions that people thought were "fundamental" but which turned out not to have any fundamental significance. For instance, up to the time of Kepler, the size of the orbits of the planets were thought to reflect something deep in creation, and people tried to find various schemes for predicting the orbits. We now know that there is nothing particularly significant about planetary orbits, they are historical accidents of the way the solar system formed. It may be that our current focus on the CC problem is similarly wrong, and that there is nothing deep about why the CC is small—it is just a historical accident of the corner of the megaverse we ended up in, where structure could have formed." [4]

"So, I don't have knee-jerk reaction against anthropic arguments; however, I do not believe they have yet come close to solving the CC problem either. The problem is that while

Weinberg's argument is beautiful if the only parameter that is anthropically varied is the cosmological constant, however, it is very unlikely, especially in a stringy landscape picture, that this is the case. More likely all parameters are varied. In that case, unless some very special properties for the landscape are assumed, there are still many many regions of the megaverse with structure and with much larger values of the CC than we observe. [4]

"(IV) Something we haven't thought of yet. I think this is the most likely possibility!" [4]

As previously discussed, the cosmological constant problem solution had been to assume that the cosmological constant was exactly zero. Thus, a fine-tuning problem for cosmology was created by Type Ia supernova measurements.

According to cosmologists, when they look at the geometry of space with regards to the Friedmann–Robertson–Walker model, the geometry of space depends on whether the mass/energy density is greater than, equal to, or less than a "critical value." Thus, the cosmological constant contributes to *the ratio of the mass/energy density of the universe to this critical value.* This clearly shows that the type Ia supernova data was consistent with a flat universe. In others words, the contributions from ordinary matter (protons and neutrons), from dark matter, and from the cosmological constant to the ratio of the mass/energy density of the universe to this critical value, were adding up to something close to one. Inflation was no longer in jeopardy at least as far as its prediction for the ratio of the mass/energy density of the universe to this critical value was concerned.

It turns out, however, that the value of the cosmological constant needed is theoretically unnaturally small. Many scientists thought that there was some systematic error in the supernova data because of this. For example, perhaps unseen intergalactic dust was making the supernova appear dimmer.

But hold on! Some astronomers and physicists believe that an inconstant constant may be the only way to explain the accelerating expansion of the universe phenomena.

Inconstant Constant

Mount Everest may crumble, the Leaning Tower of Pisa may tumble, but some physical constants are set in granite, aren't they? According to astronomers, the speed of light is the same everywhere in the universe and has been since the Big Bang. The same is true for the gravitational attraction between two masses and the charge of an electron. In fact, astronomers and physicists can only talk with any certainty about the Big Bang and what they see in other galaxies. Of course, they can only do this if they assume the physics they see through their telescopes is the same as what they see in their labs.

But that assumption was challenged recently by a team of astronomers and physicists in Australia, England, and the United States. As mentioned earlier in the chapter, the researchers found evidence that over the past ten billion years, the strength of the bond between an atomic nucleus and its surrounding electrons has changed by one part in 100,000. If the finding holds up, the astronomers and physicists are going to have to put a lot of physics in the garbage can.

The speed of light ten billion years ago is impossible to measure directly, so researchers focused on what is known as the fine-structure constant. The constant is actually made up of three other constants—the speed of light, the charge of the electron, and Planck's constant (the ratio between the energy and frequency of radiation). When all three are combined so that their units of measurement cancel one another out, what's left is a number: approximately 1/137 or .0072992.

Next, the researchers gathered measurements of light frequencies from 72 distant quasars, in order to see if the fine-structure constant (or any of its constituent parts) had changed.

They then looked for instances where intervening clouds of gas between Earth and the quasar absorbed some of the light. The results were unsettlingly clear: The more distant the gas cloud, the more the usual pattern of light absorption was altered. Clearly, the atoms in those clouds behaved differently from those closer to home.

On the other hand, researchers from Hawaii had looked for just the same change in absorption in the mid-1990s and did not find it. The researchers, therefore, hesitate to read too much into the new results: They are so surprising that you really have to nail it to the wall before you believe it. Measurements of such precision, stretching over so many light-years, can be thrown off by the smallest errors. But if that's the case, why aren't the differences more random? Why did they all change in the same direction?

"I believe the question is about the apparent variation of the fine structure constant as a function of cosmic time," says Dr. Gaudi. "This variation is measured by looking for changes in the wavelength of absorption lines, which depend on the value of the fine structure constant, as a function of look-back time. A few years ago, there was a report of the positive detection of a change of this constant via the measurement of a large number of absorption lines. This result was considered extremely surprising, to say the least, and therefore was treated with some skepticism. On the other hand, the authors worked quite hard to test the reality of this signal, and found that no more mundane reasons could explain the result. However, it's important to realize that the measurement was hard, and was made using a somewhat unconventional technique that is more prone to problems. Therefore, there were reasons to believe that some unaccounted-for error was resulting in a false detection. Furthermore, a recent independent study that did not find this variation seems to support this interpretation." [3]

An inconstant constant may seem contradictory. But some physicists believe that an inconstant constant may be the only way to explain puzzling phenomena—such as the accelerating expansion of the universe. However, it will be tricky to find out which component of the fine-structure constant has changed, even if the findings hold up. Presently, physicists can't verify whether it's the speed of light, or the charge of the electron, or all of them. Physicists are scrambling to find measures with which to check each component independently, but there are a lot of theories out there!

For instance, during the past few years (1999–2003), exciting, new, detailed measurements of the cosmic microwave background radiation also reveal that the universe is accelerating. It appears, though, that the cosmological problem exists and that the cosmological constant is non-zero. But wait! What about quintessence?

Quintessence

Theorists recently developed a new way of driving the acceleration of the universe called quintessence that may solve the preceding problem. The idea uses a scalar particle field. A particle field is the quantum source of an elementary particle. Basically, the vibrational modes of the scalar particle field produce one or more (possibly moving) elementary particles of a particular type. For example, the quantum electromagnetic field is the source of all photons. The quintessence scalar field produces both a mass/energy density and a cosmological constant: something that is sometimes called "dark energy." (See sidebar, "Dark Energy Explained.") So that the contribution to the cosmological constant changes with time, the dynamics are adjusted. It is then possible to arrange things so that the cosmological constant is close to its large, natural value in the early universe, but diminishes as the universe evolves to a value that is small and acceptable today. The idea of quintessence is speculative, but is the only reasonable way theorists have found to explain the conclusions drawn from the Type Ia supernova data discussed earlier.

In other words, cosmologists now believe that roughly two-thirds of the stuff that makes up the universe consists of mysterious "dark energy" (not to be confused with the equally puzzling, though less abundant, dark matter). As previously mentioned, a prime candidate for this dark energy (which pushes the universe apart ever faster) is a hypothetical form of matter that permeates all space, called "quintessence."

Dark Energy Explained

Two recent discoveries from cosmology prove that ordinary matter and dark matter are still not enough to explain the structure of the universe. There's a third component out there, and it's not matter but some form of dark energy.

The first line of evidence for this mystery component comes from measurements of the geometry of the universe. Einstein theorized that all matter alters the shape of space and time around it. Therefore, the overall shape of the universe is governed by the total mass and energy within it. Recent studies of radiation left over from the Big Bang show that the universe has the simplest shape—it's flat. That, in turn, reveals the total mass density of the universe. But, after adding up all the potential sources of dark matter and ordinary matter, astronomers still come up two-thirds short.

The second line of evidence suggests that the mystery component must be energy. Observations of distant supernovae show that the rate of expansion of the universe isn't slowing as scientists had once assumed; in fact, the pace of the expansion is increasing. Unless a pervasive repulsive force constantly pushes outward on the fabric of space and time, this cosmic acceleration is difficult to explain.

Why dark energy produces a repulsive force field is a bit complicated. Quantum theory, according to cosmologists, says virtual particles can pop into existence for the briefest of moments before returning to nothingness. That means the vacuum of space is not a true void. Rather, space is filled with low-grade energy created when virtual particles and their antimatter partners momentarily pop into and out of existence, leaving behind a very small field called vacuum energy.

According to cosmologists, that energy should produce a kind of negative pressure, or repulsion, thereby explaining why the universe's expansion is accelerating. Consider a simple analogy: If you pull back on a sealed plunger in an empty, airtight vessel, you'll create a near vacuum. At first, the plunger will offer little resistance, but the farther you pull, the greater the vacuum and the more the plunger will pull back against you. Although vacuum energy in outer space was pumped into it by the weird rules of quantum mechanics (not by someone pulling on a plunger), this example illustrates how repulsion can be created by a negative pressure.

Like inflation, quintessence is thought by cosmologists to have somehow originated when the universe was just 10^{-35} seconds old, and it is also driven by a scalar field whose energy varies gradually. (See sidebar, "Inflationary Scalar Fields Model.") According to cosmologists, the resemblance is not coincidental, because the idea was modeled after inflation. The big difference is the energy and time scale. Inflation occurred quickly at very high energies, whereas the scalar field responsible for quintessence operates at much lower energies over a much longer time frame. It looks like the universe is sandwiched between two periods of cosmic acceleration.

Inflationary Scalar Fields Model

As is typical in most cosmological models, what if inflation is driven by two scalar fields instead of just one? The idea makes sense to physicists, who indicate that scalar fields often operate in concert.

According to physicists, previous inflationary models don't do a good job of ending inflation. A switching mechanism was the main thing lacking. A dramatic improvement is the idea of having a separate field to turn off inflation.

Today's physicists have devised a so-called "inflationary scalar fields model." One field in this model provides the bulk of the energy density needed to drive inflation, while the second field acts as a trigger to turn it off. According to Dr. Alan Guth, the model works like a seesaw. In other words, the energy field holds down one end while the second field presses on the other end with increasing force, as if sand were being dropped on it. Eventually the whole thing tips over and inflation ends. [5]

Dr. Guth acknowledges that in reality there may be even more than two scalar fields at play. He indicates that supersymmetry models usually have a large number of scalar fields. In that case, the inflationary scalar fields model might be an improvement over the status quo, but it is still an oversimplification. [5]

So, if cosmologists look for a sharp peak at short wavelengths in the spectrum of the universe's microwave background radiation, they will find that this is one of many ways to find out if the model is right. Unfortunately, that peak will be hard to detect for technical reasons. But, according to cosmologists, the inflationary scalar fields model predicts the creation of primordial black holes in the early universe, long before other structures formed. Therefore, studies that indicate galaxy formation is "seeded" by primordial black holes (an issue yet to be resolved) might provide support for the inflationary scalar fields model, long before the completion of painstaking microwave measurements.

According to Alan Guth, the currently observed acceleration of the universe (if it's real, and it certainly appears so) provides direct evidence that gravity can act repulsively. [5]

That gives cosmologists more confidence right now that inflation is correct.

Finally, some researchers believe there may be an even stronger link between inflation and quintessence. They have proposed that the same scalar field could drive both inflation and quintessence. According to cosmologists, the premise of this model is that the field responsible for inflation operates in a high-energy regime. But the field's effects (masked by other more energetic phenomena) do not become apparent until much later in the universe's history, as its intensity drops off quickly.

NOTE

Cosmologists are not ready to say that this is better than other models, but at least it's a testable scenario.

Conclusion

The parameters that govern the cosmology of our universe are now determined with unprecedented precision, especially when all the recent astronomical observations are taken into account. The current estimated age of the Universe is about 15 billion years, with an uncertainty of less than 500 million years or so! The contributions to *"the ratio of the mass/energy density of the universe to this critical value"* from the cosmological constant, from dark matter and from proton/neutron, are respectively about 0.7, 0.25 and 0.045. The Hubble expansion rate is 0.69 kilometers per second per megaparsec (to an accuracy of 0.02). Results are consistent with inflation, with a flat universe (regular flat space) to within 10%, and the spectrum of density fluctuations is the predicted one within experimental errors.

Finally, the earlier 20th century painted a simple picture of our world as a de-accelerating, expanding universe, driven by the mass of ordinary matter. This has been supplanted by a new cosmology in the early 21st century: There is now some evidence to support all of this, the evidence being that inflation took place at less than a tiny fraction of a second during which space underwent an enormous expansion. During the past 15 billion years or so, space has been expanding, but at increasing rates: The universe is accelerating, in other words. According to theoretical cosmologists, most of this expansion is driven by a cosmological constant and by mysterious, unknown dark matter and not by the mass in protons and neutrons. It's a weird, weird world after all!

References

[1] "Astronomy Picture of the Day," [http://apod.gsfc.nasa.gov/apod/archivepix.html], NASA Headquarters, 300 E Street SW, Washington DC 20024-3210/Jet Propulsion Laboratory, 4800 Oak Grove Drive, Pasadena, California 91109/NASA/Goddard Space Flight Center, Greenbelt, MD 20771, USA, 2003.

[2] Dr. Richard Muller, Physicist, Physics Department, University of California, Berkeley, CA 94720.

[3] Dr. Scott Gaudi, Astronomer, Harvard-Smithsonian Center for Astrophysics, 60 Garden Street, Cambridge, MA 02138.

[4] Nima Arkani-Hamed, Physicist, Harvard University Department of Physics, Jefferson Laboratory of Physics, 570 Jefferson, 17 Oxford Street, Cambridge, MA 02138.

[5] Alan Guth, physicist, (MIT, Cambridge, MA) *The Inflationary Universe: The Quest for a New Theory of Cosmic Origins*, published by Perseus Books, 1997 [quoted with permission of the author].

Physics and Astrophysics

Gravity: The Construction of a Consistent Quantum Theory of Gravity

"Gravity—It's not just a good idea, it's the Law!"
—NASA briefing slide

Current theories of gravity are based on the geometric curvature of space. Current theories of other fundamental forces in the universe are "quantum field theories," where particles pass other particles back and forth among themselves to interact. Physicists and astrophysicists know that geometric gravity theories conflict with quantum field theories, and that this conflict means that they don't know what happens under extreme conditions.

What Is Gravity?

Gravity is the odd force out when it comes to small particles and the energy that holds them together. When Einstein improved on Newton's theory, he extended the concept of gravity by taking into account both extremely large gravitational fields and objects moving at velocities close to the speed of light. These extensions lead to the famous concepts of relativity and space–time. But Einstein's theories do not pay any attention to quantum mechanics, the realm of the extremely small, because gravitational forces are negligible at small scales, and discrete packets of gravity, unlike discrete packets of energy that hold atoms together, have never been experimentally observed.

Nonetheless, there are extreme conditions in nature in which gravity is compelled to get up close and personal with the small stuff. For example, near the heart of a black hole, where huge amounts of matter are squeezed into quantum spaces, gravitational forces become very powerful at tiny distances. The same must have been true in the dense primordial universe around the time of the Big Bang.

What Is a Black Hole?

A black hole is an object whose gravitational pull is so great that nothing (not even light) can escape from it. At one time, black holes were entirely theoretical. But now, their existence, both in our own galaxy and elsewhere in the universe, has been confirmed. A black hole forms when a massive star dies and collapses under its own mass. (See sidebar, "True Identities.") Some black holes, billions of times more massive than the sun, lie at the center of distant galaxies. Others may have been forged in the first fiery moments of the universe.

NOTE

Black hole theory is far from new. The idea began in 1783 when the Rev. John Michell used Newton's theory of gravity to predict the possibility of "dark stars."

True Identities

When giant stars die, what happens to them? Do they crumble to nothingness, forming black holes that slowly gnaw away at their neighbors?

The answers hang high overhead in the constellation Cygnus, the Swan. There, a distant blue star, viewable through binoculars, whirls furiously around a seemingly empty spot. X-ray telescopes have revealed that the vacancy is actually a brilliant source of radiation, called Cygnus X-1. Most astronomers believe they are watching gas from the visible star falling toward a black hole, emitting a blare of X-rays before it reaches the point of no return.

Day by day, researchers are learning more about these ultradense objects whose extreme gravity folds space–time like a taco. Recently, the orbiting Rossi X-ray Timing Explorer observed odd flickerings from another likely black hole with the catchy name GRO J 1655-40, as shown in Figure 4.1.[1] Like many distant quasars, this object sends geysers of atom fragments spewing at right angles to a disk of hot, spinning gas. But GRO J1655-40 is much smaller and lies just 10,000 light-years away, inside our own galaxy.

A black hole seven times the mass of the sun seems responsible for this sci-fi-like scene. One clue is GRO J1655-40's X-ray emission, which varies in brightness 300 times each second. That's just what physicists would expect from debris orbiting 40 miles above a black hole's event horizon, the point where even light cannot escape. The big news is that Rossi's sensors picked up a second tone at 450 cycles per second, indicating the presence of additional flotsam just 30 miles from the horizon. For gas to orbit so close to a black hole, the entire space–time continuum in that area must be rotating, which means the black hole must be rotating as well. This is the first sign that black holes don't just make our heads spin, they do it themselves.

In an earlier effort, researchers teamed Rossi with three other space observatories (the Hubble Space Telescope, the Extreme Ultraviolet Explorer, and the Chandra X-ray Observatory) and aimed all four at another presumed black hole, XTE J1118+480, which pivots around a sun-like star. The results showed that the surrounding disk of spiraling gas halts mysteriously far from the

Figure 4.1 In the center of a swirling whirlpool of hot gas is likely a beast that has never been seen directly: a black hole. Studies of the bright light emitted by the swirling gas frequently indicate not only that a black hole is present, but also likely attributes of the black hole. The gas surrounding GRO J1655-40, for example, has been found to display an unusual flickering at a rate of 450 times a second. Given a previous mass estimate for the central object of seven times the mass of our sun, the rate of the fast flickering can be explained by a black hole that is rotating very rapidly. What physical mechanisms actually cause the flickering (and a slower quasi-periodic oscillation [QPO]) in accretion disks surrounding black holes and neutron stars remains a topic of much research.

black hole. The disk's edge lies 600 miles out, instead of circling at a point 25 miles from the event horizon, as scientists expected. Intense radiation apparently heats the disk, swelling it into a huge bubble. Such observations let researchers visualize something nobody will ever see in person: the physical violence surrounding these cosmic rat traps.

How Stars Become Black Holes

Isolated black holes cannot be seen because nothing escapes from them, including radiation. If, however, a black hole accretes matter, such as from a close stellar companion, radiation is emitted from an accretion disk surrounding the black hole that can give physicists clues to the black hole's existence.

Care must be taken not to confuse radiation coming from a stellar black hole with that coming from neutron stars. The differences can be subtle, particularly since these stellar systems lie at great distances and the radiation physicists receive from them is faint. Besides, physicists' interpretations of the observations rely on theories that are not well tested. Here, physicists describe three characteristics that should help them identify stellar black holes:

1. The most reliable method of ascertaining that an X-ray binary system contains a black hole is by measuring the mass of the compact star. This requires, however, knowing a number of parameters about the binary system that are difficult to pin down with confidence—the separation of the two stars, their orbital period, the inclination of the orbit to our line of sight to the system, and the mass of the secondary star. If the mass of the compact star turned out to be approximately three solar masses or greater, the compact star would almost certainly be a black hole. If it turned out to be less, it would be a neutron star.
2. A stellar black hole has a deeper gravitational potential well than a neutron star, and matter falling toward it should form an accretion disk that is heated to higher temperatures than those observed in the case of neutron stars. Therefore, one expects the radiation coming from a black hole to consist of very high-energy X-rays and, possibly, gamma rays.
3. The X-ray flux from some X-ray binary systems flickers irregularly on time scales as short as 1/1,000 of a second. Such flickering is not observed in

> systems in which the compact star is known to be a neutron star. The flickering could come from the inner regions of accretion disks surrounding black holes. [1]

Furthermore, many black holes have similar masses to regular stars. They are too dark to be seen directly, but astronomers infer their presence by monitoring their gravitational effects on nearby bright stars. Sometimes, a black hole's gravitational pull "captures" material from a companion star (see Figures 4.2 and 4.3). [1] This material becomes agitated and heats up,

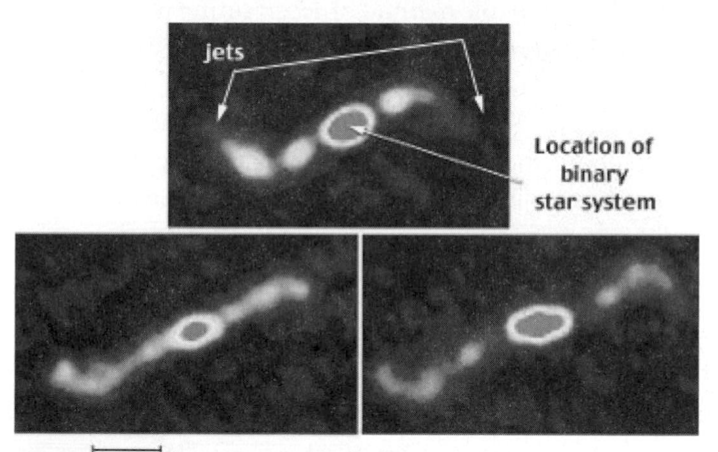

Figure 4.2 Three radio images of jets of high-speed gas from black hole candidate SS433 taken on December 7 and 22, 1991, and on January 4, 1992. SS433 is a binary star system consisting of a massive star (probably more than 10 solar masses) and a black hole candidate. A strong stellar wind blows from the massive star. Some of the wind's matter is pulled toward the black hole and forms an accretion disk around it. Two narrowly focused jets of gas are ejected outward in opposite directions from the inner edge of the accretion disk. The speed of the jets is approximately one-fourth that of light. For reasons not well understood, the jets of SS433 precess once every 162.5 days (they wobble like the axis of a spinning top). The continued ejection of new material and the precession give the jets their S-shaped morphology. (Note: Some astronomers think the compact star in SS433 is not a black hole, but a neutron star.)

Figure 4.3 SS433 is one of the most exotic star systems known. Its unremarkable name stems from its inclusion in a catalog of stars which emit radiation characteristic of atomic hydrogen. Its very remarkable behavior stems from a compact object, a black hole or neutron star, which has produced an accretion disk with jets. As illustrated in this artist's vision of the SS433 system based on observational data, a massive, hot star (left) is locked in a mutual orbit with a compact object. Material transfers from the massive star into an accretion disk surrounding the compact object blasting out two jets of ionized gas in opposite directions—at about 1/4 the speed of light! Radiation from the jet tilted toward the observer is blue shifted, while radiation from the jet tilted away is red shifted. The binary system itself completes an orbit in about 13 days while the jets precess (wobble like a top) with a period of about 164 days. Are the jets from SS433 related to those from black holes at the centers of galaxies?

causing it to emit X-rays that astronomers can also detect (see Figures 4.4 and 4.5). [1] For example, as shown in Figure 4.4, Cygnus A pumps out 10 million times more energy than our neighboring galaxy, Andromeda. How is that possible?

According to Physicist and Astronomer Dr. Andrea Ghez, "The jets are powered by accretion of material toward the central black hole. Other galaxies can be much quieter (lack jets), if their central black holes are not be fed much material." [4]

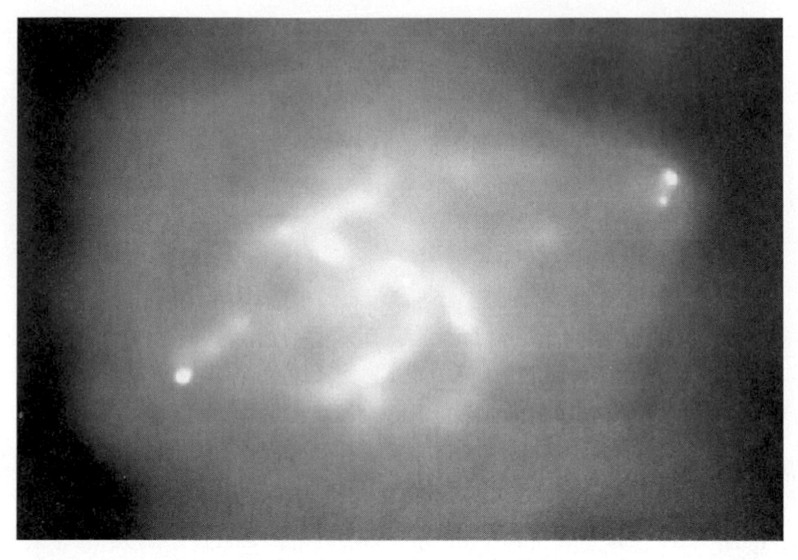

Figure 4.4 Amazingly detailed, this false-color X-ray image is centered on the galaxy Cygnus A. Astronomers believe that the two lobes (located lower right) of the active galaxy Cygnus A are produced by jets of material blasted from the poles of a supermassive black hole located at its center. Recorded by the orbiting Chandra Observatory, Cygnus A is seen here as a spectacular high-energy X-ray source. But it is actually more famous at the low-energy end of the electromagnetic spectrum as one of the brightest celestial radio sources. Merely 700 million light-years distant, Cygnus A is the closest powerful radio galaxy and the false-color radio image shows remarkable similarity to Chandra's X-ray view. Central in both pictures, the center of Cygnus A shines brightly while emission extends 300,000 light-years to either side along the same axis. Near light speed jets of atomic particles produced by a massive central black hole are believed to cause the emission. In fact, the X-ray image reveals "hot spots" suggestive of the locations where the particle jets are stopped in surrounding cooler, denser gas. The X-ray image also shows that the jets have cleared out a huge cavity in the surrounding gas. Bright swaths of emission within the cavity likely indicate X-ray hot material ... swirling toward the central black hole. Cygnus A pumps out 10 million times more energy than our neighboring galaxy, Andromeda.

Figure 4.5 The center of nearby giant galaxy M87 is a dense and violent place. In this 1994 photograph from the Hubble Space Telescope, a disk of hot gas was found to be orbiting at the center of this massive elliptical galaxy. The disk is evident at the lower left of the picture. The rotation speed of gas in this disk indicates the mass of the object the gas is orbiting, while the size of the disk indicates an approximate volume of the central object. These observations yield a central density so high that the only hypothesized object that could live there is a black hole. The picture also shows a highly energetic jet emanating from the central object like a cosmic blowtorch. The jet is composed of fast moving charged particles and has broken into knots as small as 10 light-years across. Furthermore, by studying the speed of the material rotating around the center of M87, astronomers have been able to calculate the enormous mass that must exist at its center. Again, the fact that this mass is concentrated in such a relatively small area means that it can only be a black hole.

So far, the best candidate for a stellar-mass black hole is V404 Cygni, discovered in 1989 (see Tables 4.1 and 4.2).[1]

Table 4.1 Black Hole Candidates Large and Small

Object	Mass	Radius of Event Horizon
V404 Cygni	12 Sun Masses (24,000 trillion trillion tons)	24 miles
Cygnus X-1	20 Sun Masses (40,000 trillion trillion tons)	40 miles
Center of Milky Way	1 Million Sun Masses (2 billion trillion trillion tons)	2 million miles (about 10 light-seconds)
Center of M106	36 Million Sun Masses (72 billion trillion trillion tons)	72 million miles (the radius of Venus' orbit of the sun)
Center of M87	36 Million Sun Masses (6 trillion trillion trillion tons)	6 billion miles (60 times the radius of the Earth's orbit of the sun)
Mini Black Hole*	About 100 million tons	The size of an atomic nucleus

* Existence not proved yet.

Table 4.2 If Our Earth and Sun were Black Holes, How Big would They Be?

Object	Mass	Radius of Event Horizon
The Earth	7 billion trillion tons	About a third of an inch
Our Sun	300,00 Earth masses (2,000 trillion trillion tons)	2 miles

Astronomers have identified about two dozen X-ray binary systems that satisfy one or more of the characteristics that have just been described. These systems may contain black holes, although at present, astronomers can't be sure. They are, therefore, called black hole candidates (see Tables 4.1 and 4.2).[1]

Into The Heart of a Black Hole

Throw a ball into the air and gravity will bring it back down to Earth. But, throw the ball fast enough (around 25,000 mph), and it will escape from Earth's gravity into space. More massive objects than the Earth have stronger gravitational fields that demand higher escape velocities (EVs); the sun's EV, for example, is about 100 times the Earth's.

Black holes are bodies so massive and dense that their theoretical EV is greater than the speed of light itself. This means that not even light can escape from such bodies, rendering them effectively invisible—hence the name, black hole.

NOTE

Black holes have a reputation for sucking in everything around them, but they are no more deadly than any other object of the same mass—if you keep your distance.

A black hole is formed when a very large star (at least 30 times the mass of the sun) dies and collapses under its own weight. It shrinks down to an infinitely dense point known as a *singularity,* at which point conventional mathematics can no longer cope. Around the singularity is an imaginary circle called the *event horizon,* beyond which all light from the object is turned back by the force of its own gravity. And, since Einstein proved that gravity can not only capture light, but can also distort time and space, inside the event horizon the entire concept of space and time completely breaks down.

NOTE

In 1915, Einstein's Theory of Relativity predicted so-called "Schwartzschild singularities." They were renamed "black holes" in 1967.

Inside a Black Hole

You could never see a person actually fall into a black hole: as they approached the event horizon, time would slow to the point where it would take infinitely long to reach it! In the mean time, the black hole's gravitational pull on light would make him appear to fade away. As for the person, who knows? All physicists know for sure is that in a black hole, reality as we know it ceases to exist. With that in mind, the following theoretical process (theoretical, because no one has actually observed what will happen) would take place if you fell inside a black hole and were actually able to observe what was happening. In reality, because of the intense deadly radiation near a black hole, you would be dead long before you ever reached the event horizon.

1. Event Horizon: The black hole's gravitational "boundary" beyond which light cannot escape. Once crossed, there is no return.
2. Falling In: If you fell feet first into a black hole, the difference between the gravitational pull on your head and your feet would be so powerful that it would instantly stretch you out.
3. Looking Back: If you could look back as you fell, you would see the future history of the universe flash before your eyes. But, once inside the event horizon, you would be unable to communicate anything you saw back to anyone outside of it. Here space and time become distorted.
4. Oblivion: As you neared the singularity at the center of the black hole, you would feel yourself being torn apart atom by atom.
5. Singularity: At this point, everything astronomers have ever known about the universe breaks down.

6. Black Hole Mystery: However, under certain conditions, a person could fall *through* a black hole and pop out the other side. But, from an observer's perspective, the process would take an infinite amount of time. So what would the person see on exiting?

NOTE

By the way, black holes are not completely black. According to English physicist Stephen Hawking, black holes give off faint radiation, which implies that eventually their mass must evaporate in a haze of subatomic particles.[2]

Now, suppose you could harness the energy of a black hole? What do you think would happen? Let's take a brief look.

Harnessing the Energy of a Black Hole

Far from consigning great tracts of the universe to oblivion, black holes could actually be beneficial. Mathematicians and physicists have calculated that a technologically advanced civilization could extract enough energy from a black hole to supply all of its power requirements.

The idea requires the civilization to build a structure encircling the black hole, from which a stream of fuel can be "fed" into it. At the event horizon, each particle in the fuel stream is then split into two fragments—perhaps using suitably placed explosive charges. The mathematicians and physicists have calculated that at the very moment one of these fragments falls over the black hole's event horizon, the other fragment will throw out a kick of pure energy supplied by the annihilation of its partner. This is something like the boost you get after jumping from a carousel.

The theory goes on to suggest that the "energy kicks" could be collected by the structure around the black hole, but energy extracted in this way would not be limitless. Calculations by mathematicians and physicists show that a black hole's rate of spin is steadily reduced. There comes a point when the black hole stops spinning, and its energy-producing properties vanish.

If the same civilization could build a craft capable of withstanding the enormous forces that apply in and around a black hole, scientists have speculated that it might be possible to cross the event horizon into a new region of space–time through what is known as a *wormhole.* So living next to a black hole could not only save on utilities—it could also provide the ultimate travel ticket to the rest of the universe, past, present, and future.

So, what would really happen if Earth were threatened by a black hole? Can you imagine?

What If Earth Were Threatened by a Black Hole?

Thankfully, the black holes astronomers know about are no danger to Earth, since only objects within a black hole's event horizon can actually be sucked into it. The closest stellar-sized black hole to the Earth is at least 15 light-years away, and supermassive black holes are confined to the centers of distant galaxies, so they pose no conceivable threat. Or do they? Dr. Ghez says no: "There is no danger posed from either type of black hole for the Earth." [4]

Fifteen light-years is not all that far away on a cosmological scale. Suppose there was a rogue black hole orbiting the galaxy, and that in a few hundred years was close enough to our solar system to disturb the Kuiper Belt objects that lie just beyond the orbit of Pluto. The gravitational disturbance of the black hole would send thousands of comets and asteroids into the solar system, causing chaos among the planets, as well as putting Earth in danger of being hit by multiple impacts—resulting in the mass extinction of our civilization.

NOTE

The Kuiper Belt lies beyond Pluto—a source of comets and believed to be the source of much of Earth's water and the simple chemical precursors of life. Pluto, the smallest planet, is actually a Kuiper Belt Object, a class of objects composed of material left over after the formation of the other planets. Pluto and other Kuiper Belt objects contain samples of the most ancient material in our solar system, the stuff that all the other planets, including Earth, were made of.

Even worse, suppose the black hole entered into our solar system? Could you imagine the chaos it would cause? Because of the intense gravitational pull from the black hole:

- Planets would be yanked from their orbit and flung from the solar system into a rogue orbit around the galaxy, or even flung out of the galaxy itself.
- Planets would be flung into the sun.
- Or worse yet, planets, including Earth itself, could be swallowed by the black hole. Imagine billions of terrified souls being crushed and torn apart atom by atom as the Earth swirls into the black hole and is pulverized.

On a lighter note, it has been suggested that there may be "mini black holes," formed shortly after the Big Bang, still scattered about our galaxy. As yet, none have been found; they remain purely theoretical. Or are they?

Creating Mini Black Holes in the Lab

If you are reading this part of the book, the Relativistic Heavy Ion Collider at Brookhaven National Laboratory on Long Island in New York has still not destroyed the Earth. Back in

1999, some scientists argued that the collider has the potential, at full power, to create a planet-swallowing black hole.

At the time, this was labeled as the most dangerous event in human history—and actually, it still has the potential to be the most dangerous ongoing experiment. Of course, despite the danger, this has the world's physicists tremendously excited. Scientists believe they can eventually use the collider to duplicate the conditions that prevailed milliseconds after the Big Bang, when the universe consisted of a primordial soup called the quark-gluon plasma. Brookhaven scientists continue to think that by colliding gold ions at extremely high speed, they can create a tiny, fleeting version of quark-gluon plasma to gain a better understanding of the origins of the universe.

Sounds like fun. The only problem according to physicists, is that, as theorized by Steven Hawking, from this quark-gluon plasma, other forms of matter are also produced. [2] The most dangerous being a black hole.

As previously discussed, black holes have zero volume and infinite density. They sit in deep space, trapping everything that comes near enough (crossing inside what's known as the Schwartzschild radius) and letting nothing escape, not even light.

But Brookhaven National Laboratory is not the only place where physicists are trying to create laboratory-made black holes. In 2002, Europe's CERN particle accelerator near Geneva, Switzerland, started gearing up to create black holes that could become a bona fide reality before the decade is out.

Black Hole Factories

Risky business is not new to the scientific community. During World War II, Manhattan Project scientists speculated that the detonation of an atomic bomb might ignite the atmosphere, sparking a global conflagration that would extinguish life on Earth. Later calculations and tests, however, proved such concerns to be unfounded.

Similarly, the idea of creating black holes in a laboratory might sound like a doomsday scenario. After all, wouldn't a black hole start gobbling up everything in its vicinity, like the Pac Man-like creatures gone berserk in Stephen King's novel *Langoliers*, eventually swallowing the entire Earth?

NOTE

Some scientists believe that a lab-created black hole would be microscopic at first, but would grow exponentially, eventually obliterating Earth. The black hole would first eat its way down toward the center of Earth and consume it from the inside out. It would not be a good time to be around to see this. In the end, all of Earth would be consumed.

According to new calculations by a team of physicists, black holes might become standard fare at Europe's CERN particle accelerator near Geneva, Switzerland. According to the calculations, the upgraded Large Hadron Collider (LHC) at CERN could produce as many as one black hole every second once the accelerator is up and running around 2009.

It is possible that the LHC and future accelerators will become black hole factories. At the LHC, experimenters will slam together beams of protons traveling at close to the speed of light. Some of the stupendous energy of these collisions will be converted into massive elementary particles according to Einstein's famous equation $E = mc^2$, which states that energy can be converted into mass, and vice-versa. The LHC's collisions will be so violent that they will create particles that haven't existed in the universe since a trillionth of a second after the Big Bang. The LHC will also produce particles in such a small volume of space, with such ferocity, that it will produce miniature black holes a thousand times smaller in size than an atomic nucleus. That is if (and this is a big if) gravity is a stronger force than scientists currently believe.

Gravity might operate in these extra dimensions as well, if the universe contains extra spatial dimensions beyond the three that we experience in our everyday lives. Consequently, extra dimensions will dilute the strength of gravity, just as extra water will dilute a cup of coffee. Scientists have only been able to hypothesize their existence, because particle accelerators built before the LHC have been unable to create enough energy to open up these hidden dimensions. However, if extra dimensions do exist, the LHC will be powerful enough to probe them, generating mini black holes in the process.

Nevertheless, before you go out spending your life's savings, figuring the world will end in 2009, note that about 100 times per year cosmic rays slam into Earth's atmosphere with enough energy to create mini black holes. Yet these black holes quickly evaporate into a stream of elementary particles known as Hawking radiation (named after the celebrated British physicist Stephen Hawking, who first predicted black hole evaporation in the early 1970s).

Similarly, the LHC's black holes will also spontaneously evaporate so quickly (in 10^{-26} seconds), that they won't have time to gobble up any surrounding material. According to CERN physicists, the very fact that cosmic rays have not made us extinct in the billions of years of Earth's existence is the best proof that the LHC is completely safe to operate.

Now, why should we all not be reassured by this? The short answer is that the experiment is conducted by human beings—the same folks who brought you the internal combustion engine, which threatens to destabilize the planet's climate, and powerful antibiotics, which ultimately created an invincible staphylococcus bacterium.

In other words, human progress has always had a nasty habit of producing unintended consequences—usually because the prideful progenitors of progress insist on pooh-poohing any possibility of danger. Now, in recreating the beginning of the universe, scientists are essentially playing at being God—an unforgivable offense, punishable, as tragedians in the Bible and other literature have prophesied for centuries, by annihilation.

This doomsday scenario dovetails creepily with the speculation put forth by the late Carl Sagan in his book *Cosmos.* Sagan believed that we could never find evidence of life anywhere else in the universe, because the pattern of evolution has been the same everywhere: Life begins and evolves through millions of years to the moment when it destroys itself. The nature of consciousness is such that evolution itself is a doomsday machine.

Sagan also considered nuclear war the likeliest cause of destruction, but the creation of an annihilating black hole is more plausible. Not only does it explain the apparent absence of life anywhere else in the universe, it also explains the absence of any ruins of past civilizations. A black hole removes all traces of everything—including, the creating civilization's planet.

So why is this doomsday scenario of major importance? Basically, there's a moral obligation here by scientists to bring this type of scenario out into the general public's view. For once, maybe once in the history of the universe, we can avoid the end and live happily ever after.

Nonetheless, if CERN physicists are eventually able to detect Hawking radiation from mini black holes, the discovery will pave the way for the unification of the twin pillars of 20th century physics. This of course would be Einstein's general theory of relativity and quantum mechanics—while at the same time revealing the nature of any extra hidden dimensions.

Are There Additional Dimensions?

Wondering about the real nature of gravity leads eventually to wondering whether there are more than the four dimensions that physicists can easily observe. To get to that place, physicists might first wonder if nature is, in fact, schizophrenic: Should they accept that there are two kinds of forces that operate over two different scales—gravity for big scales like galaxies, the other three forces for the tiny world of atoms?

Poppycock, say unified theory proponents—there must be a way to connect the three atomic-scale forces with gravity. Maybe, but it won't be easy. In the first place, gravity is odd. Einstein's general theory of relativity says gravity isn't so much a force as it is an inherent property of space and time. Accordingly, Earth orbits the sun not because it is attracted by gravity, but because it has been caught in a big dimple in space–time caused by the sun and spins around inside this dimple like a fast-moving marble caught in a large bowl. Second, gravity, as far as physicists have been able to detect, is a continuous phenomenon, whereas all the other forces of nature come in discrete packets.

All this leads physicists to the string theorists and their explanation for gravity, which includes other dimensions. The original string theory model of the universe combines gravity with the other three forces in a complex 11-dimensional world. In that world (our world) seven of the dimensions are wrapped up on themselves in unimaginably small regions that escape our notice. One way to get your mind around these extra dimensions is to visualize a single strand of a spider web. To the naked eye, the filament appears to be one-dimensional, but at high magnification, it resolves into an object with considerable width, breadth, and depth. String theorists argue that we can't see extra dimensions because we lack instruments powerful enough to resolve them. We may never see these extra dimensions directly, but we may be able to detect evidence of their existence with the instruments of astronomers and particle physicists.

Bridging Quantum Mechanics and Gravity

Physicist Stephen Hawking identified a specific problem about black holes that requires a bridging of quantum mechanics and gravity before we can have a unified theory of anything. [2] "Even without worrying about the problem of unification of all the forces," explains Theoretical Physicist Lee Smolin (Perimeter Institute of Theoretical Physics, Waterloo, Canada), "we likely need the unification of just general relativity and quantum physics to understand what happens at black hole singularities, and why their horizons have entropy and temperature." [5]

Hawking maintains that the assertion that nothing, even light, can escape from a black hole is not strictly true according to some theoretical physicists. [2] Weak thermal energy does radiate from around black holes. Hawking theorized that this energy is born when particle–antiparticle pairs materialize from the vacuum in the vicinity of a black hole. [2] Before the matter–antimatter particles can recombine and annihilate each other, one that may be slightly closer to the black hole will be sucked in, while the other that is slightly farther away escapes as heat. This release does not connect in any obvious way to the states of matter and energy that were earlier sucked into that black hole; and, therefore, according to some theoretical physicists, violates a law of quantum physics by stipulating that all events must be traceable to previous events. New theories may be needed to explain this problem.

Quantum Theory of Gravity

A quantum theory of gravity would involve particles passing "gravitons" back and forth among themselves. This quantum theory would probably be a more accurate description of gravity, and might be accurate enough to describe the extreme conditions found at the center of a black hole as has been previously discussed. For now, let's first briefly look at how in a rather mechanical way Isaac Newton understood gravity:

- Objects with mass simply attract each other. The more mass they have, the stronger the attraction.
- The force weakens as the separation between bodies increases. This, by the way, explains why distant stars do not produce a significant pull on us on Earth—they are too far away.
- Gravity is a very weak force. Hence, only very massive objects such as the Earth and the sun can produce a sizeable gravity.

On the other hand, Einstein understood gravity in a more natural way. The essential idea is that space is dynamic. This means that space can stretch, bend, and be deformed. But what causes space to change its shape?

According to theoretical physicists, astrophysicists, and cosmologists, the answer is mass. Massive objects cause space to curve. Now, a second body moving in this curved splice will not move in a straight line. It must be accelerating by definition. Acceleration is the change in direction of motion or the change in speed, but the first object must be producing a force, since forces are things that cause accelerations. In this way, the Einsteinian viewpoint reproduces Newtonian gravity. In other words, gravity is the consequence of the curving of space. (See sidebar, "Listening for Gravity Waves in the Fabric of Space and Time.")

Listening for Gravity Waves in the Fabric of Space and Time

Physicists from around the world have been listening for gravity waves from two giant, laser-packed tuning forks that, after nine years of construction, are monitoring the most violent events in the universe: stars colliding, merging, collapsing, and exploding, and even the Big Bang itself. More than 82 years ago, Albert Einstein predicted the quarry of this $593 million project, near-twin detectors at Hanford (in south-central Washington state) and in Louisiana, collectively known as the Laser Interferometer Gravitational-Wave Observatory (LIGWO).

Recently, physicists have calculated that violent motions by massive objects generate quivers racing at light speed through the fabric of space and time—subtly altering geometry itself as they pass. So far, gravity waves have not been convincingly detected. Einstein's general theory of relativity would triumph if proof could be found. Of major importance would be the opening of a clear channel to titanic events difficult or impossible to see by light, radio waves, X-rays, or by the other signals of conventional astronomy.

Whispers

Though born in vast cataclysms, gravity waves would be but faint whispers when passing the Earth. According to physicists, the detectors need to be extraordinarily sensitive—and hence are easily misled by more mundane vibrations. The worst thing physicists could do is say they've seen a gravity wave and find out they're wrong.

Stretching across Hanford's treeless plain and skewering the bogs and piney woods near Livingston, LA, LIGWO's antennae are easy to see from passing airliners. Forming a letter "L" 2-1/2 miles long on a side, two concrete tunnels wider than the height of a man extend at right angles from a central laboratory. Protected inside are 4-foot-wide stainless steel pipes pumped to a near vacuum. Down their lengths shoot infrared laser beams that bounce at the ends off exquisite mirrors of fused quartz and reverberate in the tubes up to 100 times. The lasers ultimately merge back in the central facility.

According to physicists, relativity says that as a gravity wave passes, one of LIGWO's legs will stretch while the other shrinks—but not by much. Even the wave from a cataclysm in or near our own galaxy will make the legs oscillate just a few billionths of a trillionth of a percent in length. This is much less than the width of an atomic nucleus. However, the ripples will throw the beams detectably out of sync, if the mirrors, lasers, and all the optics work just right.

It's very difficult to keep the laser beams on target. Also, many other things besides gravity waves can knock them off kilter. A heavy footstep nearby can set the light waves trembling, even with delicate suspensions to isolate the optics from vibrations. Earthquakes knock the detectors offline entirely. At Hanford, physicists can tell how far apart the axles are on trucks out on Route 240 (the freeway a mile away).

Such delicacy is why two machines were built. No one will believe that a gravity wave, rather than some local disturbance, created a signal that is not seen at both places. Physicists have also used computers to predict the exact patterns of vibration that would reveal gravity waves. So far, the machines are swamped by outside noise. The physicists expected that. Overseas, somewhat smaller detectors are joining the effort and should further reduce false alarms.

Waves of Thunder

Few scientists doubt the waves are there to be seen. Astronomers already have indirect evidence for them: Radio signals from super dense neutron stars closely orbiting each other show they are spiraling inward exactly as fast as expected if energy is leaking away in gravity waves. When these stars (each more massive than the sun, but only six miles across) finally crash together, they should generate a thunderclap of waves. So should stars exploding or taking swan dives into gargantuan black holes at the centers of galaxies. Lingering shudders of the Big Bang may still shake the cosmos, and their patterns may winnow competing theories of the first moments of creation.

Thus, physicists are looking into regimes of nature that have never been looked at.

LIGWO's first-generation lasers and mirrors should pick up any neutron stars or black holes that happen to merge in the Milky Way and nearby galaxies, but by 2007, better mirrors made of sapphire and more powerful lasers will multiply its sensitivity thousands of times. That may be enough to detect such cataclysms throughout much of the universe, where they could occur daily. Plans for an even more sensitive orbiting laser array would get a boost, if Earth-based detectors work. It would detect shifts in distance among satellites millions of miles apart.

Yet veterans of the gravity-wave hunt have not forgotten what happened to the late physicist Joe Weber of the University of Maryland, who pioneered the field in the 1960s. Weber built instrument-be-decked metal bars weighing a ton or more, believing gravity waves would get them vibrating. He soon claimed lots of gravity waves, but similar detectors turned up nothing. Weber went to his grave in lonely certainty he was right, but the field faded into near disrepute. In the late 1980s, a few stubborn physicists persuaded the National Science Foundation to give it another, full-hearted try. LIGWO is the result.

How Gravity Is Produced by General Relativity

One analogy of how all of this works is what happens when you take a bowling ball and place it on a bed. As the ball is lowered, the surface around the bowling ball will depress (you can try this at home if you have a bowling ball and marble).

NOTE

Mass/Energy Curves Space–Time: A massive body such as the Earth creates a significant curvature of space–time. The situation is analogous to setting a bowling ball on a bed; the heavy ball depresses a nearby region of the bed's surface.

Now, if you then toss a marble onto the bed, the marble will roll toward the bowling ball. It is as though the marble is attracted to the bowling ball. In this analogy, the bowling ball is like the Earth, the marble is like an apple (the apple was used as part of the initial experiment by Isaac Newton) and the surface of the bed is like physical space. The marble goes toward the bowling ball when tossed, just as the apple heads toward the Earth.

NOTE

Curvature of Space–Time Leads to Acceleration: When another object moves through space–time, it feels the curvature created by the massive body. A marble thrown onto the bed moves toward the bowling ball along a curved trajectory and hits the ball. Because its motion is not along a line and its speed changes, the marble is accelerating.

So, with that in mind, what are some of the unusual features of quantum mechanics? Let's take a look.

Unusual Features of Quantum Mechanics

Quantum mechanics is an important discovery of the early 20th century. It governs the behavior of microscopic entities such as atoms and electrons. A feature of those tiny worlds is uncertainty. An electron circling an atom is never at a definite position. Rather, there is a "probability cloud" associated with it. The electron has a certain chance of being at a particular position, another chance at being at another position, and so on. The denser regions of the probability cloud represent regions where the electron is most likely to be found. Another feature of quantum mechanics is fluctuation. This is closely related to the preceding uncertainty. Like the motion of a leaf in a wind, the position of an electron or atom in the microscopic world seems to fluctuate, being here and there.

So, if the theory of quantum mechanics weren't valid, no one would be walking around with the communications technology that we have today. Furthermore, why do so many scientists ignore the far larger implications of quantum mechanics, like black holes, worm holes, time travel, parallel universes, and multi-universes? Let's take a look.

The Uncanny Implications of Quantum Mechanics

Although physicists might not agree with various interpretations of the uncanny laws of quantum theory, few of them deny their validity. The laws insist that the fundamental constituents of reality, such as protons, electrons, and other subatomic particles, are not hard and indivisible. They behave like both particles and waves. Furthermore, they can disappear and reappear out of nothing (a pure void). Physicists have even managed to teleport atoms, to move them from one place to another without passing through any intervening space. On the quantum scale (as if created by a besotted god), objects seem blurred and indistinct. A single particle occupies not just one position, but exists here, there, and many places in between. Most physicists agree that quantum theory is outlandish. It seems completely in conflict with the world of big physics according to Newton and Einstein.

Most physicists have chosen an easy way out to grapple with the contradictions: They restrict the validity of quantum theory to the subatomic world. Other physicists argue that the theory's laws must hold at every level of reality. The same weird quantum rules must apply to all of you, because everything in the world, including ourselves, is made of these particles, and because quantum theory has proved infallible in every conceivable experiment. Even if you don't realize it, you too must exist in many states at once. For example, there must be many versions of myself, Earth, and the entire universe. All conceivable variations on your lives, and all possible events, must exist. You live not in a single universe, but in a vast and rich multi-verse, or multi-universe. (See sidebar, "Shadow Universes.")

Shadow Universes

In the 1999 hit movie *The Sixth Sense,* the film's central character whispers, "I see dead people." The terrified kid, Cole Sear, is an eight-year-old boy who can see the invisible spirits of the dead. Ghosts exist alongside of us in the movie's universe, but can't interact with the living. The film raises a deep and abiding question in the realm of physics, despite the melodramatic, paranormal story line: How much of the universe do we live in? How much of it are we part of?

"If we can see a ghost," comments physicist Richard Muller, "then we are interacting with it. The ghost emits light, and we absorb it. The movie gives no insight. The 'universe' is what we live in." [3]

Every day, you live your life in a way that you consider to be normal, always believing that the world you experience is the world that is. Could there be more? Could there be a universe alongside this one, a parallel world that moves right through you?

As physicists and astronomers have pushed inward to the heart of matter and backward to the moment of creation, they have been thrown against this question. Their answers make it clear we are already missing a lot of the universe, with even more potentially unseen right under our noses.

Welcome to the invisible worlds of dark matter and shadow universes. Physicists who study matter on its most fundamental level have already found the seeds of a parallel universe. You have to go back to the beginning to understand it—all the way back. The universe was a roiling cauldron of energy, less than a billion billion billionth of a second after the Big Bang. It was a simple place, but in a sense, it was also a nasty place. Conditions were so extreme that none of the forms of matter physicists and astronomers are familiar with today could exist. Our world had yet to be born.

Physicists and astronomers know very little of what this fetal universe was like. They do know that as it expanded, the universe cooled and the forces and particles they recognize today froze out of the hyper-hot primordial stew the way ice crystals freeze out of liquid water. From formlessness, form coalesced. You might call this the "Big Freeze." The stuff we are built of, the stuff that went on to become stars and galaxies and planets, emerged at this specific moment in cosmic history. In the Big Freeze, the original simplicity of the early universe was broken, creating the potential for the rich structure of which we are so happy to be part. Our world, however, may not have been the only one created.

Good theoretical physicists never let a little thing like ignorance stand in their way. Despite the beyond-the-pale extremes of the early universe, they have sketched out how the story of the universe's birth might be told. There is one version of the story that includes a shadow world.

Two distinct families of matter are created in the Big Freeze. One kind of "stuff" constitutes the world theoretical physicists know, the protons and electrons and other denizens of the subatomic world that have combined into splendid forms like us. The other family of matter would represent the shadow world. It would exist alongside of, but separate from, our own world.

Normal and Shadow Matter

The two universes of normal and shadow matter could interpenetrate, but not interact. A rich shadow universe would evolve along with our own, in time. Shadow planets could form around shadow stars and be populated by intelligent shadow beings. Stranger still, since the two worlds wouldn't interact, shadow matter could exist alongside normal matter right here on Earth. How could two universes exist in the same place? To attack this question you first have to ask how the universe you live in exists at all.

Musing over the nature of reality can seem a pretty sterile affair. When some philosopher wannabe starts pondering the existence of a chair in front of us, there's a strong urge to take a cue from the World Wrestling Federation and hit him over the head with it. "There" you want to say. "Now how real do you think it is?"

Things are different in physics and astronomy. You can't take any answers for granted. For example, the chair that downed the philosopher is really 99% empty space. The atomic nuclei in the chair, where essentially all the mass resides, are as far apart from each other as stars are in a galaxy. So why, when you sit down, don't the atoms in your butt just sail right past the atoms in the chair? The answer gets you halfway to the shadow world. The atoms in the chair exert forces that reach across space and affect the atoms in your butt. Forces count for everything and matter is secondary, on the most fundamental level of fundamental physics.

"Throughout almost the entire history of science, states Physicist, Dr. Stephen Wolfram (Wolfram Research, Inc., Champaign, IL), "space has been viewed as something fundamental—and typically continuous. In his 'A New Kind of Science' documentation, Dr. Wolfram "suggests that space as we perceive it is in fact not fundamental, but is instead merely the large-scale limit of an underlying discrete network of connections. Models constructed on this basis then lead to new ideas about such issues as the origins of gravity and general relativity, the true nature of elementary particles and the validity of quantum mechanics." [6]

Atoms would not know that their companions exist, without some kind of push or pull (in other words, a force). Even a head-on collision between two particles would produce no effect without an exchange of forces. The two particles would pass right through each other, like a ghost walking untouched through a wall.

The universe we live in has just four ways to push or pull: gravity, electromagnetism, the strong nuclear force, and the weak nuclear force. These are the only forces that emerged from the Big Freeze. We all have direct experience with the first two forces. Gravity is what keeps your feet on the ground. Electromagnetism is the explicit reason why your butt doesn't slip through the chair (electromagnetic forces between chair atoms and butt atoms give both their appearance of solidity). The strong and weak nuclear forces are not so familiar because they operate only on subatomic scales. They are responsible for giving us the elements, allowing gold its luster and sulfur its stench.

The four forces bind the world together in its diversity of shape, structure, color, and form. However, things could have been different; there is nothing mystical about the number four. Like so much in life, the number and form of the forces that emerge from the Big Freeze has an element of randomness to it, and that's what makes the shadow world possible.

Feeling without Matter

Have you ever known unrequited love? If so, you already understand a deep principle of subatomic physics. You are attracted to someone, but that person doesn't know you exist. In the same way, some forms of matter don't "feel" all four forces. A neutron, for example, has no electric charge, so it doesn't care about electromagnetism. Create a new form of matter that doesn't feel any of the four forces and it wouldn't feel our world at all. It would exist, but be undetectable. If this exotic stuff had its own set of forces, then it could bind together in interesting forms and create a rich universe parallel to our own. Each universe would pass through the other without ever knowing of the other's existence.

In the mid 1980s, some physicists recognized the shadow world in equations describing the universe during the Big Freeze. The math hinted that as the universe cooled, all the normal forces and particles would get a shadow twin. Only gravity would be shared between the two worlds. Shadow electrons, for example, would feel only shadow electromagnetism and would be "neutral" to our world's version of that force. By having a full complement of forces available to it, the shadow world could evolve into as varied a place as the one we inhabit.

Forming Shadows

For now, unfortunately, the shadow world is only the stuff that dreams are made of. Shadow matter remains a theoretical curiosity, a shadow of a possibility existing only on paper. But, researchers are not giving up on the existence of a real parallel universe. In fact, they already know exactly where to find one. While shadow matter is simply an idea, dark matter is an inescapable fact.

As previously explained in Chapter 1, astronomers already know that most of the universe exists in a dark, invisible form. For more than 22 years, evidence has gathered steadily that well over 90% of all the stuff in the cosmos reveals itself entirely through the push and pull of a single force—gravity. Galaxies spin faster than can be accounted for by the pull of their own visible mass. Galaxies in clusters swarm around a common center at speeds too high to be gravitationally driven by the luminous matter astronomers can see. In these and countless other cases, astronomers have been forced to admit that something else is out there. The dark matter continuously provides the gravitational tugs that keep what astronomers can see in motion. As the third millennium opens, there can be little doubt that dark matter exists.

It is also clear that whatever dark matter is, we aren't part of it. At first astronomers believed dark matter was normal stuff like us, subject to all four universal forces. Dark matter was then just material not emitting much light, things like rocks or dead stars. As the sophistication of their instruments and theories progressed, astronomers have been forced to abandon that idea. Dark matter is "exotic" in a similar way to shadow matter. It does not respond to the full complement of forces born in the Big Freeze. It is not stuff like us. As you read this chapter, countless dark matter particles are streaming untouched through your body like sunlight streaming through a window.

Yet there is a big difference between dark matter and shadow matter. The dark stuff doesn't appear to have its own special forces to bind it together. All it can feel is gravity and perhaps the weak nuclear force. The weak force is, well, weak, and so for that matter is gravity. You need to gather a lot of stuff together before gravity becomes useful. You cannot, for instance, hold a person-sized lump of matter together with just gravity. With only these two forces at its disposal, no lumps, or at least no interesting lumps like trees or people, can form in the dark universe. Even making a star or planet is difficult with gravity alone—the electromagnetic force is crucial to slowing material down enough to hold it in one place. The dark matter universe can hold no dark stars or dark planets populated by dark matter aliens.

So, does a parallel universe exist? The answer is yes—but we are it.

With more than 90% of the cosmos composed of an invisible form of stuff that barely notices our world's presence, it seems that we are the outsiders. All the luminous stars and galaxies stretching across space and time, all of our kind of stuff, is no more than tiny bubbles of foam riding on a vast dark matter sea. In a very real sense, we do not even live in the universe that is. We are an afterthought.

Diamonds Are Forever

That may seem like the ultimate cosmic insult, the grandest statement yet of our insignificance. But just the opposite is true. After all, a diamond's worth lies in the fact that it is not dirt. It is uncommon. In all the dark matter world there can never be anything as wondrous as a star, a meteor, or even a blade of grass. Our share of the universe may be small, but it is, by far, the only one worth the time.

The idea of multi-universes takes some getting used to, especially when one pauses to consider what it means on an everyday level. It solves once and for all the ancient question of whether we have free will, for starters. The bottom line is that the universe is open. We have free will (see Chapter 13 for a detailed discussion of free will) in the relevant sense of the word.

According to theoretical physicists, we also have every possible option we've ever encountered acted out somewhere in some universe by at least one of our other selves. Unlike the traveler facing a fork in the road who is pondering why he or she could not travel both and be one traveler, we take all the roads in our lives. This, of course, has some unsettling consequences.

Driving a car, for example, becomes extremely hazardous, because it's almost certain that somewhere in some other universe the driver will accidentally hit and kill a child. So, should you never drive? It's really impossible to control the fate of your other selves in the multi-verse. But, if you're cautious, other copies of you may decide to be cautious. There's also the argument that because the child's death will happen in some universes, you ought to take more care when doing even slightly risky things.

These represent some very startling views by theoretical physicists. Spurred by these ideas, researchers around the globe are attempting to construct a fundamentally different type of quantum computer that is powerful almost beyond imagining.

Physicists may argue about what the theory means, but fortunately for the rest of us they have no qualms about working with it. By some estimates, 40% of the United States' gross national product is said to derive from technologies based on quantum theory. Without the insights provided by quantum mechanics, there would be no cell phones, no CD players, no portable computers. Quantum mechanics is not a branch of physics; it is physics.

And yet 103 years after it was first proposed by German scientist Max Pianck, physicists who work with the theory every day don't really know quite what to make of it. They fill blackboards with quantum calculations and acknowledge that it is probably the most powerful, accurate, and predictive scientific theory ever developed. But despite the unrivalled empirical success of quantum theory, the very suggestion that it may be literally true as a description of nature is still greeted with cynicism, incomprehension, and even anger. To understand why the theory presents a conceptual challenge for physicists, consider the following experiment, based on an optical test first performed in 1801 by Thomas Young.

In the experiment, particles of light (photons) stream through a single vertical slit cut into a screen and fall onto a piece of photographic film placed some distance behind the screen. The image that develops on the film isn't surprising—simply a bright, uniform band. But if a second slit is cut into the screen, parallel to the first, the image on the film changes in an unexpected way: In place of a uniformly bright patch, the photons now form a pattern of alternating bright and dark parallel lines on the film. Dark lines appear in areas that were bright when just one slit was open. Somehow, cutting a second slit for the light to shine through prevents the photons from hitting areas on the screen they easily reached when only one slit was open.

The pattern is usually explained by physicists as an indication that light has a dual nature. Although it consists of individual photons, it behaves like a wave. Overlapping wave crests meet at the film to create the bright lines, when light waves emerge from the two slits. Troughs and crests cancel out to produce the dark lines.

But, according to physicists, there's a problem with this explanation: The same pattern of light and dark lines gradually builds up even when photons pass one at a time through the slits, as if each photon had somehow spread out like a wave and gone through both slits simultaneously. That clearly isn't the case, because the distance between the two slits can be hundreds, thousands, or in principle, any number of times greater than the size of a single photon. And if that isn't confusing enough, consider this: If detectors are placed at each slit, they register a photon traveling through only one of the slits, never through both at the same time.

Nevertheless, the photons behave as if they had traveled through both slits at once. The same baffling result holds not just for photons, but also for particles of matter, such as electrons. Each seems able to exist in many different places at once—but only when no one is looking. As soon as a physicist tries to observe a particle (by placing a detector at each of the two slits, for example), the particle somehow settles down into a single position, as if it knew it was being detected.

When pressed, most physicists will usually say that the lesson quantum mechanics has for them is that their concepts of how a particle should behave simply don't match reality. But the implications of the theory are clear: If in every case a particle (be it a photon, an electron, or any other denizen of the quantum world) appears to occupy more than one position at a time, then it clearly does occupy many positions at once. And thus, so do we, and so does everything else in the universe.

But, drawing from a simple pattern of shadow and light, isn't that an awfully big conclusion? Well, it is a similarly huge assumption that the universe is expanding, based on subtle light and shadow observations. Yet hardly any physicist anywhere disputes it.

Physicists never encounter the multiple realities of quantum mechanics under normal circumstances. They certainly aren't aware of what their other selves are doing. Only in carefully controlled conditions (as in the two-slit experiment) do they get a hint of the existence of a multi-universe. That experiment offers a rare example of two overlapping realities, in which photons in one universe interfere with those in another. In our universe, physicists see a photon passing through one slit that seems to interact with another invisible photon traveling through the second slit. In another universe, the photon that a physicist sees is invisible to the physicist in that world, while the one that a physicist can't see in our world is the photon the otherworldly physicist detects. Peculiar?

According to physicists, there is no alternative way of looking at quantum mechanics. When it comes to a conflict about what a theory of physics says and what physicists are expecting, then physics has to win.

The multi-universe concept is not new. That credit goes to Hugh Everett, whose 1957 Princeton doctoral thesis first presented what has come to be called the "many worlds" interpretation of quantum mechanics.

Everett was trying to solve the problem of why physicists see only one of the multiple states in which a particle can exist when he created the many worlds view. Some years before Everett's work, physicists had crafted an ad hoc explanation that to this day remains the standard way of coping with quantum phenomena. In the conventional view, the very act of a physicist's observation causes all the possible states of a particle to "collapse" abruptly into a single value, which specifies the position, say, or energy of the particle. To understand how this works, imagine that the particle is an e-mail message. There are multiple possible outcomes when the message is sent: The sender might receive a notice that the message could not be delivered; the e-mail could reach its intended destination; or any number of people could get it by mistake. But all other possibilities with regard to the e-mail delivery collapse into one reality when one outcome is observed.

Tacked on to smooth over the uncomfortable fact that the theory mandates multiple states for every particle in existence, to some physicists, the notion of collapse is an unsightly addition to quantum mechanics. And the collapse model creates its own problems: Because it says our observations affect the outcomes of experiments, it assigns a central role to consciousness.

Everett argued that nothing like a quantum collapse ever occurs and that human consciousness (see Chapter 14 for a detailed discussion of consciousness) does not determine the outcome of experimental results. In other words, Everett labored to move beyond those laws. He said the collapse only seems to happen from a physicist's limited perspective. Everett believed that all quantum states are equally real and that if physicists see only one result of an experiment, other versions of themselves must see all the remaining possibilities.

Physicists who use quantum mechanics in a utilitarian way suffer from a loss of nerve (and that means most physicists working in the field today). They simply can't accept the strangeness of quantum reality. This is probably the first time in history that physicists have refused to believe what their reigning theory says about the world. Indeed, most other physicists view the many worlds route as a road best not taken.

A number of physicists believe that quantum mechanics doesn't tell them so much about the world itself as it does about a person's interaction with the world. It represents a person's interface with reality. It really doesn't go further than that. The odd properties of quantum mechanics, such as the apparent ability of particles to exist in many places at once, merely reflect a person's ignorance of the world and are not true features of reality. When a quantum state collapses, it's not because anything is happening physically, it's simply because this little piece of the world called a person has come across some knowledge, and he or she updates his or her knowledge. So, the quantum state that's being changed is just the person's knowledge of the world; it's not something existent in the world in and of itself.

So, according to physicists, quantum mechanics describes a reality that shrinks away from them when they probe it too closely. It is this extreme sensitivity of quantum systems that keeps physicists from ever knowing more about them than can be captured with the formal structure of quantum mechanics.

To some physicists, such arguments are just complex rationalizations for avoiding the most straightforward implications of quantum theory. It's a tenable point of view to say I don't know what the world is like. The obvious question, then, is: What is in fact happening in reality? If quantum theory is true, it puts heavy constraints on what the world can be like.

Physicists will never advance to a new, deeper understanding of nature. This is the most serious consequence of refusing to consider the many worlds view. What one can hope for in the long run is that a new theory will be facilitated by understanding this present theory. Once you understand the existing theory, you have a handle on what you can change in it. Whereas if you don't understand it, if it's just a set of equations, then it's astronomically unlikely that you will happen upon a better theory.

In the meantime, according to some physicists, a refined application of quantum mechanics principles will produce a tool that could bolster arguments for the existence of parallel universes. Many physicists around the world are trying to build a quantum computer that would manipulate atoms or photons and exploit the particles' abilities to exist simultaneously in more than one state. Those quantum properties would tremendously increase the speed and capacity of the computer, allowing it to complete tasks beyond the reach of existing machines. In fact, a quantum computer could in theory perform a calculation requiring more steps than there are atoms in the entire universe.

The computer would have to be manipulating and storing all that information somewhere in order to do that. Computation is, after all, a physical process. It uses real resources, matter and energy. But if those resources exceed the amount available in our universe, then the computer would have to be drawing on the resources of other universes. So, if such a computer is built, the case for many worlds will be compelling.

What if quantum theory is wrong? It's possible that quantum theory will be proved false one day, because it seems inconceivable that physicists have stumbled across the final theory of physics. But will the new theory either retain the parallel universe feature of quantum physics, or will it contain something even more weird like quantum time travel?

Quantum Time Travel

Time travel is easy: All of us do it every day, but only in one direction. For thousands of years, scientists and philosophers have talked of time as a river that flows steadily onward year after year. But what if there were a way to swim against the flow, or to run down the bank ahead of the river? Might we be able to journey back and forth in time just as we travel through space? The idea is not as far-fetched as it sounds, and the implications for the future are intriguing.

Ever since Einstein, scientists have considered three-dimensional space and time not as two different things, but as different aspects of four-dimensional "space–time." Quantum physicists, who study the world of subatomic particles, often find it easier to explain events by assuming that time runs backward as well as forward, however much it defies common sense.

At the other extreme, cosmologists looking at the universe on a grand scale have found that space and time can be warped by gravity and speed. Back in the 1940s, German mathematician Kurt Goedel proved that if we could warp and twist space–time enough (creating what he called "closed, time-like curves") then we could bore tunnels through time itself. But no one knew how to do the twisting—until black holes.

NOTE

One problem for any time traveler will be the energy bill. Even a small step through time is likely to use up the energy output of a fair-sized star.

As previously explained, the gravitational pull of a black hole is so enormous that it distorts the very fabric of space–time into what is called a *singularity.* When singularities were found to spin, it was proved that closed, time-like curves not only can occur—they must occur. The singularity forms a doughnut shape in space–time, while the hole in the middle is a perilous gateway to somewhere—or sometime. In any event, there are presently three ways to travel in time without breaking the rules:

- Wormholes
- Rotating Cylinder
- Cosmic Strings

Wormholes

Every week on the Sci-Fi Channel, the *SG-1* team uses the Stargate to travel to other worlds. The Stargate is composed of a stable wormhole that connects to another Stargate on another world when the coordinates of that world are dialed in. In the world of theoretical physics, wormholes can also be used for time travel.

Since the 1930s, physicists have speculated about the existence of "wormholes." Wormholes are essentially gateways between different parts of the universe and are made by linking a pair of black holes. This effectively creates a tunnel through time and space: A traveler entering at one end would exit at the other at a different time, as well as a different place.

The difficulty lies in keeping the wormhole open while the traveler makes his or her journey: If the opening snaps shut, he or she will never survive to emerge at the other end. For years, scientists believed that the transit was physically impossible. But recent research suggests that it could be done using exotic materials capable of withstanding the immense forces involved. Even then, the time machine would be of limited use—for example, you could not return to a time before the wormhole was created.

Using wormhole technology would also require a society so technologically advanced that it could master and exploit the energy within black holes. The trip would not be impossible—just very, very difficult!

Rotating Cylinder

Civilizations with the technology to harness black holes might be better advised to leave wormholes alone and try the time-warp method suggested by some astronomers. They have a simple recipe for a time machine: First, take a piece of material 10 times the mass of the sun, squeeze it together and roll it into a long, thin, super-dense cylinder—a bit like a black hole that has passed through a spaghetti factory. Then spin the cylinder up to a few billion revolutions per minute and see what happens.

Some astronomers predict that a ship following a carefully plotted spiral course around the cylinder would immediately find itself on a "closed, time-like curve." It would emerge thousands, even billions of years from its starting point and possibly several galaxies away.

There are problems, though. According to some astronomers, for mathematics to work properly, the cylinder has to be infinitely long. Also, odd things happen near the ends, and you would need to steer well clear of them in your timeship. However, if you make the device as long as you can and stick to paths close to the middle of the cylinder, you should survive the trip!

Cosmic Strings

As a variation on the rotating cylinders, some scientists have suggested using "cosmic strings" to construct a time machine. At the moment, these are purely theoretical objects that might possibly be left over from the creation of the universe in the Big Bang.

So, would it be possible to use "cosmic strings" to construct a time machine? According to Dr. Smolin, "So far as I know, not a realistic one. You have to distinguish carefully between cosmic strings and string theory, as a conjectured fundamental theory that unifies the different forces. They are not at all the same thing. Cosmic strings are consequences of gauge theories, including the currently successful theories of the strong, weak, and electromagnetic forces." [5]

As previously explained, a black hole contains a one-dimensional singularity—an infinitely small point in the space–time continuum. A cosmic string, if such a thing existed, would be a two-dimensional singularity—an infinitely thin line that has even stranger effects on the fabric of space and time.

According to theoretical physicists, no one has actually found a cosmic string. Astronomers have suggested that this may explain some strange effects seen in distant galaxies. By maneuvering two cosmic strings close together (or possibly just one string plus a black hole), it is theoretically possible to create a whole array of closed, time-like curves. Your best bet is to fire two infinitely long cosmic strings past each other at very high speeds, then fly your ship around them in a carefully calculated figure eight. In theory, you would be able to emerge anywhere, anytime!

Back to the Future

Many scientists are uncomfortable with time travel, not because it is impossible, but because of the paradoxes it creates. In other words, no known laws of physics rule out time travel, but venturing to another era is rife with problems.

Thus, the previously discussed "many worlds" theory solves another of the famous time travel paradoxes. For example, let's look at the "grandpa-cide" scenario: Say you went back in time and shot your grandfather before he met your grandmother. Would you never have been born? If not, you could never have traveled back in time and shot your grandfather. Which means that you *were* born, so you *could* have gone back. According to "many worlds," when you go back in time you actually emerge in another universe that develops in parallel to our own. But, with an infinite number of universes to choose from, how can time travelers ever hope to find their way back to the one they started out from?

Or, imagine this loopy scenario: You pack the complete works of Shakespeare, travel back to Elizabethan England, and sell them all to a struggling young playwright named William, who uses the mysterious gift from the future to make his fame and fortune. Who then wrote all those plays and sonnets? They seem to have sprung from nowhere.

NOTE

Scientists say building a time machine may be impossibly difficult. But time travel is not against the laws of physics.

Or, imagine another loopy scenario: You travel back in time to visit the composer Ludwig Van Beethoven and took with you a CD of his Fifth Symphony. Beethoven listens to it and writes the music down, then later his score is used to record your CD. Where does the music come from? In science, as in life, there is no such thing as a free lunch!

The many worlds interpretation of quantum theory solves all of these paradoxes. In the first example, when you make the fateful call on your grandfather, the past you visit (and change) is not the past of the universe you came from. In your home universe, the grandfather remains alive. The grandfather you murder belongs to the past of a parallel space and time, one in which you will never be born, but one in which you could remain should you so choose.

In other words, according to the many worlds theory, an infinite number of universes is constantly being created. In quantum physics, when subatomic particles have a "choice" of options (such as going through one hole or another in a screen), they select one at random. The many worlds theory says that there is a universe for each possible choice made by the particle.

So, what about the Shakespeare paradox? The plays and sonnets do not emerge from thin air. Somewhere in the multi-universe at least one version of Shakespeare created them without any help from a time traveler. But a time traveler could bestow them upon a Shakespeare in another universe, and that lucky bard would become famous through no effort of his own. In the big picture, though (and the multi-universe is the biggest picture possible), knowledge always has a creative source. The same can be said for the Beethoven scenario.

NOTE

Scientists often talk about time travel in the context of its implications for theoretical physics. But they try not to mention "time machines" in case it gets picked up by the supermarket tabloids.

In the many worlds view, time travel is no more paradoxical (although it may prove a bit more difficult) than any other form of transportation. If you got particularly angry with yourself for something you once did, or might do, you could even travel to the past (or future) and murder the other you.

WARNING

You might be advised to be leery of any visits from another time-traveling you.

So, if time travel is possible, could tourists from the future visit us? Let's take a look.

Don't Drink the Water

If time machines are possible, it is likely that someone in the future will already have constructed one. After all, in the future there is time to complete even the largest engineering project! Even if humans are not up to the task, creatures from other planets may try. So, why are we not overrun by visitors from the future?

NOTE

So far we have managed to avoid being overwhelmed by tourists from the future for good scientific reasons. In fact, time tourism is unlikely to be a problem: Theoretically, time machines would have a very limited capacity.

This is the argument used by the famous English physicist Stephen Hawking in what he called his "chronology protection conjecture." [2] Like many other scientists, Hawking is troubled by the weird paradoxes of time travel. [2] He argues that the universe simply couldn't allow time travel to happen, because its evolution since the Big Bang cannot be reversed. [2] If the universe were to contract instead of expanding, asks Hawking, would human beings "unevolve" in the same way as they have evolved over millions of years? [2]

A second explanation for the absence of visitors from the future is that none of the time machines envisaged so far lets the voyager go back before the moment the machine was first constructed. So relax. Since no one has built a time machine yet, out-of-time tourists are not a problem!

In any event, different times are nothing less than different universes. The universes we can affect we call the future. Those that can affect us we call the past.

Now that some of the unusual features of quantum mechanics (there are many) have been explained, it shouldn't be a problem to try to explain the construction of a quantum theory of gravity. Or, will it?

How Difficult Could It Be?

Finally, when theorists try to construct a microscopic extension of gravity; that is, a description of gravity at tiny, tiny distances, they must do so in obeisance of the rules of quantum mechanics. Is this true?

"No," replies Dr. Smolin, "there is also the possibility that quantum mechanics is a consequence of the microstructure of space and time, rather than the other way around. This has been explored in some recent papers." [5]

Such a theory is called a quantum theory of gravity. Dr. Smolin goes further to explain that this is "the theory that describes the unification of quantum theory with general relativity, the theory of space, time and gravitation." [5]

Although many suggestions have been made for such a theory (including the most promising candidate of superstrings), no proposal has gained universal acceptance. Almost all ideas encounter difficulties. It is easy to understand why—the marriage of quantum mechanics and Einstein's gravity theory requires space itself to fluctuate, a mind-boggling concept. How can it be that space has certain probabilities of having certain shapes? If there are a variety of spaces, which one do we live in? The answer must be all of them if the probabilistic nature of quantum mechanics is to be implemented. Although these issues seem philosophical, mathematical difficulties also arise. The construction of a quantum theory of gravity remains one of the greatest unsolved problems of theoretical physics.

Conclusion

Fresh ideas are being born to explain the final theory of everything. Physicists are now looking to higher dimensions for answers to the construction of a quantum theory of gravity, and there are preliminary indications, at least on the theoretical front, that they may be on the right track after all.

Four dimensions (three spatial and one temporal) adequately describe reality in the realm of human experience. But what if there were, say, six more spatial dimensions hidden from our worldly view? Would physicists ever be able to tell the difference if one of these extra dimensions was so large that it contained the entire visible universe, while the other five were wrapped up so small that they couldn't be seen at all? This just may be the case according to the M-theory (the M stands for the mother of all theories, magic, or matrix, depending on the source), the leading contender for the coveted title of Ultimate Theory.

But, does a theory like the M-theory reflect reality? Renowned theoretical physicist Dr. Stephen Hawking was recently asked this question by the editor/author of this book. Dr. Hawking's response follows:

"One can't ask whether a theory reflects reality, because we have no theory independent way of determining what is real. Even the everyday objects around us that we regard as obviously real, are in the positivist view, just a model we construct in our minds, to interpret the data from our optical and sensory nerves. Doctor Johnson cried, I refute it thus, and stubbed his toe on a large stone, when told Bishop Berkeley's opinion that nothing is real. [2]

"But maybe we are all linked in to a giant computer simulation, that sends a signal of pain, when we send a motor signal, to swing an imaginary foot, at an imaginary stone. Maybe we are characters in a computer game played by aliens. [2]

"Joking apart, the important point is that we can have several different descriptions of the universe, all of which predict the same observations. We can not say that one description is more real than another, just that it may be more convenient for a particular situation. So all the theories in the M-theory network, are on a similar footing. None can be said to be more real than the others. [2]

"Remarkably enough, in many of the theories in the M-theory network, space–time has more than the four dimensions we experience. Are these extra dimensions real? I must admit I have been reluctant to believe in extra dimensions. But the M-theory network fits together so beautifully, and has so many unexpected correspondences, that I feel to ignore it, would be like claiming that God put fossils in the rocks, to trick Darwin into believing in evolution." [2]

The M-theory is the latest, most fundamental version of the superstring theory as it currently stands. But, is it the name of an existing theory? Dr. Smolin replies as follows:

"M-theory is not the name for an existing theory. It is a conjecture, about the existence of a theory that would unify different string theories. Although the conjecture was made in 1996, there has not been since any definite proposal as to what the theory is. We cannot say what the fundamental principles of the theory are, what the laws are, or what the right mathematics is to express it." [5]

So, is the M-theory really a theory? Again, renowned theoretical physicist Dr. Stephen Hawking was recently asked this question by the editor/author of this book. Dr. Hawking's response follows:

"Why it should be called M-theory, is completely obscure. M-theory is not a theory in the usual sense. Rather it is a collection of theories, that look very different, but which describe the same physical situation. These theories are related by mappings, or correspondences, called dualities, which imply that they are all reflections of the same underlying theory. Each theory in the collection works well in the limit, like low energy or low dilaton, in which its effective coupling is small, but breaks down when the coupling is large. This means that none of the theories can predict the future of the universe to arbitrary accuracy. For that, one would need a single formulation of M-theory that would work in all situations." [2]

"Up to now, most people have implicitly assumed that there is an ultimate theory that we will eventually discover. Indeed, I myself have suggested we might find it quite soon. However, M-theory has made me wonder if this is true. Maybe it is not possible to formulate the theory of the universe in a finite number of statements." [2]

Thus, The M-theory is a construct that holds that elementary particles (and, by extension, the entire universe) can be described by wave-like vibrations of fundamental entities called strings. The M-theory, as a matter of fact, unites five previously disparate versions of the superstring theory into one coherent package. According to M-theorists, our universe lies on a three-dimensional membrane or "brane" that floats in a fifth dimension (with the fourth dimension being time). The theory also posits the existence of five higher spatial dimensions, which can generally be ignored, as they are wrapped up in tiny circles far smaller than the radius of a single proton. As weird as these ideas may sound, several physicists have found recent success with applying the M-theory to a host of cosmological questions.

NOTE

Though their influence may be huge, the truth is that many hidden dimensions may simply be too small for physicists to detect. And, while the usual four dimensions of space–time may be infinite, this need not be so for any extra dimensions. In fact, extra dimensions may be wrapped up in spaces millions of times smaller than subatomic particulars or perhaps as large as one millimeter. A human hair and a magnifying glass provide a good analogy. At first glance, a human hair may appear to be a two-dimensional object. Yet, with the aid of a magnifying glass, physicists can see that it actually spans a third dimension as well. At present, no gravitational studies have probed distances smaller than 0.2 mm. If our observable universe were somehow wrapped around another large (millimeter-scale) spatial dimension, this could explain the more mysterious aspects of gravity such as dark matter and why gravity is so much weaker than other fundamental forces. Some physicists think that they will be able to detect extra dimensions by 2007.

According to cosmologists, the solutions to several cosmic conundrums may stem from the interactions of branes in higher dimensions. So far, conventional physics has been unable to tell cosmologists what the Big Bang really was, what caused it, or why the expansion of the universe is accelerating. The Big Bang theory has been incredibly successful, but it's really only a theory of what happened after the Big Bang. Equally mystifying is the dark energy (see sidebar, "The Big Crunch") associated with cosmic acceleration, which is thought to comprise the bulk of energy in the universe. Getting to the bottom of these mysteries with extra dimensions may sound like a stretch, but cosmologists need crazy ideas to get to the bottom of this.

The Big Crunch

In a few billion years, a mysterious force permeating space–time will be strong enough to blow everything apart, shred rocks, animals, molecules, and finally every atom in its last, seemingly mad, instant of cosmic self-abnegation. Recently, astronomical measurements by scientists have not ruled out this possibility.

A mysterious force called dark energy seems to be wrenching the universe apart. This is called the Big Rip, which is only one of a constellation of doomsday possibilities resulting from the discovery by two teams of astronomers back in 1998.

The galaxies started speeding up about five billion years ago, instead of slowing down from cosmic gravity, as cosmologists had presumed for a century. Dark energy sounded crazy at the time, but in the intervening years, a cascade of observations have strengthened the case that something truly weird is going on in the sky. It has a name, but that belies the fact that nobody really knows what dark energy is.

Since 1998, dark energy has become one of the central and apparently unavoidable features of the cosmos. This has been the surprise question mark at the top of every scientist's list. It undermines what physicists presumed they understood about space, time, gravity, and the future of the universe.

In a quest to take the measure of dark energy by tracing the history of the universe with unprecedented precision, armies of astronomers are now fanning out into the night, enlisting telescopes, large and small, from Chile to Hawaii to Arizona to outer space. Some of them are following the trail blazed by the first two groups back in 1998. They are searching out a kind of exploding star known as a Type 1a supernova. By enabling scientists to plumb the size of the universe and how it grew over time, Type 1a supernova serve as markers in space. New efforts by astronomers intend to harvest hundreds or thousands of supernovas, where in the past, astronomers could only base their conclusions on observing a few dozen of them.

Other scientists are seeking to gain leverage by investigating how the antigravitational force of dark energy has retarded the growth of conglomerations of matter like galaxies. Some astronomers are building an array of radio telescopes to count and study clusters of galaxies deep in space–time. Others are building giant cameras that use light-bending powers of gravity itself as lenses to map invisible dark matter in space and compile a growth chart of cosmic structures or are already probing the internal dynamics of galaxies by the thousands.

The idea of an antigravitational force pervading the cosmos does sound like science fiction. But, in concert with Einstein's equations, theorists have long known that certain energy fields would exert negative pressure that would in turn, produce negative gravity. Indeed, some kind of brief and violent antigravitational boost, called inflation, is thought by theorists to have fueled the Big Bang.

As astronomers and physicists try to figure out how this strange behavior could be happening in the universe today, the ultimate prize from all the new observing projects could be as simple as a single number. That number, known as *w*, is the ratio between the pressure and density of dark energy. Knowing this number and how it changes with time (if it does), might help scientists pick through different explanations of dark energy, and thus the future of the universe.

Caused by the energy residing in empty space, one possible explanation for dark energy (perhaps the sentimental favorite among astronomers) is a force known as the cosmological constant. It was first postulated back by Einstein in 1917. A universe under its influence would accelerate forever.

As the universe grows, there would be more space and thus more repulsion, while the density of energy in space would remain the same over the eons. Within a few billion years, most galaxies would be moving away from the Milky Way faster than the speed of light and so would disappear from the sky altogether. Thus, like a black hole, the edge of the observable universe would shrink around your descendants.

However, by using the most high-powered modern theories of gravity and particle physics, attempts to calculate the cosmological constant have resulted in numbers 1,060 times as great as the dark energy astronomers have observed. This is big enough, in fact, to have blown the universe apart in the first second—long before atoms had time for form. Theorists admit they are at a loss. Some theorists now feel that Einstein's theory of gravity (the general theory of relativity) needs to be modified.

Another possibility comes from string theory (the putative theory of everything). Associated with particles or forces as yet undiscovered, the string theory allows that space could be laced with other energy fields. Those fields, collectively called quintessence, could have an antigravity effect.

For example, quintessence could even change from a repulsive force to an attractive one (as the universe expanded and diluted the field, getting weaker and eventually disappearing) or could change with time. All of this could set off a big crunch.

The density of phantom energy would go up and up, eventually becoming infinite, while the density of the energy in Einstein's cosmological constant stays the same as the universe expands. Such would be the case if the parameter w turned out to be less than minus 1. Physicists are stunned by the possibility and until recently simply refused to consider it.

This version of doomsday would start slowly. Then, as phantom energy increased its push and the cosmic expansion accelerated billions of years from now, more and more galaxies would start to disappear from the sky as their speeds reached the speed of light.

But things would not stop there. Depending on the exact value of w, the phantom force from the phantom energy will be enough to overcome gravity and break up clusters of galaxies billions of years from now. That will happen about a billion years before the Big Rip itself. After that the apocalypse speeds up. The Milky Way galaxy will be torn apart about 900 million years later, and about 60 million years before the end. The solar system will fly apart about three months before the rip. When there is half an hour left on the cosmic clock, the Earth will explode.

The dark energy surveys now under way hope to be able to measure *w* to an accuracy of 5 percent. But, it may not be sufficient to eliminate the nightmare of phantom energy, even if that can be done.

Variations in the atmosphere and gaps in astronomers' understanding of supernova explosions, add uncertainty to the dark energy measurements. As a result, some astronomers fear that the results may leave them on the razor's edge, unable to decide between a cosmological constant and the other possibilities—the big rip or quintessence. Cosmologists could then be stuck with a *standard model* of the universe that fits all the data, but which they have no hope of understanding.

The work of the dark energy hunters has been complicated by the impending loss of the Hubble Space Telescope (HST) due to cuts by the Bush administration and NASA. Recently, citing safety concerns, NASA canceled all future shuttle maintenance missions to the telescope, dooming it to die in orbit, probably by 2007 or 2008. According to astronomers, HST can see far enough out in space and time to measure how and if the dark energy parameter *w* is changing over the eons.

Similarly, plans for a special satellite that was to have been jointly sponsored by NASA and the Department of Energy, have temporarily disappeared from NASA's five-year budget plan. This was the result of the agency's presidentially ordered shift toward the Moon and Mars.

A handful of scientists have already started putting M-theory to work on the most pressing of these problems. Scientists are now drawing on M-theory to offer an alternative to inflation. In addition, scientists have offered an explanation for what caused the Big Bang and what came before. According to the ekpyrotic model (also dubbed the ekpyrotic scenario), after the Greek word "ekpyrosis," meaning conflagration—the Big Bang could have been the by-product of a fiery collision between two fiat and parallel branes drifting through the fifth dimension. (See sidebar, "Ekpyrotic Model.")

Ekpyrotic Model

In March 2001, Justin Khoury and Paul Steinhardt of Princeton University, along with Nell Turok of Cambridge University and Burt Ovrut of the University of Pennsylvania, unveiled a potentially serious alternative to inflation. Their so-called "ekpyrotic model," derived from the Greek word ekpyrosis, meaning "conflagration," is based on the so-called M-theory—the latest, most fundamental version of the superstring theory. According to M-theorists, our universe might lie on a three-dimensional sheet-like membrane, or "brane" which floats in a hidden fifth dimension (the fourth dimension being time). The theory also posits the existence of six higher spatial dimensions, all tied up in extremely tiny circles, smaller than the radius of a single proton.

In the ekpyrotic scheme, the driving force behind the universe's expansion comes not from inflation, but from a collision between our universe and another parallel "brane" universe from an imperceptible dimension. This scenario does not contradict the idea of a universe that has expanded and cooled since the Big Bang. What their model does is amend the earliest moments of the story.

This model was challenged about a week later in a 21-page paper written by Andre Linde and his wife, Renata Kailosh, both of Stanford, along with Lev Kofman of the University of Toronto. The model works better, they indicate, by simply changing the sign of one parameter. They call their "improved scenario" the "pyrotechnic universe," while acknowledging it still suffers from problems that beset the original version.

Paul Steinhardt is unconvinced by Linde critique. This is a whole different approach, rather than a specific model. It's worthwhile, in any case, to try to devise alternative approaches. Either physicists will learn that this is a viable alternative or that it's hard to come up with an alternative to inflation.

There's no restriction against the universe emerging from such a mega-dimensional head-on collision in principle. Without inflation, sufficient energy and heat would be generated in such an event to fuel the observed expansion of the universe, and the temperature and density fluctuations that astronomers see today could be attributed to quantum wrinkles on each of the branes, making some regions hotter than others.

According to physicists, what looks like a singularity in four dimensions may not be one in five dimensions. When the branes crunch together, the fifth dimension disappears temporarily, but the branes themselves don't disappear. So, the density and temperature don't go to infinity, and time continues right through. Although general relativity goes berserk, string theory does not, and what once looked like a disaster in the ekpyrotic model now seems manageable.

The ekpyrotic scenario has come under attack because of its stringent set of initial conditions, requiring that the two branes be perfectly homogenous, flat, and parallel. According to physicists, if you start with perfection, you might be able to explain what you see. But you still have to answer the question: Why must the universe start out perfect?

According to physicists, inflation rests on assumptions as well. It works over a broader range of initial conditions, but not an infinite range. Physicists indicate that there is this whole prehistory to the universe, an immense period of time before the Big Bang that is exponentially longer than the inferred age of our universe, during which these branes could flatten themselves into their most symmetric, lowest energy state. After the collision, there would also be more time for the branes to settle into equilibrium and reach a uniform temperature.

NOTE

According to M-theory, the particles that comprise the universe, including photons, are confined to the brane on which they were created, except for gravitons (the yet-undiscovered carriers of gravitational force), which can drift into extra dimensions. Therefore, with the ability to carry information to alternate dimensions, gravity may offer a unique window into megadimensional space where parallel universes roam. A nearby brane, or fold in the brane, concealed by extra dimensions, might provide the gravitational pull commonly attributed to dark matter, and the current acceleration of the universe might be due to gravity leaking into extra dimensions over cosmological distances.

But, before astronomers race off to replace inflation with colliding branes, physicists warn that the two scenarios need not be mutually exclusive. In fact, they may complement each other. In the long run, since inflation seems to be an obvious solution to the problems it was designed to address, physicists think it's inevitable that string theory and M-theory will need to incorporate inflation. This is why the universe is so uniform and flat.

According to the original so-called "branefall" scenario, physicists feel that standard inflation could still occur during the slow-moving phase when two parallel branes slowly drift toward each other. This is because the attractive tug between branes (owing to gravity and other string forces) provides a source of potential energy that stays virtually constant, even with space expanding like mad. One virtue of the ekpyrotic model is that it presents a geometrical picture of where the energy that sustains inflation comes from.

Some physicists feel that it doesn't take a genius to say that all of the ideas out there today are wrong—or incomplete, at best. Physicists are still at a very early stage of understanding the extra dimensions (assuming they exist) and their cosmological implications.

No one disputes the proposition that M-theory is a work in progress. And, since it's being billed as a "theory of everything," it could take a long time to nail down the details. Even though physicists don't know the full theory, there's still a lot they can do with confidence. In the same way, physicists know that Newton's laws are just part of the story, but they use them all the time. They work fine so long as you pick the right regime.

Physicists agree that the new physics is still taking shape, but consider the interface between cosmology and string theory to be the most fruitful place to be working today. Cosmology offers the best opportunities to test ideas from M-theory and string theory, because the energy produced in accelerator experiments is more than a quadrillion times too small. Physicists have been working on this theory for more than two decades with no new data.

Finally, by searching for specific features in the Cosmic Microwave Background, physicists hope to find observable signatures that could tell them they're working with the right theory. They also need to have the right tools to apply to questions, such as the origin of the universe, that they couldn't get at before.

Black Hole Theory Modified

Recently though, theoretical physicist and cosmologist Stephen Hawking backpedaled on the idea that nothing can escape the limitless gravitational pull of these collapsed, dark stars. He now claims that some information (matter, heat, and/or light) can break free of their once-thought-inescapable event horizons (see sidebar, "Update: Hawking's Black Hole Theory Is a Reversal in Itself," for more information). This news-making event also reflects some of the nonstatic nature of the questions answered in this book.

Update: Hawking's Black Hole Theory Is a Reversal in Itself

Stephen Hawking's new theory about black holes may defy the conventional wisdom he created: All information is lost in the gravitational collapse that creates these hungry holes in space and time. However, the fundamental laws of quantum physics say that information is conserved, and never can be destroyed—not even by a voracious black hole.

In a reversal, Hawking now argues that black holes never quite shut off completely. And, as they start to emit more heat from their gravitational pull, they eventually open up and release *qubits* or quantum bits of information.

Thus, it goes without saying that Hawking's possible solution to the black hole information paradox has caused tremendous excitement in the scientific community. But, until other physicists study his findings in detail, the jury is still out.

References

[1] "Astronomy Picture of the Day," [http://apod.gsfc.nasa.gov/apod/archivepix.html], NASA Headquarters, 300 E Street SW, Washington DC 20024-3210/ Jet Propulsion Laboratory, 4800 Oak Grove Drive, Pasadena, California 91109/NASA/Goddard Space Flight Center, Greenbelt, MD 20771, 2003.

[2] Dr. Stephen Hawking, Theoretical Physicist, CH CBE FRS, Lucasian Professor of Mathematics Department of Applied Mathematics and Theoretical Physics, University of Cambridge, Cambridge, CB3 0WA, United Kingdom.

[3] Dr. Richard Muller, Physicist, Physics Department, University of California, Berkeley, CA 94720.

[4] Dr. Andrea Ghez, Physicist and Astronomer, Department of Physics and Astronomy/IGPP, UCLA.

[5] Dr. Lee Smolin, Theoretical Physicist, Perimeter Institute of Theoretical Physics, Waterloo, Ontario, Canada.

[6] Dr. Stephen Wolfram, Physicist, "Quick Takes on Some Ideas and Discoveries in a New Kind of Science," Wolfram Research, Inc., (© 2004 Stephen Wolfram), Champaign, IL, 61820.

Particle Physics: The Mechanism That Makes Fundamental Mass

"Scientists can always be counted on for creating unintended consequences."
—Anonymous

In the world, everything has mass. The mass of everything is due to the mass of the subatomic particles of atoms which are composed of electrons, protons, and neutrons. The weight of an electron is about 2,000 times lighter than a proton. A neutron is slightly heavier than a proton.

With that in mind, physicists have constructed a theory of subatomic particles and their interactions during the last part of the 20th century. This theory is known as the Standard Model of particle physics. Three forces operate in this microscopic world:

1. Electromagnetism, which encompasses the electric and magnetic forces.
2. The strong nuclear force, which holds the protons and neutrons together in the nucleus or a heavy central core of an atom.
3. The weak subnuclear force, which causes some nuclei to decay in radioactive processes and helps to generate energy at the center of the sun.

But what really is the Standard Model? Let's take a look.

Standard Model

For more than 35 years, the Standard Model has been a pillar of fundamental physics. It is the crown jewel of 20th-century particle physics—and has taken decades to construct. In 1979, Drs. Sheldon Glashow, Abdus Salam, and Steven Weinberg won the Nobel Prize for unifying the sectors of the theory comprising the weak subnuclear and electromagnetic forces, what is now known as the electroweak theory. The remaining sector (the strong interactions) is based on a so-called non-abelian gauge theory (a theoretical invention by Drs. R. L. Mills and C. N. Yang in 1954). The strong force holds the protons and neutrons together in a nucleus. Gravity is the only fundamental force not included in the Standard Model of particle physics: Although string theory (see sidebar, "What Is the String Theory") holds some promise, physicists have been unable to construct a consistent quantum theory of gravity.

What Is the String Theory?

The idea that the microscopic constituents of matter are incredibly tiny bits of string is what is known as string theory. The string is like a guitar wire that undergoes only certain vibrations. These correspond to different musical notes for a guitar wire. Each vibrational mode is an elementary particle for the string: one vibration is an electron, another is a quark, another is the photon (the particle of light), and so on. By exchanging bits of string, the fundamental forces are created. For example, when a string in its photon mode is repeatedly "passed back and forth" between two electrically charged strings, the electric force is produced. Because strings can join and split into two strings and are able to fuse ends to form one string (similar to tying together two shoe laces), they are able to accomplish this exchange. Or, similar to cutting a long shoe lace into two shorter pieces, one string can break into two.

Because it offers the ultimate unification, some theorists are excited about string theory. All forms of matter and all types of forces are manifestations of a single structure. String theory also has the potential to solve one of the greatest problems of theoretical physics of the 20th century: the incompatibility of quantum mechanics and Einstein's general theory of gravity. See Chapter 8 for a detailed discussion of String Theory.

Two types of generic particles, *leptons* and *quarks*, make up all matter. Electrons, muons, and taus are the three types of charged leptons. Only the electrons are stable out of these three particles. The other charged leptons, which are heavier, decay in a small fraction of a second. Electrons, which flow through a wire to make electricity, form the electronic cloud of the outer part of an atom. Leptons that are electrically neutral and have very tiny masses also exist. These are called *neutrinos*. See Chapter 6 for a detailed explanation of what neutrinos are.

There are six types of quarks: down, up, strange, charm, bottom, and top. Two down quarks and one up quark compose a neutron, while two up quarks and one down quark compose a proton. The neutrons and protons fuse together in a nucleus, which is the tiny, central core of an atom. The heavier quarks (strange, charm, bottom, and top) decay into lighter quarks and other particles through weak subnuclear interactions.

The particle forces are generated by the exchange of vector gauge bosons: Passing virtual photons between charged particles yields the electromagnetic forces. Exchanging virtual Ws (also known as W Boson, a carrier particle of the weak interaction) and Zs (also known as Z Boson, a carrier particle of weak interactions and involved in weak processes that do not change flavor) between quarks and leptons produces the weak subnuclear forces, and rapid emission between gluons themselves and quarks creates the strong interactions.

Electroweak symmetry breaking generates the masses of the quarks and leptons (Ws and Zs). This process leaves the photon massless. It thereby renders the electromagnetic forces long-ranged, macroscopic, and quite different from the subnuclear weak interactions. Thus, symmetry breaking is achieved using a field called the Higgs particle (or "God Particle") in the Standard Model. Experimentalists have searched for the Higgs, but to date, have not detected it. The least appealing part is the Higgs sector of the electroweak theory (see a detailed discussion of the Higgs particle or field later in the chapter), and many theorists believe that nature might use a more elegant way to accomplish mass generation and symmetry breaking.

The Standard Model has been subjected to thorough experimental scrutiny and has survived several attacks for decades. For example, the experimental result for the fractional production of bottom anti-bottom through a Z particle gauge boson differed from the theoretical prediction in the 1990s. Eventually the discrepancy was eliminated by further experiments and analysis. The Standard Model has overcome so many of these experimental–theoretical conflicts that now particle physicists are almost complacent about deviations, and they are willing to assume that any problem is merely a statistical fluctuation that will eventually resolve itself.

The main machines testing the Standard Model have been the CERN collider near Geneva, Switzerland; the SLAC facility in Stanford, California; and the proton accelerator at Fermilab near Chicago, Illinois. Thus, a comparison of dozens of quantities revealed no discrepancy between experiment and theory in 1998. Much to the dismay of inventive theorists, the Standard Model appeared to be in perfect shape. These theorists had hoped to find small imperfections that would indicate new physics such as supersymmetry, compositeness, or undiscovered subnuclear forces.

Much of the effort has focused on Z particle production. An electron and positron annihilate to produce a Z particle, which exists virtually for less than a split second before decaying. The decay is most often into a lepton and an antilepton or into a quark and an antiquark.

So, what does all of this mean? One possibility is that the mass of the Higgs is below 113.5 Giga Electron Volt (GeV) and has somehow escaped detection. This is unlikely, but possible. Normally, to establish extraordinary evidence, a new theory is needed. Perhaps, the inconsistencies in the data are telling scientists that electroweak breaking does not make use of a Higgs field, since the Higgs sector has not been experimentally confirmed. This would be an exciting and intriguing result. The resolution to the preceding quandary remains unknown, until additional experiments are analyzed.

So, how does breaking destroy electroweak unification? Let's take a look.

How Breaking Destroys Electroweak Unification

As previously explained, the weak subnuclear force and electromagnetism have been unified into a single structure known as the electroweak theory. There is a reason why electromagnetism (which manifests itself at macroscopic distances) is so different from the weak subnuclear force (which only operates inside a nucleus). Also, as previously explained, the unified electroweak theory has undergone what is called "breaking." This breaking destroys the apparent unification of electromagnetism with the weak subnuclear force. In particular, it creates masses for two subatomic particles (the W and Z). This leaves the photon without mass. One of the great principles of particle physics is that all forces are created through the exchange of a special set of particles called gauge bosons. Electromagnetism is generated through the exchange of photons. In the meantime, the weak subnuclear forces are created through the exchange of Ws and Zs. It is difficult to exchange particles of mass over significant distances. Subatomic particles W and Z are very heavy, and can only be exchanged over distances much smaller than a nucleus since the particles are about 90 times heavier than a proton. This explains why the electromagnetic force and the weak subnuclear force are so different.

So, with this in mind, what's the big problem here? When are scientists likely to solve this problem?

What Is the Unsolved Problem?

The "breaking" mechanism also produces the masses of the electron, the proton, and the neutron in the Standard Model. It would seem that scientists have discovered the mechanism that makes all mass. However, there are several theoretically attractive ways to implement the breaking mechanism, but physicists do not know which is correct. In a way that theorists have not imagined, it is also possible that nature has found a way to break the electroweak theory. One important purpose of the Superconducting Super Collider that was going to be built in Texas was to determine the breaking mechanism. The cancellation of that project shattered the dreams of many scientists, who had hoped to discover one of nature's most fundamental secrets. Physicists have now turned to the Large Hadronic Collider (LHC) now being built near Geneva, Switzerland. They hope that it will have sufficient power to discover how "breaking" works. The LHC particle accelerator will be completed around 2006, and results should be available by the end of the next decade.

Now, let's look at one of the most important discoveries of particle physics in the last 32 years. It's called the "God Particle."

The Possible Discovery of the "God Particle"

One of the most important discoveries of particle physics of the last 32 years has possibly just been made by experimentalists at CERN (the giant laboratory just outside of Geneva on the border of Switzerland and France). Scientists there think that they have discovered the Higgs field, also nicknamed the "God particle" by Nobel laureate Dr. Leon Lederman [1], who wrote a book with that title.

The author/editor of this book recently had a very brief interview with Nobel laureate Dr. Leon Lederman. His comments follow:

"I have had the great privilege and joy of leading Fermilab and engaging in the most powerful microscope ever constructed. Fermilab's particle accelerator completed a hundred-year effort to identify the basic constituents of matter—the quarks and leptons that make us and our galaxy—only to learn, with chagrin and fascination, that all of this is only 4% of the matter/energy in our Universe. Still so much to do!" [1]

The Higgs should have a mass of about 125 times the mass of the proton when the result is verified. This makes Higgs as heavy as a medium-sized nucleus. If the results are true, it will "fill in" the last missing piece of a puzzle involving the solution of one of the great outstanding problems in physics of the 21st century: the origin of all mass. If the properties of the Higgs are confirmed, the picture of fundamental particle forces will have been completed. (See sidebar, "Unsolved Mystery: Higgs Discovery Update.") As previously explained, that picture is known as the Standard Model.

Unsolved Mystery: Higgs Discovery Update

CERN ended its one-month extension of the Large Electron Positron (LEP) experiment in early November, 2000. With the hope of clearing up the somewhat ambiguous situation concerning the Higgs discovery, an extension had been granted. The final results are in, and the situation has not been clarified. There are four clean candidate events leading to about 95% confidence that the Higgs has been seen. CERN experimentalists are unable to definitively say that they have uncovered the particle, because of the extraordinary importance of the Higgs. If they have succeeded, the Higgs mass is around 115 GeV/c^2 (for comparative purposes the mass of the proton is 0.94 GeV/c^2).

CERN is now constructing the LHC (large hadronic collider), which could not be delayed further. As previously explained, the LHC should be completed by 2007. It should allow scientists to probe new short-distance properties of nature, to explore higher energies, and to make new discoveries (including that of the Higgs particle).

Again, as previously explained, the Standard Model of particle physics provides a description of microscopic matter and their fundamental interactions. All matter is comprised of quarks and leptons. Three quarks bind to form the proton and neutron. The neutrons and protons stick together to form nuclei—the tiny, heavy, central "hearts" of atoms. (See sidebar, "Making Heavy Elements.") Leptons appear in nature in two types: electrically charged and neutral. Neutral leptons are called neutrinos and hardly interact with matter at all. There are three known charged leptons, the lightest of which is the electron. Electrons, which are negatively charged, are attracted to nuclei, which are positively charged, to form atoms. Much the way bees might swarm around a queen that has left her hive, a good pictorial representation of all atom is a cloud of electrons swarming around a tiny nucleus. Quarks and leptons are the fundamental building blocks of nature, since atoms make up everything in the world.

Making Heavy Elements

How were the heavy elements from iron to uranium made? Both dark matter and possibly dark energy originate from the earliest days of the universe, when light elements such as helium and lithium arose. Heavier elements formed later inside stars, where nuclear reactions jammed protons and neutrons together to make new atomic nuclei. For instance, four hydrogen nuclei (one proton each) fuse through a series of reactions into a helium nucleus (two protons and two neutrons). That's what happens in our sun, and it produces the energy that warms Earth.

But when fusion creates elements that are heavier than iron, it requires an excess of neutrons. Therefore, astronomers assume that heavier atoms are minted in supernova explosions, where there is a ready supply of neutrons, although the specifics of how this happens are unknown. More recently, some scientists have speculated that at least some of the heaviest elements, such as gold and lead, are formed in even more powerful blasts that occur when two neutron stars (tiny, burned-out stellar corpses) collide and collapse into a black hole.

Fundamental Forces

There are three fundamental forces:

1. Gravity
2. Magnetism, Electric Force, and Weak Subnuclear Interaction
3. Strong Nuclear Force

Gravity

As explained in Chapter 4, the most familiar fundamental force is gravity. Gravity holds humans and other objects to the Earth; makes the moon go around the Earth, thereby leading to tides, lunar phases, and eclipses; and causes the Earth to orbit the sun, thereby leading to seasons. Again, as explained in Chapter 4, gravity is generated by objects with mass. But, because gravity is such a weak force, only bodies of huge mass, such as the Earth and sun, create a significant effect. Gravity plays no role in the subatomic world, where protons, neutrons, and electrons are extremely light.

Magnetism, Electric Force, and Weak Subnuclear Interaction

The second fundamental force is a combination of three forces previously thought to be insignificant to one another: magnetism, the electric force, and the weak subnuclear interaction. Leading to electromagnetism, the unification of the electric and magnetic forces was achieved in the 19th century. The source of all macroscopic forces except those created by gravity is the electromagnetic force. Friction, spring forces, air pressure, the forces in collisions, and so on, originate from the electromagnetic force. The weak subnuclear interaction is responsible for certain decays of nuclei and plays a role in generating the energy of the sun and other stars in their cores. As its name implies, it operates at distances smaller than a nucleus and is very weak. It is difficult to observe for this reason. At the end of the 1960s, a theory was proposed that unified the weak subnuclear force with electromagnetism. Experiments in the 1970s and 1980s confirmed the electroweak theory. In 1979, Drs. Steven Weinberg, Sheldon Glashow, and Abdus Salam received the Nobel Prize in physics for unifying the weak subnuclear interaction with electromagnetism.

Strong Nuclear Force

The third fundamental force is called the strong nuclear force. It binds three quarks together to form the proton and neutron. It is also responsible for causing protons and neutrons to stick to one another in a nucleus. But can protons become unstable in the nucleus? (See sidebar, "Unstable Protons," for further details.)

Unstable Protons

Are protons unstable? In case you're worried that the protons you're made of will disintegrate, transforming you into a puddle of elementary particles and free energy, don't sweat it. Various observations and experiments show that protons must be stable for at least a billion trillion trillion years. However, many physicists believe that if the three atomic forces are really just different manifestations of a single unified field, the alchemical, supermassive bosons previously described will materialize out of quarks every now and then, causing quarks, and the protons they compose, to degenerate.

At first glance, you'd be forgiven for thinking these physicists had experienced some sort of mental decay on the grounds that tiny quarks are unlikely to give birth to behemoth bosons weighing more than 10,000,000,000,000,000 times themselves. But there's something called the Heisenberg uncertainty principle, which states that you can never know both the momentum and the position of a particle at the same time, and it indirectly allows for such an outrageous proposition. Therefore, it's possible for a massive boson to pop out of a quark making up a proton for a very short time and cause that proton to decay.

Basic Particle Forces

The basic particle forces are generated through the exchange of vector gauge bosons. This is one of the key ideas in physics. These are particles that incorporate an enormous amount of symmetry and spin with one fundamental unit. The electromagnetic force is generated when charged particles exchange photons (spin-one particles of "light"). As previously explained, the weak subnuclear interactions are generated by exchanging heavy vector bosons known as Ws and Zs, while the strong force is produced by eight gluons.

The fundamental constituents (quarks and leptons) along with the two fundamental particle interactions, the electroweak interaction and the strong nuclear force, constitute the Standard Model of particle physics. Because a good quantum theory of gravitation is not available, gravity has not yet been incorporated.

It would seem that scientists know everything there is to know about microscopic matter and its interactions. However, the mechanism that produces fundamental mass is one aspect of the Standard Model that has remained a mystery. Because the W and Z particles have very heavy masses of 90 to 100 times the mass of a proton), the weak subnuclear interactions are feeble and short-ranged. The photon is massless in contrast. It is this great mass difference that makes electromagnetism so different from the weak subnuclear force. So, how are masses for the W and Z particles created?

Theorists have proposed various mechanisms during the past 37 years, many of which can only be verified by experiment. The use of particles known as Higgs fields is one way to give masses to the W and Z particles. Four such particles are needed; one is left over, and three are absorbed by the W and Z particles. There should also be one spin-zero, electrically neutral particle observed in nature in this mechanism. This is the "God particle" or Higgs particle. It also suggests that the W and Z masses are created by the Higgs mechanism, which is the potential breakthrough discovery at CERN. By respectively absorbing positively charged, negatively charged, and neutral Higgses, the positively charged W particle, the negatively charged W particle, and the neutral Z particle obtain mass.

The Higgs field also has the ability to generate masses for quarks and leptons. Thus, the origin of all mass will be understood if the expected properties of the Higgs field are confirmed.

The generation of mass proceeds through a process known as spontaneous symmetry breaking. If rotating an object does not change its appearance, it has symmetry. For example, if a rod is rotated counterclockwise, its appearance is unchanged. Because a sphere can be rotated in many ways without changing its shape, it has even more symmetry than a rod.

When a system or object naturally loses its rotational invariance, spontaneous symmetry breaking occurs. Suppose pressure is exerted on the rod, say, by pushing on it with a finger. Then the rod will buckle if sufficient force is applied. One can tell that a buckled rod has been rotated when it is rotated. Thus, the symmetry has been broken.

By the way, one might think that it is impossible for the rod to buckle in a particular direction if the force is exerted exactly along the axis of the rod from the top. The buckling will occur, but the direction is unpredictable. This is a feature of spontaneous symmetry breaking. The breaking takes place, but one does not know in which direction.

The Higgs field thus acts as the finger applying pressure to the system to cause it to buckle and lose its symmetry. It is a deep result of the quantum field theory, that when this happens, the W and Z particles (which are messengers of the symmetries broken by the Higgs) acquire masses. In other words, without spontaneous symmetry breaking, the W and Z particles would be massless.

The Higgs field produces masses for the quarks and the electrically charged leptons through its interactions with these fields. These masses are proportional to the strength with which the Higgs couples with the particles. The top quark weighs the most (about 200 times the mass of a proton), because the Higgs interacts most strongly with the top quark. The electron interacts very weakly with the Higgs, and that is why it is the lightest particle (about 2000 times lighter than a proton). Physicists do not understand why the Higgs couplings differ so greatly. Certain features of the Standard Model will remain a mystery, even if the CERN experiment is confirmed and the Higgs mechanism is realized.

So, if one of the most important scientific discoveries has been made, why has there been so little news about it? The answer is that because of its great importance, experimentalists must be sure of the result before announcing it.

Finally, there are four experimental detectors at CERN. Of these four, only Aleph is seeing convincing evidence of Higgs production. That detector sees three Higgs-candidate events. Another detector, Delphi, also has possibly produced one Higgs in a single positron-electron collision. Although Aleph detects that the Higgs has been seen with better than 99% confidence, no strong claims can be made with so few events.

Conclusion

In 2002, researchers at the Brookhaven National Laboratory on Long Island, NY announced that they had accomplished something unprecedented: the creation of a form of matter not seen since the first tick of the history of the universe. That form of matter, a super-dense state called a quark-gluon plasma, has long been a goal of particle physicists. In this plasma, the protons and neutrons that make up atomic nuclei are shattered into a cloud of quarks and gluons (particles that carry the force that normally keeps quarks together). The last time quarks had been rolling around loose (rather than bound up in protons, neutrons, or other subatomic particles) was only one millisecond after the Big Bang, when the whole universe was a toasty 1.8 trillion degrees Fahrenheit. But the evidence showing they had created the plasma was not conclusive, and it may be until 2005 before Brookhaven physicists are confident enough to declare success.

Inaugurated in 2000, the collider was designed to create novel states of matter by smashing together larger bits of matter at higher speeds than had been achieved before. These big bits of atoms (such as gold or silicon atoms stripped of their electrons) are much more complex than the protons that physicists have been accustomed to studying. That complexity means physicists aren't terribly sure what they will see when they smash things together.

After a shakedown period of several months, the collider began running at full energy for the first time in July of 2001, with a series of experiments. But, even at two-thirds power in January 2002, physicists indicated that the collider provided evidence of phenomena that are difficult to explain. The number of particles spawned by a collision of two atomic nuclei of gold is larger than what one would get by smashing together the individual particles that make up those nuclei. According to physicists, where those additional particles come from is anyone's guess.

Finally, for now, the big prize is the quark-gluon plasma. Physicists running a competing experiment at the CERN lab outside Geneva announced that they had created such a plasma in 2000. Their report was quickly criticized as premature, and Brookhaven National Laboratory does not want to repeat that mistake. Before Brookhaven National Laboratory physicists declare they have proof of a quark-gluon plasma, they want to make sure every potential source of error is found and accounted for. It's an important announcement. And though physicists might not have bulletproof evidence, they want it to be bullet-resistant.

Reference

[1] Leon M. Lederman, Pritzker Professor of Physics, 1988 Nobel Laureate in Physics, Fermilab Director Emeritus, Batavia, IL 60510.

Particle Physics and Astrophysics: The Solar Neutrino Problem

"The mathematician lives long and lives young; the wings of his soul do not early drop off, nor do its pores become clogged with the earthy particles blown from the dusty highways of vulgar life."
—James Joseph Sylvester (1814–1897)

Most people are under the impression that scientific discoveries happen rather suddenly: On a particular day, all experiments lead to a remarkable result, or perhaps, after a week of analyzing a years worth of data, a radical new effect is uncovered. Indeed, many scientific breakthroughs transpire in this manner. But a few occur gradually, taking years or even decades to unravel. The realization that neutrinos have mass is an example of the latter. But what are neutrinos?

Neutrinos: What Are They?

Neutrinos are fundamental, subatomic particles that are similar to electrons, but are neutral—meaning that they possess no electric charge (see Figure 6.1).[1] The lack of charge renders the behavior of neutrinos very different from the behavior of electrons. Because of electromagnetism, an electron is attracted to or repelled by another object most of the

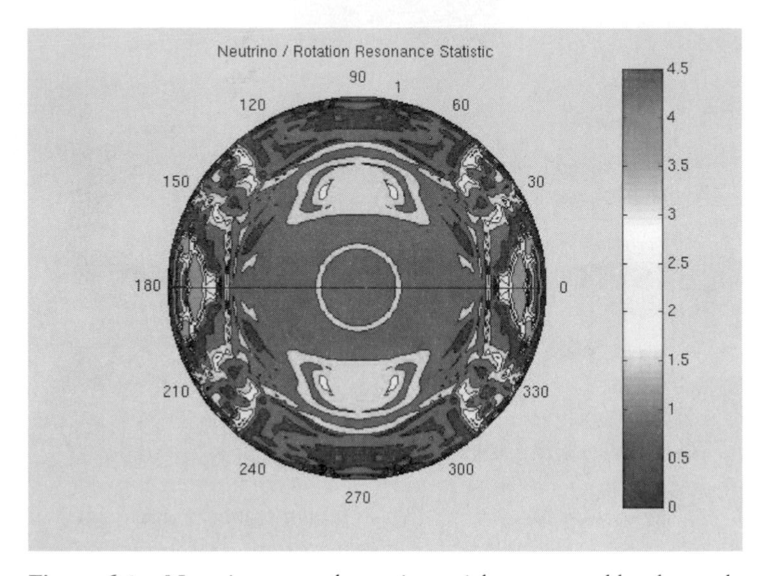

Figure 6.1 Neutrinos are subatomic particles generated by the nuclear reactions that power stars like our sun. Flying outward from the sun's core, they easily pass through the sun (and almost anything else!) unimpeded and should be detectable by earth-based neutrino "telescopes." Still, to the long-standing consternation of astrophysicists, the observed flux of solar neutrinos is less than expected. In a new twist to this solar neutrino saga, an analysis of data from the GALLEX/GNO neutrino detector finds a solar neutrino flux that varies over about 27 days—approximately matching the sun's rotation period. In fact, since different parts of the sun rotate at different rates, the neutrino flux variations match most exactly the rotation rates of the areas shown on this cross-sectional map of the solar interior rotation. So how could solar rotation affect the neutrino flux? Some theoretical models say that neutrinos can change quantum properties when they interact with tangled solar magnetic fields and become particles that the neutrino experiments were not designed to detect. Then, as the sun rotates, the neutrinos sometimes come to us unaffected and sometimes come through magnetic fields, diminishing the flux that can be measured.

time, but a neutrino experiences no electromagnetic force because it is neutral. Neutrinos have extremely weak interactions for this reason—they hardly deflect off other matter. When they do deflect, it is due to the subnuclear weak force, a feeble interaction that is, among other things, responsible for certain decay of nuclei. For example, consider the following in order to get an idea of how weakly interacting neutrinos are. A neutrino released in a nuclear decay will travel on average a distance of roughly one light-year (the length of time it would take for you to travel a year if you were going at the speed of light) through lead (neutrinos can travel through solid matter) before changing its direction of motion! There is only about one chance in ten that a neutrino will bounce off a nucleus in your body, despite the fact that about 100 cosmic-ray-generated neutrinos pass through your body every second!

There are three types, or "flavors," of neutrinos:

- One that is associated with the electron
- One that is associated with the muon
- One that is associated with the tau lepton

The muon and tau are heavier versions of the electron.

NOTE

The mass of the muon and the tau is about 200 and 3,500 times, respectively, the mass of an electron.

Therefore, only electron-neutrinos are initially produced when electrons are involved in a high-energy scattering process. (Ditto for tau- and muon-neutrinos.)

Neutrinos spin in a very specific way when they are initially produced: If you take your left hand and point your thumb in the direction that the neutrino is moving and let your fingers curl naturally, the fingers indicate the spinning direction. For this reason, neutrinos are called left-handed particles. The Standard Model of particle physics has only left-handed neutrinos. Adding in right-handed neutrinos is the simplest way to incorporate massive neutrinos in the Standard Model.

NOTE

Neutrinos, like electrons, can possibly spin in only two directions; such particles are called spin-one-half fermions.

Except through the effects of neutrino masses, right-handed neutrinos do not experience any forces. If neutrinos have mass, then it is expected that they should also oscillate: Electron-neutrinos can transform into muon-neutrinos or tau-neutrinos and then go back into electron-neutrinos. Similarly, muon-neutrinos and tau-neutrinos can undergo oscillatory flavor changes. The evidence is now convincing that the solar neutrino problem and the atmospheric neutrino anomaly are solved by neutrino oscillations (see sidebar, "Solving the Solar Neutrino Problem").

But, do neutrinos really have mass? Let's take a look.

Do Neutrinos Have Mass?

Nuclear reactions such as those that create heavy elements also create vast numbers of ghostly subatomic bits known as neutrinos. These belong to a group of particles called leptons, such as the familiar electron and the muon and tau particles. Because neutrinos barely interact with ordinary matter, they can allow a direct look into the heart of a star. This works only if physicists are able to capture and study them, something they are just now learning to do.

Not long ago, physicists thought neutrinos were massless, but recent advances indicate that these particles may have a small mass. Any such evidence would also help validate theories that seek to find a common description of three of the four natural forces—electromagnetism, strong force, and weak force. Even a tiny bit of heft would add up because a staggering number of neutrinos are left over from the Big Bang.

Neutrinos Do Have Mass

The first indications that neutrinos have mass arose about 36 years ago. In 1968, Dr. Ray Davis [2] and his colleagues published the initial results of an experiment that measured neutrinos from the sun using a large container of cleaning fluid as a neutrino detector in the Homestake Mine in South Dakota. The number of electron-neutrinos detected was less than the figure predicted from theoretical solar models. The experiment was continued for two decades and the theory of the sun was improved, but the deficit in the observed number of neutrinos persisted, thereby creating the solar neutrino problem. (See sidebar, "Solving the Solar Neutrino Problem.")

Solving the Solar Neutrino Problem

Two of the three winners of the Nobel Prize in Physics for 2002 worked on the Solar Neutrino Problem. What is this problem? Has it been solved?

According to astronomer Dr. Scott Gaudi, "The essence of the solar neutrino problem is that astrophysicists can predict the number of neutrinos produced by the nuclear reactions that power the sun, but the number of neutrinos directly detected here at the Earth are smaller than this. This is only a problem under the assumption of the standard model of particle physics that the neutrinos remain unchanged between the sun and the Earth. It has not been solved, per se. It has been shown conclusively that the problem exists, and this has led to the conclusion that the neutrinos must change their properties between the sun and Earth, and therefore that the standard model of physics is incomplete." [3]

NOTE

Those two Nobel Prize winners are Raymond Davis [2], who started measuring neutrinos from the sun with a huge container of dry cleaning fluid in a mine in South Dakota, and Masatoshi Koshiba, who led the Kamiokande neutrino experiment in Japan that confirmed and extended the solar neutrino measurements.

The sun, like most stars, is ultimately powered by fusion reactions gained by the energy when simpler atomic nuclei like hydrogen combine to form heavier nuclei like helium. Fusion processes, which take place in the core of the sun, produce a huge flux of neutrinos. These neutrinos can be detected on Earth using large underground detectors, such as those built by Davis and Koshiba. [2] The neutrino flux has been measured to see if it agrees with theoretical calculations based upon physicists' understanding of the workings of the sun and the details of the Standard Model of particle physics. The measured flux is roughly one half of that expected from theory (especially the work of John Bahcall, although checked by many others). This discrepancy between theory and observation is the "Solar Neutrino Problem."

The cause of the deficit was a mystery. Therefore, is the Standard Model of particle physics wrong? Is the model of the solar interior wrong? Are the experiments in error? "It has been proven," explains Dr. Gaudi, "in a model independent way, that in fact the models of the Sun are correct, and it is the Standard model of particle physics that is wrong (or actually, just incomplete)." [3]

Thanks to recent measurements (particularly those with the Super-Kamiokande detector in Japan and the Sudbury Neutrino Observatory in Canada, there is probably an answer to how the sun's nuclear fusion source of energy operates as predicted, but the neutrinos themselves seem to behave in a way unexpected by the Standard Model of particle physics. In the Standard Model, neutrinos are massless particles that travel at the speed of light. The three types of neutrino, called electron, muon, or tau neutrinos were thought to be different particles, completely independent of each other. The new results strongly suggest that this part of the Standard Model is incorrect. Neutrinos have a small mass and can "oscillate" or change from one type of neutrino into another. The change actually occurs before the neutrinos escape from the sun. The reason the earlier underground experiments were seeing a deficit of neutrinos is that they were not detecting all three types. The Sudbury results do measure all three types of neutrinos, and their results match the theoretical predictions as well.

As so often happens in science, solving one mystery reveals a new one. The new puzzle is how to correct the Standard Model of particle physics to include these new neutrino results.

Correcting the Standard Model of Particle Physics

Scientists may soon have to go back to the drawing board to develop a new Standard Model of physics that includes extra forces, particles, or dimensions above and beyond those already accounted for. According to a report by a 45-member collaboration of American physicists, they began suspecting problems with the current model when neutrinos generated in the lab didn't behave as predicted.

Using the Tevatron accelerator (the world's most energetic particle accelerator) at the Fermi National Accelerator Laboratory in Batavia, Illinois, the team of physicists generated billions of neutrinos with a 100-million-watt proton beam, slammed them into a series of target nuclei, and observed the aftermath. According to theory, when neutrinos would strike a target nucleus, they could either pass on through or turn into a muon, a heavier cousin of the electron. To their surprise, the team found a one percent discrepancy between the results and the theoretical predictions.

It might not sound like much, but a room full of physicists fell silent when the discrepancy was first revealed. The team eventually reached the conclusion (after spending a year evaluating the data) that the neutrinos acted according to prediction in only one out of every 400 trials. The team isn't absolutely sure how to explain the discrepancy, but it could be explained by unknown particles or forces that are altering the expected neutrino interactions.

Physicists around the world are developing future accelerator experiments that will be capable of solving this riddle for good. Others, in the meantime, are using accelerators to search for evidence of missing particles and forces.

If some of the electron-neutrinos that are created in the energy-producing core of the sun mutate into one of the two other known types of neutrinos, the muon-neutrino and the tau-neutrino, this discrepancy could be explained. Because of the eventuality that the transformed neutrino will change back into its original form only to undergo the transformation again and again, such a process is called neutrino oscillation.

Therefore, unless neutrinos have mass, it is a fundamental result that neutrino oscillations cannot occur. The radiochemical experiment of Ray Davis indicated that neutrinos are massive, but this finding was not conclusive. Many astrophysicists did not believe that the solar models were sufficiently accurate and they were predicting too large a solar output. Others felt that Davis and his co-workers were not measuring all the neutrinos that they should have been. Because neutrinos are extremely weakly interacting particles, rending them the most difficult particles in nature to detect, such an error in the experiment would not be surprising. So, not knowing whether there was a problem (1) with the theory, (2) with the experiment, or (3) with the Standard Model of particle physics that assumes neutrinos to be massless, for several decades scientists lived with the solar neutrino problem.

The elusive nature of the neutrino and the inherent difficulty in its detection has created many false signals. In their experiments involving nuclear beta decay, a team of Russian physicists announced that electron-neutrinos had a mass in the several electron-Volt range. However, subsequent beta decay experiments did not confirm the result.

NOTE

An electron-Volt, or eV, is actually a unit of energy. Particle physics measure masses in terms of energy using $E = mc^2$. An electron-Volt is 1.78×10^{-36} kilograms.

A new neutrino mystery arose in 1985. Cosmic rays are elementary particles such as protons and electrons or electromagnetic radiation such as X-rays and gamma rays that travel throughout space. They are produced by many kinds of astrophysical objects such as stars, including our sun, supernovae, black holes, neutron stars, active galactic nuclei, and quasars. When such cosmic rays strike the Earth's atmosphere, pions are often produced.

NOTE

A pion is the lightest or least massive type of mesons. They are copiously produced in high energy particle collisions. Pions can have electric charges of +1, –1, or 0.

The charged pion decays into a muon and a muon-neutrino. The muon then decays into an electron, a muon-neutrino and an electron-neutrino. Thus there should be about twice as many muon-neutrinos as electron-neutrinos streaming down on Earth from cosmic rays. (See sidebar, "Frozen Cosmic Ray Neutrinos.")

Frozen Cosmic Ray Neutrinos

A field of sensors buried deep in Antarctica has produced a novel view of the sky: a map that shows not light but neutrinos, ethereal particles emitted by star-munching black holes, supernova explosions, and other violent phenomena. Neutrinos have no charge and almost no mass, so they zip unflinchingly through space and matter. That quality makes them ideal for probing the most inaccessible corners of the universe, but also makes them almost impossible to detect.

An international team of physicists built the Antarctic Muon and Neutrino Detector Array II (AMANDA II) in order to find the astronomical neutrinos. This is an array of 677 bowling ball-sized light detectors strung along 19 cables and sunk nearly half a mile down into the polar ice. A high-energy neutrino from the northern sky burrows through Earth and crashes into a water molecule near one of the sensors several times a day. Called Cerenkov radiation, the resulting subatomic chain reaction produces brief trails of blue light zigging through the ice. Those signals are picked up and amplified by the detectors. The original course of the incoming neutrino is indicated by the path of the radiation.

AMANDA II picked up 1,000 neutrino collisions during its first year of operation. The team of physicists processed the data into the first-ever neutrino sky map. In order to track down the brightest neutrino emitters, the researchers are now combing through a second and third year of data. Some may be the long-sought sources of the astoundingly energetic particles called cosmic rays that rain down on Earth from parts unknown. The physicists have no clue where they come from because their paths are bent and scrambled by Earth's magnetic field. Why? "Neutrinos travel along straight lines because they don't interact with things very well (this is also the reason they are so hard to detect!)," explains Physicist and Astronomer Dr. Andrea Ghez.[4]

Neutrinos, however, would fly on an arrow-straight path from their point of origin—perhaps places where hot gas falls into supermassive black holes.

NOTE

According to the physicists, one exciting possibility is that the neutrino sky map will reveal the sources of cosmic rays as well.

Earlier measurements of atmospheric neutrinos had error bars sufficiently large as to be compatible with this 2–1 ratio. However, in 1985, two experiments in deep mines (one in the United States called Irvine Michigan Brookhaven (IMB) and one in Japan called Kamiokande) observed that the ratio was less than two-to-one. Either more electron-neutrinos or fewer muon-neutrinos were being detected. This became known as the atmospheric neutrino anomaly. It again could be explained by neutrino oscillations; however, the oscillations could not be of the same type as those that would explain the solar neutrino problem. Furthermore, two other solutions to the atmospheric neutrino anomaly were possible: The experimental difficulty in detecting neutrinos might be producing an incorrect result, or the theory of cosmic ray showers could be wrong.

Then an experiment in England in the late 1980s announced that a neutrino had a mass of 17,000 KeV. This became known as the 17 KeV neutrino (kilo-electron-Volts). Eventually the experiment was found to be faulty, because other physicists were unable to reproduce the result.

In the early 1990s, two new solar neutrino experiments (the Soviet–American Gallium Experiment [SAGE] in Russia and the Gallium Experiment [GALLEX] in Italy) that used gallium as a target confirmed the experimental deficit in solar neutrinos. At this point, it seemed clear to most physicists that an error in the experiment could be ruled out. Because the possibility still existed that the theoretical solar models were in error in regard to solar neutrino production, few physicists were ready to announce that neutrinos had masses.

An experiment called Liquid Scintillator Neutrino Detector (LSND) at Los Alamos in 1996 surprised the world by announcing that it had seen the oscillation of muon-anti-neutrinos to electron-anti-neutrinos: 22 events had been observed whereas four had been expected. Already ruled out by experiments in Europe, however, the result was near the border of a region of neutrino parameters. In addition, subsequent other accelerator experiments did not support the LSND findings. Now, a majority of scientists believe that the Los Alamos experiment was a fluke.

The "official" values for the masses of the three neutrinos were zero until the late 1990s. However, the combined analysis of a variety of experiments actually favored this bizarre result: The masses squared were negative! Because this is not physical, few scientists believed that the neutrinos could really be tachyonic (that is, have a negative m^2). In violation of Einstein's special theory of relativity, tachyonic particles should travel faster than the speed of light.

Quite a few theoretical extensions of the Standard Model of particle physics predicted that neutrinos had masses. Many theorists believed the neutrinos oscillated, since there were numerous theoretical suggestions and hints from experiments. For neutrinos to be massless particles, they felt that too many unusual things were happening.

Super-Kamiokande (Super K) was completed and began taking data in 1996. Super K is a scaled-up version of Kamiokande. It began detecting atmospheric neutrinos at rates much higher than its predecessors. After analyzing more than 500 days of data in 1998, the experimentalists at Super-K announced that the atmospheric neutrino anomaly was not a statistical aberration: The deficit in muon-neutrinos is due to the upward traveling ones. Oscillations easily explain this: Those muon-neutrinos raining down on the mine do not have sufficient time to oscillate, while those traveling through the Earth do. The scientific press reported that neutrinos had mass in early June of 1998. A majority of physicists became convinced that neutrino oscillations were a reality at this point. However, there remained a small chance that neutrinos did not have mass, given all the false signals of the past, the difficult nature of neutrino experiments, and the remote possibility that the theory of cosmic ray neutrino production was incorrect.

In 2001, that lingering uncertainty was eliminated. In a paper submitted on June 18, 2001, a team of almost 200 scientists released the first results from the Sudbury Neutrino Observatory (SNO), an experiment which uses a kiloton of heavy water (that is, water containing significant amount of deuterium) that is located in Sudbury, Ontario. (See sidebar, "SNO Status.")

SNO Status

In April 2002, the SNO group announced an improvement of its earlier June 2001 measurement that indicated that neutrinos have mass. Previously, SNO had to rely on data from the Super K to be better than 99% sure that electron-neutrinos were changing into muon- or tau-neutrinos on their way from the sun to the Earth. Such a metamorphosis, which is known as a neutrino oscillation, is only possible if the masses of neutrinos are non-zero. Now, an additional three year's worth of data has allowed SNO to eliminate its dependence on Super K. This is significant in that it is sometimes difficult to calibrate and compare results from different experiments. SNO was able to measure the total number of neutrinos being emitted by the sun and show agreement with standard solar models. If neutrinos were massless, all the sun's neutrinos should be of the electron type. However, SNO observes that only about one third are electron-neutrinos. This means that two thirds of the neutrinos are of the muon or tau type and that they changed into non-electron neutrinos sometime after being created in the core of the sun (most probably as they passed through the sun). The experimental results are in nice agreement with theoretical considerations and are so well established with data that there can be little doubt that neutrinos have mass.

With greater than 99% confidence, scientists believe that solar neutrinos are undergoing changes on their way from the sun to the Earth. When combined with data from Super K, SNO was able to demonstrate that the electron-neutrinos are changing into muon-neutrinos or tau-neutrinos. Therefore, the solar neutrino flux is consistent with the theoretical models of the sun, if neutrino oscillations are occurring. Of the three possible solutions to the solar neutrino problem previously mentioned, only one was viable: Neutrinos had to undergo oscillations and had to have masses.

In order to establish this result, it has taken more than 35 years to incorporate neutrino masses in the standard model of Particle Physics. Also, it is easy to modify the Standard Model of Particle Physics, but to determine precisely how this is done, additional experimental information is needed.

The most likely scenario is that solar electron-neutrinos from the sun oscillate into muon-neutrinos and perhaps tau-neutrinos, and some atmospheric muon-neutrinos oscillate into tau-neutrinos. The mass of the heaviest neutrino is roughly 0.05 eV; the mass of the next is somewhere between 0.01 eV and 0.0001 eV; and the third has a mass significantly less than these two values. But, for comparative purposes, the lightest non-neutrino particle is the electron, with a mass of 511,000 eV.

Thus, neutrinos can only account for a small fraction of the dark matter that permeates the Universe, because neutrino masses are less than a few electron Volts. As discussed in Chapter 1, the mystery of dark matter thus remains unresolved.

Finally, a major discovery has taken place in physics in a short time. It has taken 35 years to achieve the result, however.

Conclusion

The year 2002 went down as the year of the neutrino for physicists. In October of that year, Raymond Davis, Jr. of the

University of Pennsylvania and Brookhaven National Laboratory, shared a Nobel Prize for detecting solar neutrinos and discovering that the sun emits far fewer than expected of these ghostly subatomic particles. This was a finding that exposed a serious flaw in physicists' understanding of fundamental natural laws. Earlier in 2002, an international team of physicists conducted an elegant experiment that finally solved the enigma Davis had uncovered nearly 30 years earlier.

Standard models of how the sun shines tell exactly how many neutrinos should be created by nuclear reactions in the solar core. Checking those models proved quite a challenge. Neutrinos are so inert, they mostly pass right through Earth, but Davis managed to capture and count a few in an enormous underground detector. He was shocked: He found just one-third as many as theory had predicted. Repeated tallies have since confirmed the solar neutrino deficit. More recently, physicists at the Sudbury Neutrino Observatory in Ontario, Canada, and at the Super K detector in Japan, have provided a possible explanation. Neutrinos are known to exist in three varieties, called flavors, each of which is associated with another subatomic particle. Until recently, physicists could effectively detect only one flavor, the electron-neutrino. According to theory, that is the kind that should be generated by the nuclear fusion of hydrogen in the sun. Some physicists have speculated however, that certain solar neutrinos might transform en route into the other flavors, making them extremely difficult to find.

In 2002, preliminary evidence from the Sudbury and Super K detectors showed tentative hints of such neutrino transformations, but with limited statistical accuracy. Then in April of 2002, physicists working at Sudbury announced the results of a challenging new study that compared the total flux of all three neutrino types with the flux of the electron-neutrinos alone. The data showed conclusively that the bulk of the neutrinos had transformed into one or both of the other flavors, muon-neutrinos and tau-neutrinos. Physicists observed clearly that there are significantly more than just electron-neutrinos reaching Earth.

The results imply that, contrary to physicists' assumptions, neutrinos are not massless; otherwise, such transformations would not be possible. That finding is forcing researchers to revamp the standard model of physics, which describes the interactions of all the fundamental particles in the universe. This is the first major extension of the model in more than 25 years. Mass-bearing neutrinos would also account for some of the invisible matter thought to hold galaxies and galaxy clusters together. Neutrinos are a mystery that physicists are just beginning to understand.

Overall, no one doubts the reality of neutrinos now. To understand the phenomenon, you have to think of a neutrino not only as both particle and wave, which is typical quantum mechanics, but also as a combination of waves. Each wave has a slightly different frequency, and each frequency corresponds to a certain mass. When the neutrino is born, the waves are in step but, because of their different frequencies, the waves gradually drift out of step as the neutrino travels through space and time. The waves start to interfere with each other; at times one frequency dominates, at times another. At times the neutrino has one mass, at times another. At times it has one flavor, at times another. When you pin down one quality, the other blurs: It's the old quantum mechanical uncertainty again, but this is the very essence of a neutrino, as it is of no other fundamental particle.

Finally, physicists are still trying to figure out the basic properties of neutrinos. You don't have many constituents of matter. You only have 12. Neutrinos are one fourth of the population. How they interact with other particles, and whether they have a mass or not, are not trivial questions.

References

[1] "Astronomy Picture of the Day," [http://apod.gsfc.nasa.gov/apod/archivepix.html], NASA Headquarters, 300 E Street SW, Washington, DC 20024-3210; Jet Propulsion Laboratory, 4800 Oak Grove Drive, Pasadena, CA 91109; NASA/Goddard Space Flight Center, Greenbelt, MD 20771, 2003.

[2] Dr. Raymond Davis, Jr., physicist, University of Pennsylvania and Brookhaven National Laboratory, recently shared a Nobel Prize for detecting solar neutrinos and discovering that the sun emits far fewer than expected of these ghostly subatomic particles—a finding that exposed a serious flaw in our understanding of fundamental natural laws.

[3] Dr. Scott Gaudi, Astronomer, Harvard-Smithsonian Center for Astrophysics, 60 Garden Street, Cambridge, MA 02138.

[4] Dr. Andrea Ghez, Physicist and Astronomer, Department of Physics and Astronomy/IGPP, UCLA.

Astrophysics: The Source of Gamma-Ray Bursts

"Love is a matter of chemistry, but sex is a matter of physics."
—Anonymous

Gamma-ray bursts (GRBs) pose one of the greatest mysteries of modern astronomy. About once a day, the sky lights up with a spectacular flash, or burst, of gamma rays. More often than not, this burst outshines all of the other sources of cosmic gamma rays added together (see Figure 7.1). [1] The source of the burst then disappears altogether. Astrophysicists can't predict when the next burst will occur or from what direction in the sky it will come. At present, astrophysicists don't even know what causes these flashes or how far away they are!

The first gamma-ray burst was seen in the year 1967 (although it was not reported to the world until 1973) by satellite-borne detectors intended to look for violations of the Nuclear Test Ban Treaty. What was seen in the data by Ray Klebesadel and others at the Los Alamos National Laboratory was an unexpected, large increase in the number of gamma rays seen by detectors on several widely separated satellites. The data indicated that the source of the gamma-ray emission was not from the vicinity of Earth, but from an outer region of space! In their discovery paper, they reported the detection of 16 such events during the time from July 1969 to July 1972. There was no doubt that the phenomena they reported were

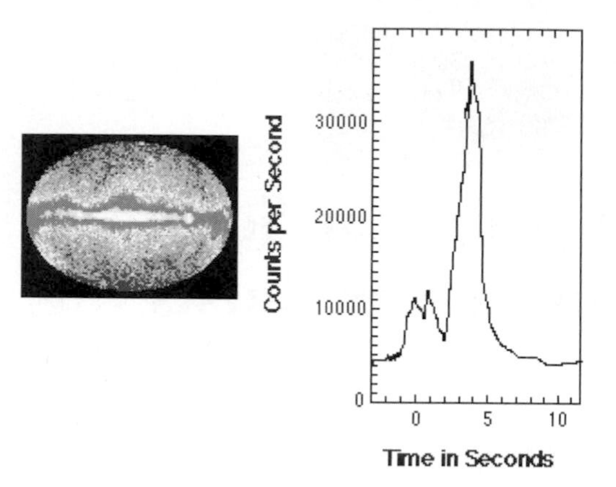

Figure 7.1 Using graphics and data from NASA's Compton Gamma Ray Observatory, this illustration shows one of the most exciting mysteries of modern astrophysics—gamma-ray bursts. Incredibly, gamma-ray bursts, sudden flashes of radiation with over 100,000 times the energy of visible light photons, occur several times a day. They typically last from fractions of a second to many minutes and appear from random directions, unexpectedly triggering space-based gamma-ray instruments. At left a burst suddenly appears, flickers, and fades in a false-color gamma-ray all-sky map, briefly overwhelming all other sources of celestial gamma-rays. The graph at right shows the corresponding response of an orbiting gamma-ray detector as its counting rate suddenly climbs and falls recording the passage of the mysterious burst. Originating far across the universe, gamma-ray bursts are now known to be the most powerful explosions since the Big Bang and may yet prove to be useful tools for exploring the distant cosmos. Future space and ground-based observatories will also work to discover the nature of the bursters and the source of their extreme energy.

real, and that no one had a reasonable explanation for it. Now, 25 years and data from several thousand bursts later, scientists still have no clear explanation for the strange celestial events they call gamma-ray bursts.

NOTE

Gamma rays are a highly penetrating type of nuclear radiation, similar to x-rays and light, except that it comes from within the nucleus of an atom, and, in general, has a shorter wavelength.

Gamma-rays emission is a decay mode by which excited state of a nucleus de-excite to lower (more stable) state in the same nucleus.

Why Scientists Study Gamma-Ray Bursts

Scientists are understandably excited about gamma-ray bursts. They are among the brightest light sources in the sky in any wavelength region. They may come from the most distant reaches of our galaxy—or they may represent an entirely new type of object near the edge of the observable universe. If the former is true, then astrophysicists are seeing something in the distant parts of the galaxy that is presently unobservable by any other means. If the latter is true, then gamma-ray bursts could help them to understand the form and evolution of the universe as a whole; they would be a new observational tool for cosmologists. Either way, the resolution of the mystery of gamma-ray bursts will result in a major breakthrough for astronomy (see Figure 7.2).[1]

What Scientists Know and What They Don't Know

Perhaps the most striking feature of the time profiles of gamma-ray bursts is the diversity of their time structures. Some bursts' light curves are spiky with large fluctuations on all time scales, while others show rather simple structures with few peaks (see Figure 7.3).[1] However, some bursts are seen with both characteristics present within the same burst! In a few cases, burst sources have repeated their appearance. However, no persistent, strictly periodic behavior has been seen from gamma-ray bursts.

The durations of gamma-ray bursts range from about 30 ms to over 1000 ms. However, the duration of a gamma-ray burst is difficult to quantify since it is dependent upon the sensitivity and the time resolution of the experiment that observes the

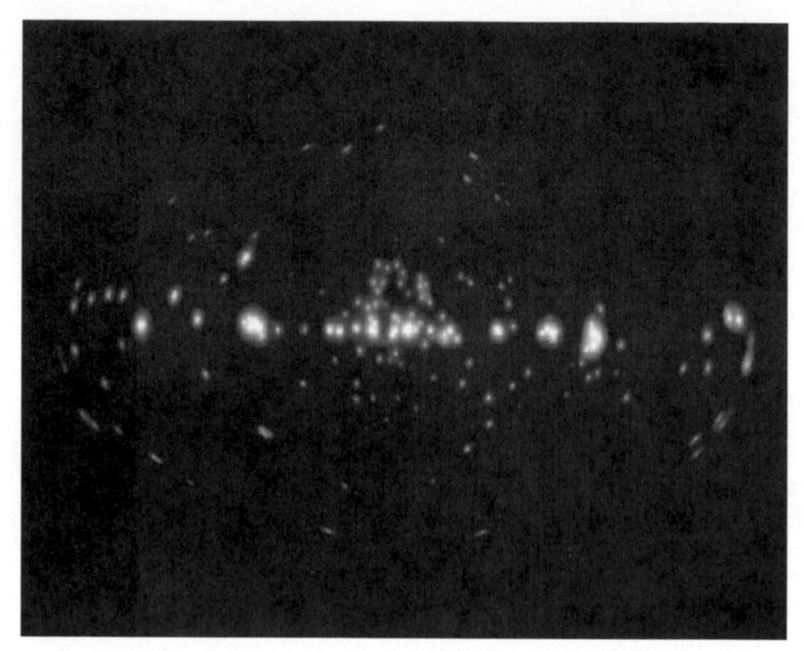

Figure 7.2 Gamma rays are the most energetic form of light, packing a million or more times the energy of visible light photons. If you could see gamma rays, the familiar skyscape of steady stars would be replaced by some of the most bizarre objects known to modern astrophysics—and some which are unknown. When the EGRET instrument on the orbiting Compton Gamma-ray Observatory surveyed the sky in the 1990s, it cataloged 271 celestial objects of high-energy gamma rays. Researchers identified some in relation to exotic black holes, neutron stars, and distant flaring galaxies. But 170 of the cataloged objects, shown in the above all-sky map, remain unidentified. Many objects in this gamma-ray mystery map likely belong to already-known classes of gamma-ray emitters and are simply obscured or too faint to be otherwise positively identified. However, astronomers have called attention to the ribbon of objects winding through the plane of the galaxy, projected here along the middle of the map, which may represent a large unknown class of galactic gamma-ray emitters. In any event, the unidentified objects could remain a mystery until the planned launch of the more sensitive Gamma-ray Large Area Space Telescope in 2005.

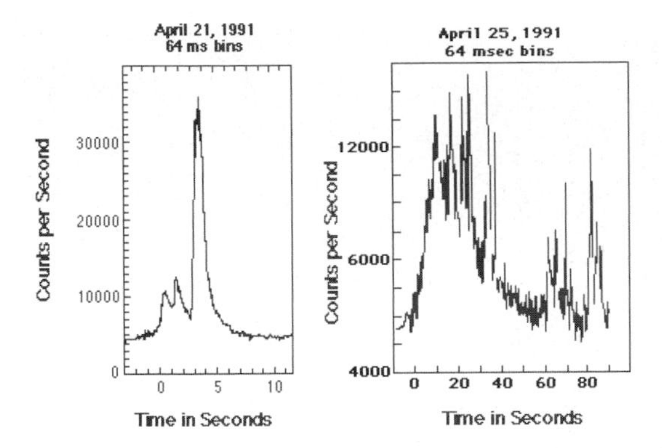

Figure 7.3 Astrophysicists know what gamma-ray bursts look like in time.

event. The "tip of the iceberg" effect tends to cause weaker bursts to be observed as shorter, since only the higher parts of the peak emission are observable.

Astrophysicists Know That They Come from Every Direction in the Sky

For a long time, astronomers thought that the source of a gamma-ray burst would be in our Milky Way Galaxy. If the source was in our own galaxy, it would be easier to explain how it produced the amount of energy it does (the farther away it is, the more energy it would have to contain for astrophysicists to still see it as bright as they see it). But being in our own galaxy would then cause the distribution of gamma-ray burst locations (that is, how astrophysicists see them spread out in the sky) to be concentrated along the galactic plane. This would be so because the galactic plane is where most of the stars are located in our galaxy, and it was believed that a gamma-ray burst had to be related to some stage of the life of a star.

However, this is not what astrophysicists see! Thanks to data primarily from the Burst and Transient Source Experiment

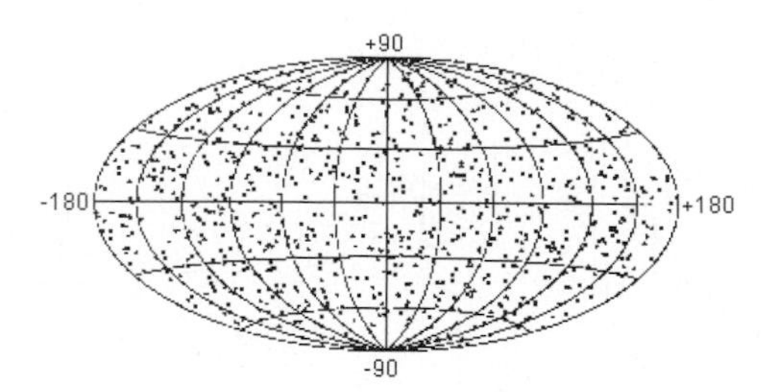

Figure 7.4 BATSE GRBs in Galactic Coordinates.

(BATSE) on the Compton Gamma-Ray Observatory, they know that bursts occur randomly all over the sky, as shown in Figure 7.4.[1] This makes it very hard for scientists to figure out what is causing the gamma-ray burst and where to look for the next burst. After all, the whole sky is a very big place to try to watch all at once!

In fact, this is one of the biggest problems with gamma-ray bursts. Since astrophysicists never know from where the next burst will come, they can't bring all of the different telescopes in the world (or in orbit) to observe the burst location and thus (hopefully) identify the object that emits the flash of gamma rays.

Astrophysicists Don't Know What Kinds of Objects Emit the Radiation

In fact, astrophysicists have absolutely no idea what kind, or kinds, of objects are responsible for gamma-ray bursts! As previously mentioned, astronomers first thought that the events were probably related to an activity of a star in some part of its life cycle. Then they saw enough bursts to understand that their initial ideas had to be essentially wrong. Now, astronomers just don't know. But scientists are never at a loss for ideas! Several ideas, or theories, about what kinds of objects are responsible for gamma-ray bursts that are being discussed these days are:

- Supernovae: The death explosions of massive stars (see Figures 7.5 and 7.6).[1]
- Neutron stars or black holes in a big halo that surrounds the outside of the part of our galaxy that we can see with our eyes.
- Some undefined type of very, very powerful object that is not necessarily in our galaxy in great abundance, but which can be found in all galaxies in the universe.

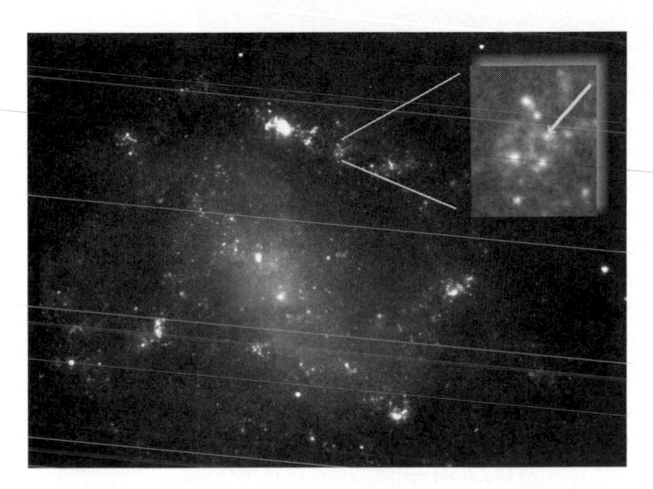

Figure 7.5 Modern astronomers keep a long list of things that go bump in the night. Near the top are supernovae—the death explosions of massive stars, and gamma-ray bursts—the most powerful explosions seen across the universe. Intriguingly, the galaxy in the above Hubble Space Telescope image may have been host to both a supernova and a gamma-ray burst, which were one and the same event. ESO 184-G82 is a spiral galaxy with a prominent central bar and loose spiral arms dotted with bright star-forming regions. The inset shows an expanded view of one of the star-forming regions, about 300 light-years across. Indicated is the location of an extraordinarily powerful supernova explosion whose light first reached planet Earth on April 25, 1998. That location and date also correspond to the detection of an unusual gamma-ray burst, which may be representative of a peculiar class of these cosmic high-energy flashes. So far, this combination is unique and makes barred spiral ESO 184-G82, at a distance of only 100 million light-years, the closest known gamma-ray burst host galaxy.

Figure 7.6 New evidence has emerged that a mysterious type of explosion known as a gamma-ray burst is indeed connected to a supernova of the type visible in the above image. On April 1, 2003, the orbiting HETE satellite detected gamma-ray burst GRB 030329. The extremely bright burst was found hours later to have an extremely bright afterglow in visible light, and soon set the record for the closest measured distance at redshift 0.17. The afterglow brightness allows unprecedented coverage of its evolution. On April 14, 2003, as many astronomers suspected would happen, the afterglow began to appear as a fading type II supernova. Type II supernovae might not appear coincident with gamma-ray bursts, however, when the gamma-ray beam goes in another direction. The above spiral galaxy, NGC 3184, was home to a type II supernova in 1999 at the position of the arrow. Astronomers are currently pressing hard to find the host galaxy for GRB 030329.

One way astrophysicists may find the answer to the question of what kind of objects are responsible for gamma-ray bursts is to find what is called a "counterpart" to the burst. The "counterpart" is an object that is connected to the object that emits the gamma-ray burst in a close way—for example, in a binary system. It is called the counterpart because it is an object that astrophysicists can use to study the gamma-ray burst-emitting object in an indirect way. Hopefully, they can see the counterpart in another part of the electromagnetic spectrum. This would then allow astrophysicists to bring a whole range of science tools to bear on what is causing the gamma-ray burst. These tools might include spectra, photometry, distance estimates, comparisons with other objects, and so on. Recently, such counterpart searches have paid off.

How the Mystery Will Be Solved

Until recently, the big problem was that, with the exception of soft gamma-ray repeaters (which may be a distinct class of objects), no gamma-ray burst had been seen in wavelengths longer than X-ray and no quiescent counterpart had been seen (see Figure 7.7).[1] This greatly hampered the study of gamma-ray bursts, since we didn't know what objects to concentrate on theoretically.

Now All of That May Have Changed!

Astronomers may have solved one of the longest standing puzzles in science. Two optical counterparts to recent gamma-ray bursts (GRBs) are being scrutinized by the most powerful telescopes on Earth and in space. The results of these observations seem to show the bursts originate billions of light-years from Earth.

The discoveries are a result of new instruments and new cooperation between gamma-ray, X-ray, and optical astronomy. The Gamma-Ray Observatory (GRO) detects on average one burst per day. Unfortunately, GRO cannot pinpoint the position of the object or objects causing the burst to a small region of the sky. This makes searches at other wavelengths more difficult. The launch of the BeppoSAX satellite changed that. Now, more precise positions of GRBs can be determined at X-ray wavelengths and relayed to astronomers using powerful optical telescopes on the ground and the HST. Data from these telescopes show the GRBs to originate far outside our Milky Way galaxy.

Figure 7.7 A fading afterglow from one of the most powerful explosions in the universe is centered in this false-color image from the space-based Chandra X-ray Observatory. The cosmic explosion, an enormously bright gamma-ray burst (GRB), originated in a galaxy billions of light-years away and was detected by the BeppoSAX satellite on February 22, 2001. GRB 010222 was visible for only a few seconds at gamma-ray energies, but its afterglow was followed for days by x-ray, optical, infrared, and radio instruments. These Chandra observations of the GRB's x-ray glow hours after the initial explosion suggest an expanding fireball of material moving at near light speed has hit a wall of relatively dense gas. While the true nature of gamma-ray bursters remains unknown, the mounting evidence from GRB afterglows does indicate that the cosmic blasts may be hypernovae—the death explosions of very massive, short-lived stars embedded in active star-forming regions. As the hypernova blasts sweep up dense clouds of material in the crowded star-forming regions, they may also trigger more star formation.

The first break in the case of the mysterious bursts came on February 28, 1997. Within hours of the detection of a burst designated GRB 970228 by the BeppoSAX satellite (see Figure 7.8) [1], astronomers began searching the sky near the source

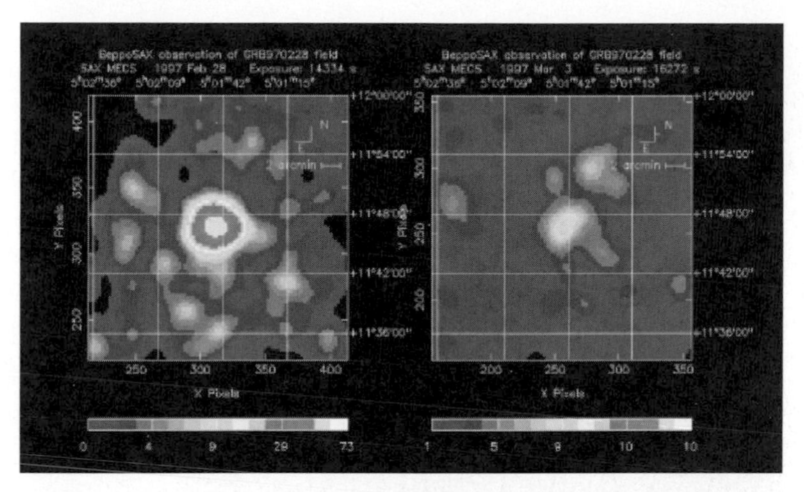

Figure 7.8 BeppoSAX observations of GRB970228. On the left is the original observation of the burst on February 28th 1997. On the right is the same region of the sky, in the constellation of Orion, March 3rd, 1997 showing that the burst object has dimmed.

of the X-rays with optical telescopes. They were rewarded with the first optical images of a GRB object.

Analysis of the HST images of the February 28, 1997 burst reveals that the burst object is associated with a faint, fuzzy patch of light dwarfed by the bright emission of the GRB source as shown in Figure 7.9. [1] This faint extended emission is presumed by many scientists to be a "host" galaxy from within which some cataclysmic event led to the GRB. The unmatched resolution of the Hubble telescope allowed astronomers to determine that the source of the burst does not lie at the center of the faint galaxy, but is offset, most likely in the disk population of normal stars. This would seem to rule out the possibility that the bursts are powered by massive black holes at the center of galaxies and suggests the products of typical stellar evolution like colliding neutron stars as GRB candidates. A galaxy like our own Milky Way could produce a bursting object every few million years, an explosion that for a few seconds outshines the entire galaxy.

Figure 7.9 Hubble Space Telescope image of the February 28th, 1997 burst location. The fuzzy patch of light to the lower left of the bright burst has been identified as a distant galaxy.

Within a week of the February outburst, the optical component had faded nearly out of sight, leaving only the faint smudge of its host galaxy as shown in Figure 7.10.[1] Of the more than 2,000 bursts detected by satellites in the past six years, only one burst location ever produced a second GRB. In all likelihood the source of the February 28 burst will not shine again.

Further insight to the source of GRBs came after another BeppoSAX observation, this time of a burst on May 8, 1997 as shown in Figure 7.11.[1] Again, within a few hours of the burst detection, telescopes all over the world were pointed towards the elusive source. Scientists at the Palomar Observatory identified an optical counterpart to the GRB that exhibited unusual variability in its brightness.

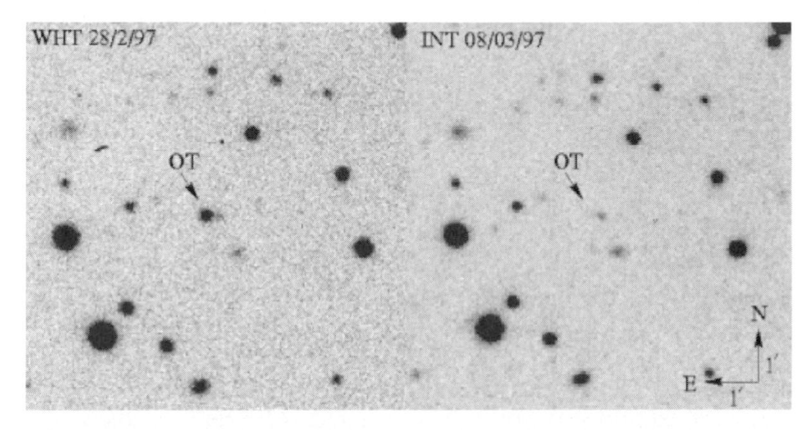

Figure 7.10 By March 8th, 1997, the GRB of February 28th has faded from sight.

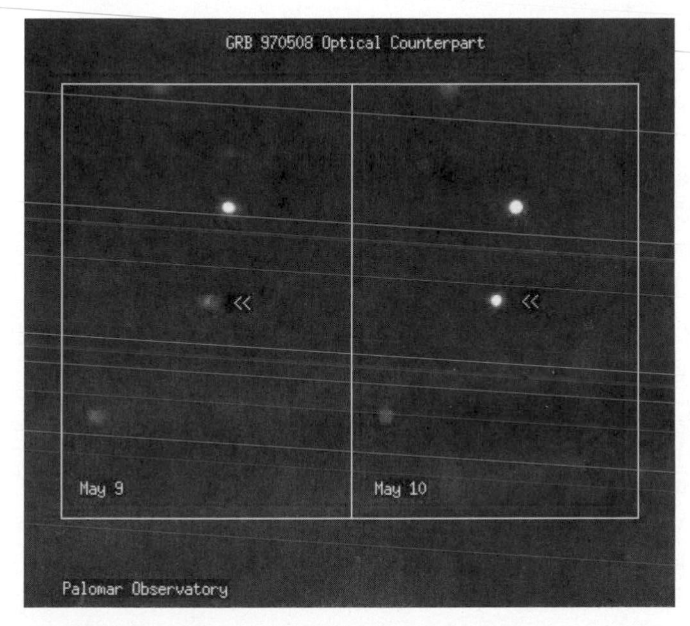

Figure 7.11 The variable optical component to the May 8th GRB.

The astronomers then turned to the world's most powerful ground-based telescopes, the 10-meter Keck pair in Hawaii. By examining the wavelengths of specific spectra features present in the light from the variable source, the scientists were able to determine the distance to the source object. The GRB host was found to lie at a redshift of at least 0.8, or several billion light-years from Earth, over half-way across the observable universe. This is definitive proof that at least some if not all of the observed gamma-ray bursts are due to objects far beyond our own galaxy. For a few seconds the burst was over a million times brighter than an entire galaxy.

GRO recorded a bright GRB event, GRB 970616, in June 1997. The position of the burst, accurate to about 2 degrees, was derived and promptly disseminated. The region of sky was scanned by the Rossi X-Ray Timing Explorer (RXTE) within four hours of the GRO discovery. RXTE discovered an X-ray source several times brighter than those detected previously by SAX. Meanwhile, a greatly improved position was derived by combining the GRO measurement with a separate detection of GRB 970616 by the Ulysses spacecraft. The RXTE X-ray source lies within this region of the sky. Follow up observations with RXTE one day later revealed that the source had faded below the detection limit. Thus, the source detected by RXTE was, in all likelihood, the X-ray counterpart to GRB 970616, and follow up studies at optical wavelengths can be performed.

Finally, the three NASA satellites (GRO, RXTE, and Ulysses), working in tandem, offer the possibility of greatly expanding the number of gamma-ray burst counterpart detections. No other events produce so much energy in so short a time. Gamma-ray bursts will continue to be studied in an attempt to discover just what can cause such violent and spectacular displays.

Conclusion

All over the world, telescopes have helped provide the data that gamma-ray bursts are produced when massive stars explode and their cores collapse to form black holes. An international team of astronomers from the California Institute of Technology (Caltech) got their proof from a gamma-ray burst (GRB 011121) that occurred in November 2001.

NOTE

The exact form of the changing brightness and color of the transient allowed astronomers at Caltech to show the presence of a supernova, which has now been designated SN 2001 ke by the International Astronomical Union.

Since supernovae are known to be exploding stars, the implication is that this gamma-ray burst originated in the explosive death of a massive star. At the same time, the Caltech team was able to determine the nature of the environment surrounding the explosion site: a cocoon of material from a stellar wind similar to those seen around massive stars in our own Milky Way Galaxy. The data were obtained from December 4, 2001 to May 5, 2002 using the Hubble Space Telescope WFPC2 instrument and its F555W, F702W, and F814W filters.

With that in mind, and as previously explained, Gamma-ray bursts are enormous blasts of gamma rays, often accompanied by an afterglow of light, X-rays, and radio waves. Their cause has been unknown, although evidence has been building since 1997 that massive stars are the culprits.

For example, careful sleuthing with the Hubble Space Telescope, CSIRO's Australia Telescope, Compact Array radio telescope, the Anglo-Australian Telescope, and telescopes in Chile showed that gamma-ray burst GRB 011121 had in fact been accompanied by the explosion of a massive star—a supernova. The Hubble Space Telescope picked up the telltale light "signature" of a supernova, while the ground-based telescopes showed that the explosion had taken place in a cocoon of matter shed by the star before its demise.

"Very recently," explains astronomer Dr. Scott Gaudi, "a very bright gamma-ray burst occurred, that enabled scientists from all of over the world to obtain precise, spectroscopic measurements of the optical afterglow. These measurements showed conclusively that the gamma-ray bursts are related to supernovae, the death of massive stars. Therefore, one of the primary mysteries about gamma-ray bursts, namely what are the progenitors, has been solved." [2]

By comparing the infrared and optical observations, astrophysicists were able to determine the amount of dust between them and the gamma-ray burst—something that hasn't been done before. The dust data confirmed that the gamma-ray burst occurred inside matter shed by the parent star.

The idea that gamma-ray bursts and supernovae are linked goes back more than a decade, but it got a boost in 1998 when Australian telescopes (the Australia Telescope, the Anglo-Australian Telescope, and telescopes of the ANU) saw for the first time a supernova that appeared to coincide with a gamma-ray burst. It was this supernova (SN 1998bw) that really pushed people to think about the link between gamma-ray bursts and supernovae.

GRB 011121 was detected in the southern sky on November 11, 2001 by the Italian-Dutch satellite BeppoSAX. Its position was further pinned down by a network of satellites. Optical telescopes in Chile identified the glowing embers of the gamma-ray burst and determined that the burst was located in a relatively nearby galaxy, only five billion light-years from Earth.

Gamma-ray bursts can be seen for vast distances across the universe. If they are caused by the core collapse by massive stars, it may be possible to use them to trace star formation in the early universe, perhaps even back to the first generation of stars.

All is not yet known about gamma-ray bursts. It may be that some bursts are produced by other exotic phenomena, such as two colliding neutron stars or a neutron star colliding with a black hole.

References

[1] "Astronomy Picture of the Day," [http://apod.gsfc.nasa.gov/apod/archivepix.html], NASA Headquarters, 300 E Street SW, Washington DC 20024-3210; Jet Propulsion Laboratory, 4800 Oak Grove Drive, Pasadena, CA 91109; NASA/Goddard Space Flight Center, Greenbelt, MD 20771, 2003.

[2] Dr. Scott Gaudi, Astronomer, Harvard-Smithsonian Center for Astrophysics, 60 Garden Street, Cambridge, MA 02138.

Theoretical High-Energy Physics: The Unification of the Basic Forces

"When I investigate and when I discover that the forces of the heavens and the planets are within ourselves, then truly I seem to be living among the gods."
—Leone Battista Alberti (1404–1472)

The unification field theory, in theoretical high-energy physics, is an attempt to describe all fundamental or basic forces and the relationships between elementary particles in terms of a single theoretical framework. Basic forces can be described by fields that mediate interactions between separate objects. So, what does unification mean?

- Two seemingly different forces are unified when a theory is found that combines them.
- The two forces then become manifestations of a single theoretical structure.

Examples of Unification

In the 19th century, scientists were able to unify the electric and magnetic forces into one structure called electromagnetism. James Clerk Maxwell formulated the first field theory in electromagnetism. Then, in the early part of the 20th century, Albert Einstein developed general relativity, a field theory of gravitation. Later, Einstein and others attempted to construct a unified field theory in which electromagnetism and gravity would emerge as different aspects of a single fundamental field. They failed, and to this day gravity remains beyond attempts at a unified field theory.

Forces That Have Not Been Unified

The two other fundamental forces, gravity and the strong nuclear force, remain as independent isolated forces. However, theorists suspect, but do not know for certain, that all the fundamental forces will emerge from one theory. This theory is called the Uni-Law.

Unification of Other Forces

At subatomic distances, fields are described by quantum field theories, which apply the ideas of quantum mechanics to the fundamental field. In the 1940s, quantum electrodynamics (QED), the quantum field theory of electromagnetism, became fully developed. In QED, charged particles interact as they emit and absorb photons (minute packets of electromagnetic radiation), in effect exchanging the photons in a game of subatomic "catch." This theory works so well that it has become the prototype for theories of the other forces.

All remaining forces (friction, string tension, support forces, the forces in springs, collision forces, atomic forces, molecular forces, etc.) are produced by the fundamental forces. Therefore, it is only necessary to unify the four fundamental interactions of electromagnetism, gravity, the weak subnuclear force, and the strong nuclear force.

Also, during the 1960s and 1970s, particle physicists discovered that matter is composed of two types of basic building blocks—the fundamental particles known as quarks and leptons. The quarks are always bound together within larger observable particles, such as protons and neutrons. They are bound by the short-range strong force, which overwhelms electromagnetism at subnuclear distances. The leptons, which include the electron, do not "feel" the strong force. However, quarks and leptons both experience a second nuclear force, the weak force. This force, which is responsible for certain types of radioactivity classed together as beta decay, is feeble in comparison with electromagnetism. So what is the best candidate for unification?

The Best Candidate for Unification

In the 1960s, the weak subnuclear force was united with electromagnetism to form the electroweak theory, which was subsequently verified in particle accelerator experiments. Sheldon Glashow, Abdus Sitlam, and Steven Weinberg independently proposed a unified "electroweak" theory of these forces, based on the exchange of four particles: the photon for electromagnetic interactions, two charged W particles, and a neutral Z particle for weak interactions.

At the same time that the picture of quarks and leptons began to crystallize, major advances led to the possibility of developing a unified theory. Theorists began to invoke the concept of local gauge invariance, which postulates symmetries of the basic field equations at each point in space and time. Both electromagnetism and general relativity already involved such symmetrics, but the important step was the discovery that a gauge-invariant quantum field theory (see sidebar, "Grand Unified Theory") of the weak force had to include an additional interaction; namely, the electromagnetic interaction.

Grand Unified Theory

During the 1970s, a similar quantum field theory was developed for the strong force, called quantum chromodynamics (QCD). In QCD, quarks interact through the exchange of particles called gluons. The aim of researchers now is to discover whether the strong force can be unified with the electroweak force in a Grand Unified Theory (GUT). GUTs unite electromagnetism, the weak subnuclear interaction, and the strong nuclear force into a single theory. Gravity is not included. It is possible that a GUT theory will emerge as part of the structure of string theory.

There is evidence, though, that the strengths of the different forces vary with energy in such a way that they converge at high energies. However, the energies involved are extremely high (ultrahigh), more than a million million times as great as the energy scale of electroweak unification, which has already been verified by many experiments.

Where Do Ultrahigh-Energy Particles Come From?

The most energetic particles that strike us from space, which include neutrinos as well as gamma-ray photons and various other bits of subatomic shrapnel, are called cosmic rays. They bombard Earth all the time; a few are zipping through you as you read this chapter. Cosmic rays are sometimes so energetic, they must be born in cosmic accelerators fueled by cataclysms of staggering proportions. Scientists suspect some sources: the Big Bang itself, shock waves from supernovae collapsing into black holes, and matter accelerated as it is sucked into massive black holes at the centers of galaxies. Knowing where these particles originate and how they attain such colossal energies will help physicists understand how these violent objects operate.

Is a New Theory of Light and Matter Needed to Explain What Happens at Very High Energies and Temperatures?

All of that violence cited in the previous question leaves a visible trail of radiation, especially in the form of gamma rays—the extremely energetic cousins of ordinary light. Astronomers have known for three decades that brilliant flashes of these rays, called gamma-ray bursts, arrive daily from random directions in the sky. Recently, astronomers have pinned down the location of the bursts and tentatively identified them as massive supernova explosions and neutron stars, colliding both with themselves and black holes. But, even now, nobody knows much about what goes on when so much energy is flying around. Matter grows so hot that it interacts with radiation in unfamiliar ways, and photons of radiation can crash into each other and create new matter. The distinction between matter and energy grows blurry. Throw in the added factor of magnetism, and physicists can make only rough guesses about what happens in these hellish settings. Perhaps current theories simply aren't adequate to explain them.

Are There New States of Matter at Ultrahigh Temperatures and Densities?

Under extreme energetic conditions, matter undergoes a series of transitions, and atoms break down into their smallest constituent parts. Those parts are elementary particles called quarks and leptons, which, as far as physicists know, cannot be subdivided into smaller parts. Quarks are extremely sociable and are never observed in nature alone. Rather, they combine with other quarks to form protons and neutrons (three quarks per proton) that further combine with leptons (such as electrons) to form whole atoms. The hydrogen atom, for example, is made up of an electron orbiting a single proton. Atoms, in turn, bind to other atoms to form molecules, such as H_2O. As temperatures increase, molecules transform from a solid such as ice, to a liquid such as water, to a gas such as steam.

That's all predictable, known science, but at temperatures and densities billions of times greater than those on Earth, it's possible that the elementary parts of atoms may come completely unglued from one another, forming a plasma of quarks and the energy that binds quarks together. Physicists are trying to create this state of matter, a quark-gluon plasma, at a particle collider on Long Island. At still higher temperatures and pressures far beyond those scientists can create in a laboratory, the plasma may transmute into a new form of matter or energy. Such phase transitions may reveal new forces of nature.

These new forces would be added to the three forces that are already known to regulate the behavior of quarks. The so-called strong force is the primary agent that binds these particles together. The second atomic force, called the weak force, can transform one type of quark into another (there are six different "flavors" of quark—up, down, charm, strange, top, and bottom). The final atomic force, electromagnetism, binds electrically charged particles such as protons and electrons together. As its name implies, the strong force is by far the most muscular of the three, more than 100 times as powerful as electromagnetism and 10,000 times stronger than the weak force. Particle physicists suspect the three forces are different manifestations of a single energy field in much the same way that electricity and magnetism are different facets of an electromagnetic field. In fact, physicists have already shown the underlying unity between electromagnetism and the weak force.

Finally, some unified field theories suggest that in the ultrahot primordial universe just after the Big Bang, the strong, weak, electromagnetic, and other forces were one, then unraveled as the cosmos expanded and cooled. The possibility that a unification of forces occurred in the newborn universe, is a prime reason particle physicists are taking such a keen interest in astronomy, and why astronomers are turning to particle physics for clues about how these forces may have played a role in the birth of the universe. For unification of forces to occur, there must be a new class of supermassive particles called gauge bosons. If they exist, they will allow quarks to change into other particles, causing the protons that lie at the heart of every atom to decay. And, if physicists prove that protons can decay, the finding will verify the existence of new forces.

The quantum field theory, upon which inflation now rests, may not be adequate to explain the universe's sudden expansion. Instead, physicists may need a theory of so-called "quantum gravity" to really understand the mechanism driving inflation. Quantum gravity, which is still in its formative stages, is an attempt to develop a "theory of everything" (TOE)—one that unites the intricacies of quantum mechanics with the force of gravity as described by Einstein's general relativity.

"Quantum gravity is no longer in its formative stages," explains Theoretical Physicist Dr. Lee Smolin. "There is a well developed approach to quantum gravity called loop quantum gravity, which after 17 years of work appears to be a consistent unification of the principles of general relativity and quantum theory." [2]

The underlying hope is to combine the strengths of both particle physics and cosmology into a single, powerful tool. According to physicist Dr. Alan Guth (MIT, Cambridge, MA), "We know that the universe is incredibly large: The visible part of the universe contains about 1,090 particles. Since we have all grown up in a large universe, it is easy to take this fact for granted: Of course the universe is big, it is the whole universe! In 'standard' Friedmann–Robertson–Walker cosmology, without inflation, one simply postulates that about 1,090 or more particles were here from the start. Many of us hope, however, that even the creation of the universe can be described in scientific terms. Thus, we are led to at least think about a theory that might explain how the universe got to be so big. Whatever that theory is, it has to somehow explain the number of particles, 1,090 or more. One simple way to get such a huge number, with only modest numbers as input, is for the calculation to involve an exponential. The exponential expansion of inflation reduces the problem of explaining 1,090 particles to the problem of explaining 60 or 70 *e*-foldings of inflation. In fact, it is easy to construct underlying particle theories that will give far more than 70 *e*-foldings of inflation. Inflationary cosmology therefore suggests that, even though the observed universe is incredibly large, it is only an infinitesimal fraction of the entire universe." [1]

String theory, which describes a universe with more dimensions than physicists can readily perceive, represents one approach to establishing a quantum gravity theory. Ultimately, physicists may have to describe inflation in terms of strings.

At the moment, there's a large variety of string theory models, which is a sign physicists don't really know enough to converge on one soon. The same holds for inflationary models, and physicists now think the two things are connected. If they're lucky, one day, physicists will converge on a unique theory, but they don't see that happening anytime soon. For the time being, though, string theory can potentially unify not only all the fundamental forces, but the elementary particles as well.

Hanging on a String

As it turns out, nature has been dropping hints over the past few decades that nothing may indeed be far richer than even Einstein and the quantum mechanics thought. This new geometry of emptiness curls, folds, loops (even rips and tears) in eleven (or perhaps twelve) dimensions of space and time.

As previously explained, the more complex landscape comes with a set of ideas generally known as "string theory" (short for superstring theory) and more recently M-theory, which can stand for magic, mother, mystery, matrix, or membrane theory, almost according to taste. As described by M-theory, the entire universe arises from the harmonics of vibrating strings, membranes, and blobs in twelve dimensions.

These unseen dimensions curl around one another in strange, convoluted shapes, forming holes, knots, and handles, leaving pieces of the universe oddly isolated, perhaps stranded on islands or hanging onto the rest of the cosmos by tenuous threads. Our entire tangible universe may be trapped on a nine-dimensional membrane attached to a larger, ten-dimensional universe only by gravity.

String theory strikes some as strangely beautiful, others as nothing short of bizarre. Ignored by all but a few smitten followers until a few years ago, it has suddenly become part of the common parlance. Recently, its success at solving certain long-standing puzzles has won over even some of its most vocal critics.

Physicists first became attracted to string theory because it, and it alone, resolves the glaring mismatch between the laws that rule the large-scale cosmos (Einstein's theory of gravity) and those that run the microcosmos (quantum mechanics). In string theory, not only do the two sets of rules get along together, they need each other; they require each other to exist. On close inspection, even the unruly quantum space–time "foam" becomes the raucous song of vibrating strings.

The trick to making this difficult marriage work seems deceptively simple: The main reason that quantum mechanics and gravity don't mesh is that the two together produce infinite solutions to equations. Everywhere gravity meets the quantum, infinity pops up, and infinite solutions are nonsense. They are the dead canaries that tell physicists the laws of nature are breaking down.

The infinities appear in part because present-day theoretical physics allows particles to be infinitely small, and space and time to squeeze down to infinitely small specks. Properties of space, time, and particles approach the dreaded zero with impunity, and zero always opens the door to infinity.

String theory solves that most fundamental of problems by doing away with infinitely small particles. It doesn't allow infinitely small anything; the loop of string is the smallest allowable size. Like glasses that blur the view, strings erase the problem. The incompatible partners, gravity and quantum theory, fit snuggly together out of sight, under the covers of size. Indeed, because the strings prevent anything from getting infinitely small, they smear out many previously troubling properties of space and time at infinitely small scales. You never get to the point where the disasters happen. String theory prevents it.

And, if string theory is right, it would do a great deal more than mend the rift between gravity and the quantum: It could explain almost every outstanding question in fundamental physics—from how the quarks are trapped inside the vacuum, to why the vacuum is the way it is in the first place. Because string theory brings together all the laws compatibly under one happy roof, it could well reveal that every tick of a clock, every barking dog, every dying star flows from a single master principle, a single grand equation. Fittingly, the power from string theory arises at least in part from the fact that it embodies a more perfect symmetry than general relativity or quantum theory alone; from its higher-dimensional perspective, string theory naturally embraces a multitude of seemingly disparate realities.

So, what is string theory exactly? Let's take a close look.

"At present," explains Dr. Smolin, "string theory is a large collection of approximate calculations, responds which suggest, but do not prove, that there exists a fundamental theory that unifies the different forces and particles in nature. At present there is some evidence for a huge number of different such theories. They are conjectured to be part of one big theory, called M-theory, but there is so far no evidence for that. String theory, as it stands, also makes no experimental predictions, by which it could be checked or tested." [2]

The Stuff Strings Are Made Of

String theory says simply that the building blocks of nature are no longer particles, but vibrating strings of some fundamental, unknown "stuff." Depending on the exact geometry of the vibrating string, it will produce different harmonic chords, just as a piano produces a different sound than a flute.

In turn, the way the string vibrates determines a particle's mass, its electric charge, its spin, and any other properties. By changing its mode of vibration, the string can masquerade as an electron or a photon or a graviton or any other particle of nature. In a sense, the resonant patterns are equivalent to the particles, the universe, and the symphony they sing.

While some strings have free ends, others curl into loops. They interact with one another by joining and splitting off again, branching into new strings and loops, growing and shrinking. If a universe built up from point-like particles resembles a city made of LEGO blocks, the universe of strings looks a lot more like a tree.

Physicists don't see the strings, in part, because they are unimaginably small. Consider that it takes about a million atoms to cross the period at the end of this sentence, and that it takes about 100,000 protons to equal the size of an atom. Then think how many protons it would take to cross Ohio. That's how many strings could, theoretically, fit inside a single proton. For the more quantitatively minded, the string scale is 10^{-33} centimeters (a fraction with 1 in the numerator and the number 1 followed by 33 zeros in the denominator).

But size is only one reason physicists can't see the strings. They also curl around dimensions unfamiliar to the physicists' senses. Consider a bug living on a beach ball. To the bug, the beach ball seems a flat, infinite space—like the Earth did to our forebears. Or, think of a fire hose: Seen at a great distance, it looks like a one-dimensional line. In fact, it curls around a second (from our perspective unseen) dimension.

If string theory is right, we are all bugs living on fire hoses that curl and twist into at least six unseen dimensions. Sometimes, dimensions wind around one another creating extra-energetic (that is, massive) particles. Sometimes, the dimensions flatten out, or deflate, like a three-dimensional beach ball flattened by a truck. These transformations can produce astonishing effects, such as turning something that looks like an elementary particle into something that looks more like a black hole. And vice versa.

The exact shapes of these extra dimensions are critical. Just as the exact shape of a tuba or guitar determines the sounds it can make, so the topology of the extra dimensions determines the possible particles produced by the strings. The topology of the extra dimensions, in which the strings vibrate, in other words, determines the ingredients available to make our universe.

Every time you sweep your hand through space it also sweeps through all the extra curled-up dimensions. But they are so small, and your hand circles through them so fast, that it gets back to its starting point before anyone can notice.

How do physicists know they're there if they can't see them? "Again," responds Dr. Smolin, "you have to distinguish between cosmic strings and string theory. They are completely different. The answer in both cases is that before being taken seriously, a theory must make predictions for new observations or experiments, which are then tested. Fundamental string theory has made no such predictions, so far. The theory of cosmic strings did make predictions which were tested. These were about the distribution of energy with frequency in the cosmic microwave radiation. The predictions of the theory of cosmic strings disagreed with the data. As such, few cosmologists now believe the theory of cosmic strings." [2]

Why do so many physicists put so much faith in something so far removed from the palpable, testable world? The reason is, the theory seems to work. Physicists have trust in string theory for the same reason most people have trust in computers or jet planes and other things they don't understand—these things work for you; they take you where you want to go, and all things considered, they seem to be fairly reliable.

On the other hand, Dr. Smolin feels that the preceding question "is a sociological question. The scientific community contains people with a wide range of attitudes and motivations. Some think critically and independently. Many do not and just follow fashion or do what they think is most likely to win them the approval of powerful older scientists, and hence further their careers. In this way, science is not much different than other ways of earning a living. Some scientists want to be independent and take risks, others prefer to do what is safe. But it is important to mention that not all experts in quantum gravity and theoretical physicists believe that string theory is relevent for nature. A great many experts work on other approaches to fundamental physics. Loop quantum gravity is one of them, but it is not the only one. Some of these theories do much better than string theory at answering questions about how gravity and quantum theory are to be combined in one theory. Because of this, many experts put more faith in them than string theory." [2]

Modifying the Concept of Space–Time

Perhaps the most basic thing physicists have learned about string theory is that it modifies the concepts of space–time that Einstein developed. This doesn't come as a complete surprise: Einstein based his theory of gravity on his ideas about space–time, so any theory that modifies Einstein's gravitational theory to reconcile it with quantum mechanics has to incorporate a new concept of space–time.

String theory actually gives a "fuzziness" to all physicists' familiar notions of space and time, roughly as Heisenberg's uncertainty principle gave a basic fuzziness to classical ideas about the motion of particles. In ordinary quantum mechanics, interactions among elementary particles occur at definite points in space–time. In string theory, things are different: Strings can interact just as particles do, but you cannot say quite when and where this occurrs.

Even to a theoretical physicist, this kind of explanation raises more questions than it answers. String theory involves a jump in physical concepts that's large, even when compared with previous revolutions in physics. There's no telling when humans will succeed in crossing the chasm.

But physicists really do understand one aspect of how string theory changes the concepts of space–time. This involves a key part of string theory called supersymmetry. Finding supersymmetry gives physicists their best hope to prove that the things that are being explained here have something to do with nature and are not just armchair theorizing.

In your everyday life, you measure space and time by numbers. For example, you say, "It is now 4 o'clock," "We are 300 feet above sea level," or "We live at 50 degrees north latitude." This idea of measuring space and time by numbers is one bit of common sense that Einstein preserved. In fact, in his time, quantities that could be measured by numbers were all physicists knew about.

But quantum mechanics changed all that. Particles were divided into bosons (like light waves) and fermions (like electrons or neutrinos). Quantities like space, time, and electric field that can be measured by numbers are "bosonic." Quantum mechanics also introduced a new kind of "fermionic" variable that cannot be measured by ordinary numbers. Fermionic variables are infinitesimal and inherently quantum-mechanical, and as such are hard to visualize.

Finally, according to the idea of supersymmetry, in addition to the ordinary familiar dimensions (the three spatial dimensions plus time), space–time has additional, infinitesimal, or fermionic dimensions. If supersymmetry can be confirmed in nature, this will begin the process of incorporating quantum mechanical ideas in the description of space–time. But how can physicists ever know if supersymmetry is right?

Conclusion

No one knows, as yet, exactly what string theory will turn out to be, but because it has performed so many amazing feats, many physicists are willing to wait to see where it takes them. And, at least until experiments prove them wrong, they are willing to trust that something more real than little green people lies hidden in the ultra-small world of extra dimensions. Nobody in this field is clever enough to have invented something like this. It's clearly something that physicists have discovered.

Finally, what really was the Big Bang, and how could there have been a beginning of time (see sidebar, "The End of the Beginning")? This question certainly involves quantum gravity, because quantum mechanics and general relativity were both important near the Big Bang. That makes this another question for the string theorists, even though physicists do not seem close to an answer. A plausible guess springs from the way string theory and quantum gravity impart a fuzziness to physicists' concepts of space–time. Under ordinary conditions, time seems like a well-defined notion, but as you get closer to the Big Bang, quantum mechanics and stringy fuzziness become more significant. The very notion of time seems likely to lose its meaning when one gets back to the beginning—and that, quite likely, will prove to be a key to understanding what the Big Bang really was.

According to Dr. Smolin, "There are reliable calculations that show that the initial cosmological singularity, which is a prediction of classical general relativity, is removed by quantum effects. These calculations have been done using the methods of loop quantum gravity. Thus, the effect of quantum gravity is to remove the evidence that the big bang was the beginning of time. [2]

"Time may have a fundamentally different nature from space," explains Physicist Dr. Stephen Wolfram, "The standard mathematical formulation of relativity theory suggests that—despite our everyday impression—time should be viewed as a fourth dimension much like space. Time as we perceive it, may instead emerge from an underlying process that makes it quite different from space. And through the concept of causal invariance, the properties of time seem to lead almost inexorably to a whole collection of surprising results that agree with existing observations in physics–including the special and general theories of relativity, and perhaps also quantum mechanics." [3]

The End of the Beginning

Recently, Astronomers using the Hubble Space Telescope, reported that they had reached far enough out in space and back in time to be very close to the beginning of the Big Bang itself. Astronomers at the Space Telescope Science Institute (Johns Hopkins University), revealed what they believe was the deepest telescopic view into the universe ever obtained.

The million-second exposure of a small patch of dark sky in the constellation Fornax captured objects a quarter as bright as previous surveys, as well as detecting roughly 10,000 galaxies. According to astronomers, when stars were just beginning to form, the several dozen faint reddish spots could even be infant galaxies just emerging from the dark ages that prevailed in the first half million years after the Big Bang.

Before astronomers know if their surmises are correct, they have cautioned that more work will be required. Until the James Webb Telescope goes into orbit in 2012, astronomers will not be able to take a deeper picture.

Dating back as far as when the universe was only a billion years old, the ultra deep survey surpasses two earlier surveys, known as the Hubble Deep Fields, which revealed thousands of new galaxies. The farther away a detected object is, the longer it has taken the light to get here, because light travels at a finite speed.

Without the aid of any gravitational amplification, the ultra deep field has the sensitivity to reach back to galaxies when the universe was only 300 million years old. Thus, the Hubble opens to exploration the period of time from 300 million to 700 million years of age. This is when theorists suggest that the first galaxies were burning themselves out of the murk that descended when the fires of the Big Bang cooled.

The ultra deep field shows that the early universe was littered with galaxies of oddball shapes and colors, like the earlier deep fields. But, astronomers homed in on a few dozen soft reddish dots, which appear on the Nicmos image, but not on the visible-light image.

Some astronomers were excited about the possibility that they might be the most distant objects ever seen (galaxies just emerging from the dark ages). But, the objects just might turn out to be dim nearby stars known as brown dwarfs.

Galaxies at great distances are being carried away at high speeds and their visible light is lengthened and stretched (redshifted in the jargon) to higher wavelengths, because the universe is expanding. The more their light becomes infrared radiation, the farther away they are. When the universe was only 300 million years old, galaxies would be moving so fast that all their light would be invisible infrared radiation. Thus, the Hubble might finally have hit the wall, if the red dots are distant galaxies. Finally, in order to pursue such galaxies, Hubble's successor, the Webb telescope, is being built to observe infrared radiation.

References

[1] Dr. Alan Guth, physicist, (MIT, Cambridge, MA) "Inflation," published in Carnegie Observatories Astrophysics Series, Vol. 2, Measuring and Modeling the Universe ed. W.L. Freedman (Cambridge: Cambridge Univ. Press, 2003), quoted with permission of the author.

[2] Dr. Lee Smolin, Theoretical Physicist, Perimeter Institute of Theoretical Physics, Waterloo, Ontario, Canada.

[3] Dr. Stephen Wolfram, Physicist, "Quick Takes on Some Ideas and Discoveries in a New Kind of Science," Wolfram Research, Inc., (© 2004 Stephen Wolfram), Champaign, IL, 61820.

Solid State Physics: The Mechanism Behind High-Temperature Superconductors

"Every body continues in its state of rest or uniform motion in a straight line, except insofar as it doesn't."
—Sir Arthur Stanley (1882–1944)

Solid-state physics deals with the properties of solids, from the atomic level upwards. It is closely linked to materials science (which also explores the chemical and engineering aspects of materials) and to electronic device technology (which has had a profound influence on our way of life). This chapter discusses the unsolved problem of solid-state physics as the mechanism behind high-temperature superconductors (HTS) and some of the technological revolutions that have relied on HTS for the basic discoveries and inventions. But in order to lay the foundation for an in-depth discussion of HTS, a general discussion of superconductor basics and history must be conducted first.

Superconductors

The phenomenon of superconductivity was discovered by Onnes in 1911, when he reduced the temperature of solid mercury below 4.3 kelvin and found that its resistance vanished. (See sidebar, "Superconductor History.") Since then, about half of the common chemical elements have been found to exhibit superconductivity (see sidebar, "Superconductivity: What Is It?") if cooled below a critical temperature (Tc). The values of Tc vary from 0.0003 K (for rhodium) to 9.5 K (for niobium).

Superconductor History

Superconductors are composed of materials that have no resistance to the flow of electricity. They are one of the last great frontiers of scientific discovery. Not only have the limits of superconductivity not yet been reached, but the theories that explain superconductor behavior seem to be constantly under review.

In 1911, superconductivity was first observed in mercury by Dutch physicist Heike Kamerlingh Onnes of Leiden University. When he cooled mercury to the temperature of liquid helium, 4 kelvin (–452° F, –269° C), its resistance suddenly disappeared. The kelvin scale represents an "absolute" scale of temperature. Thus, it was necessary for Onnes to come within 4 degrees of the coldest temperature that is theoretically attainable to witness the phenomenon of superconductivity. Later, in 1913, he won a Nobel prize in physics for his research in this area.

In 1933, the next great milestone in understanding how matter behaves at extreme cold temperatures occurred. Walter Meissner and Robert Ochsenfeld discovered that a superconducting material will repel a magnetic field. A magnet moving by a conductor induces currents in the conductor. This is the principle upon which the electric generator operates. But in a superconductor, the induced currents exactly mirror the field that would have otherwise penetrated the superconducting material—thus causing the magnet to be repulsed. This phenomenon is known as diamagnetism and is today often referred to as the "Meissner effect." The Meissner effect is so strong that a magnet can actually be levitated over superconductive material.

Other superconducting metals, alloys, and compounds were discovered in subsequent decades. For instance, in 1941, niobium-nitride was found to superconduct at 16K. Then, in 1953, vanadium-silicon displayed superconductive properties at 17.5K. And in 1962, scientists at Westinghouse developed the first commercial superconducting wire (an alloy of niobium and titanium). In the 1960s (at the Rutherford-Appleton Laboratory in the UK), high-energy, particle-accelerator electromagnets made of copper-clad niobium-titanium were then developed. In 1987, they were first employed in a superconducting accelerator at the U.S. Fermilab Tevatron.

The first widely accepted theoretical understanding of superconductivity was advanced in 1957 by American physicists John Bardeen, Leon Cooper, and John Schrieffer. Their *theories of superconductivity* became known as the BCS theory (derived from the first letter of each man's last name). This won them a Nobel prize in 1972. For elements and simple alloys, the mathematically complex BCS theory explained superconductivity at temperatures close to absolute zero. However, the BCS theory has subsequently become inadequate to fully explain how superconductivity is occurring at higher temperatures (and with different superconductor systems).

Another significant theoretical advancement came in 1962 when Brian D. Josephson (a graduate student at Cambridge University) predicted that electrical current would flow between two superconducting materials (even when they are separated by a non-superconductor or insulator). His prediction was later confirmed and won him a share of the 1973 Nobel Prize in Physics. This tunneling phenomenon is today known as the "Josephson effect." It has been applied to electronic devices such as the superconducting quantum interference device (SQUID)—an instrument capable of detecting even the weakest magnetic fields.

In the field of superconductivity, the 1980s were a decade of unrivaled discovery. In 1964, Bill Little of Stanford University had suggested the possibility of organic (carbon-based) superconductors. The first of these theoretical superconductors was successfully synthesized in 1980 by Danish researcher Klaus Bechgaard of the University of Copenhagen, and three French team members. It had to be cooled to an incredibly cold 1.2K transition temperature and subjected to high pressure to superconduct. But its mere existence proved the possibility of "designer" molecules. These are molecules that are fashioned to perform in a predictable way.

Next, in 1986, a true breakthrough discovery was made in the field of superconductivity. Alex Muller and Georg Bednorz (researchers at the IBM Research Laboratory in Ruschlikon Switzerland) created a brittle ceramic compound that superconducted at the highest temperature then known: 30K. What made this discovery so remarkable was that ceramics are normally insulators. They don't conduct electricity well at all, so researchers had not considered them as possible high-temperature superconductor candidates. The lanthanum, barium, copper, and oxygen compound that Muller and Bednorz synthesized behaved in a not-as-yet-understood way. The discovery of this first of the superconducting copper-oxides (cuprates) won the two men a Nobel Prize the following year. Due to a small amount of lead having to be added as a calibration standard (making the discovery even more noteworthy), it was later found that tiny amounts of this material were actually superconducting at 58K.

Muller and Bednorz' discovery triggered a flurry of activity in the field of superconductivity. In a quest for higher and higher Tc's, researchers around the world began "cooking" up ceramics of every imaginable combination. A research team at the University of Alabama-Huntsville then substituted yttrium for lanthanum in the Muller and Bednorz molecule in January of 1987. They achieved an incredible 92K Tc. For the first time, a material (today referred to as yttrium barium copper oxygen [YBCO]) had been found that would superconduct at temperatures warmer than liquid nitrogen (a commonly available coolant). Using exotic (and often toxic) elements in the base perovskite ceramic, additional milestones have since been achieved. Now, the current class (or "system") of ceramic superconductors with the highest transition temperatures are the mercuric-cuprates. The first synthesis of one of these compounds was achieved in 1993 by Prof. Dr. Ulker Onbasli at the University of Colorado and by the team of A. Schilling, M. Cantoni, J. D. Guo, and H. R. Ott of Zurich, Switzerland. The world record Tc of 138K is now held by a thallium-doped mercuric-cuprate comprised of the elements mercury, thallium, barium, calcium, copper, and oxygen. Then, in February of 1994, the Tc of this ceramic superconductor was confirmed by Dr. Ron Goldfarb at the National Institute of Standards and Technology–Colorado. Its Tc can be coaxed up even higher (approximately 25 to 30 degrees more at 300,000 atmospheres) under extreme pressure.

The first company to capitalize on high-temperature superconductors was Illinois Superconductor (today known as ISCO International). Formed in 1989, this amalgam of government, private-industry, and academic interests introduced a depth sensor for medical equipment that was able to operate at liquid nitrogen temperatures (< 77K).

While no significant advancements in superconductor Tc's have been achieved in recent years, other discoveries of equal importance have been made. In 1997, researchers discovered that at a temperature very near absolute zero, an alloy of gold and indium was both a superconductor and a natural magnet. Conventional wisdom held that a material with such properties could not exist! Since then, over a half-dozen such compounds have been found.

Recent years have also seen the discovery of the first high-temperature superconductor that does not contain any copper (in 2000), and the first all-metal perovskite superconductor (in 2001). Startling (and serendipitous) discoveries like these are forcing scientists to continually re-examine longstanding theories on superconductivity and to consider heretofore unimagined combinations of elements.

Superconductivity: What Is It?

Superconductivity is a phenomenon observed in several metals and ceramic materials. These materials have no electrical resistance when they are cooled to temperatures ranging from near absolute zero (–459 degrees Fahrenheit, 0 kelvin, –273 degrees Celsius) to liquid nitrogen temperatures (–321°F, 77K, –196°C). The Tc is the temperature at which electrical resistance is zero, and it varies with the individual material. Critical temperatures are achieved by cooling materials with either liquid helium or liquid nitrogen for practical purposes.

These materials can carry large amounts of electrical current for long periods of time without losing energy as heat because they have no electrical resistance (meaning electrons can travel through them freely). With no measurable loss, superconducting loops of wire have been shown to carry electrical currents for several years. This property has implications for electrical power transmission, for electrical-storage devices, and for transmission lines if they can be made of superconducting ceramics.

Another property of a superconductor is that once the transition from the normal state to the superconducting state occurs, external magnetic fields can't penetrate it. This effect is called the Meissner effect and has implications for making high-speed, magnetically levitated trains. It also has implications for making powerful, small, superconducting magnets for magnetic resonance imaging or MRI.

How do electrons travel through superconductors with no resistance? "At the beginning of the twentieth century," explains physicist Dr. Steven Weinberg (University of Texas at Austin), "several leading physicists, including Lorentz and Abraham, were trying to work out a theory of the electron. This was partly in order to understand why all attempts to detect effects of Earth's motion through the ether had failed. We now know that they were working on the wrong problem. At that time, no one could have developed a successful theory of the electron, because quantum mechanics had not yet been discovered. It took the genius of Albert Einstein in 1905 to realize that the right problem on which to work was the effect of motion on measurements of space and time. This led him to the special theory of relativity." [3] Let's look at this more closely.

The atomic structure of most metals is a lattice structure, much like a window screen in which the intersection of each set of perpendicular wires is an atom. Metals hold on to their electrons quite loosely, so these particles can move freely within the lattice. This is why metals conduct heat and electricity very well. As electrons move through a typical metal in the normal state, they collide with atoms and lose energy in the form of heat. In a superconductor, the electrons travel in pairs and move quickly between the atoms with less energy loss.

As a negatively charged electron moves through the space between two rows of positively charged atoms (like the wires in a window screen), it pulls inward on the atoms. This distortion attracts a second electron to move in behind it. This second electron encounters less resistance, much like a passenger car following a truck on the freeway encounters less air resistance. The two electrons form a weak attraction, travel together in a pair and encounter less resistance overall. In a superconductor, electron pairs are constantly forming, breaking, and reforming, but the overall effect is that electrons flow with little or no resistance. The low temperature makes it easier for the electrons to pair up.

One final property of superconductors is that when two of them are joined by a thin insulating layer, it is easier for the electron pairs to pass from one superconductor to another without resistance (the Josephson effect). This effect has implications for superfast electrical switches that can be used to make small, high-speed computers.

> The future of superconductivity research is to find materials that can become superconductors at room temperature. Once this happens, the whole world of electronics, power, and transportation will be revolutionized.

Careful measurements show that the resistivity of a superconductor really is zero. For example, a persistent current circulating in a superconducting loop of wire showed no decrease after several years. Joule heating would dissipate electrical power as heat and reduce the current flow if the resistance were non-zero. Quantum mechanics provides an explanation for this remarkable behavior: Below Tc, the electrons in a superconductor form Cooper pairs (with opposite spin) which are not scattered by atoms within a solid. In fact, the motion of all Cooper pairs is correlated. They form a single quantum-mechanical state. The thermal energy of the solid disrupts each Cooper pair, and normal conduction (with resistance) resumes if the superconductor is warmed to a temperature exceeding Tc.

Without consuming electrical power, superconductors are useful because the persistent current can generate a constant magnetic field. Unfortunately, the magnetic field tends to destroy the superconductivity. There is a critical field (Bc) above which the superconductivity vanishes. The value of Bc depends on the temperature (T). Its value is at a maximum at T = 0K, and decreases to zero at T = Tc. Therefore, the windings of superconducting magnets must be cooled to a temperature considerably below Tc in order to generate a useful large magnetic field. Alloys containing niobium have been developed that have Tc as high as 23K. They also support magnetic fields of more than 20 Tesla at T = 4.2K (the boiling point of liquid helium). Helium is used as the refrigerant in large magnets, such as those used for MRI scanning.

However, liquid helium is expensive (about $10/liter), whereas liquid nitrogen (boiling point 77K) is cheap and easier to handle. This situation created an incentive to develop superconductors with a higher Tc. As previously explained, the first major progress in this direction came in 1986. This is when two scientists at the IBM Zurich laboratories created a superconductor containing lanthanum, barium, copper, and oxygen, with a Tc being greater than 30K. In 1987, scientists at the University of Texas replaced lanthanum with the element yttrium to give a superconductor (Y Ba2 Cu3 07, generally known as YBCO) a Tc equaling 93K. That same year, a superconducting compound containing bismuth, strontium, calcium, copper, and oxygen (known as BSCCO) was found to form a phase having a Tc equaling 110K. Related compounds have been discovered with a Tc of up to 133K. Collectively, these materials are known as high-temperature superconductors.

High-Temperature Superconductors

Few technologies ever enjoy the sort of rock-star celebrity that superconductors received in the late 1980s. Headlines the world over trumpeted the discovery of HTS, and the media and scientists alike gushed over the marvels that we could soon expect from this promising young technology. (See sidebar, "Why Are They Called HTS?") Levitating 300-mph trains, ultra-fast computers, and cheaper, cleaner electricity were to be just the beginning of its long and illustrious career.

Why Are They Called HTS?

As previously explained, the first superconductors discovered in 1911 were simple metals like mercury and lead. They were ordinary conductors at room temperature, but they became superconductors when the temperature dropped to only a few degrees (3K) above absolute zero. These superconductors were too cold for many practical applications. Ever since, researchers have been trying to figure out how to make substances superconduct at room temperature. High temperature superconductors operate at 100K to 150K. That's very cold compared to the air around you, but much warmer than the original superconductors of 1911. Hence, scientists call them high-temperature superconductors.

Today you might ask: what ever happened to the high-temperature hype? It was the hottest potato of its time, but it all fizzled out.

The problem was learning to make wire out of it. These superconductors are made of ceramics—the same kind of material in coffee mugs. Ceramics are hard and brittle. Finding an industrial way to make long, flexible wires out of them was going to be difficult.

Indeed, the first attempts were disappointing. The so-called "first-generation" HTS wire was relatively expensive: five to ten times the cost of copper wire. Furthermore, the amount of current it could carry often fell far short of its potential: only two or three times that of copper, versus a potential of more than 100 times.

But now, thanks to years of research involving experiments flown on the space shuttle, this is about to change. The NASA-funded Texas Center for Superconductivity and Advanced Materials (TcSAM) at the University of Houston is teaming with Houston-based Metal Oxide Technologies, Inc. (MetOx) to produce the "smash hit" that scientists have been seeking since the 1980s. Of course, this would be a "second-generation" HTS wire that realizes the full 100-fold improvement in current capacity over copper, yet costs about the same as copper to produce. The once-famous superconductors may be about to step back into the limelight.

An Awaiting Audience

The special "talent" of superconductors is that they have zero resistance to electric current. Absolutely none. In theory, a loop of HTS wire could carry a circling current forever without even needing a power source to keep it going.

In normal conductors, such as copper wire, the atoms of the wire impede the free flow of electrons, sapping the current's energy and squandering it as heat. Today, about 6 to 7% of the electricity generated in the United States gets lost along the way to consumers, partly due to the resistance of transmission lines, according to U.S. Energy Information Agency documents as shown in Table 9.1.[2] Replacing these lines with superconducting wire would boost utilities' efficiencies, and would go a long way toward curbing the nation's greenhouse gas emissions.

Of course, with the Bush Administration's present energy policy and a Congress that is heavily influenced by special interest groups from coal-burning electric power plants (utilities) and oil companies, it is rather doubtful at this time that these lines will ever be replaced with superconducting wire to boost utilities' efficiencies to produce clean cheap energy and help curb greenhouse gas emissions. But one can hope! Nevertheless, please see Chapter 19, "Electrical Energy: Free Energy—The Quantum Mechanical Vacuum," for a very detailed discussion on free energy.

On a hopeful note, the fledgling "maglev" train industry would also welcome the availability of higher-quality, cheaper HTS wire. Economic realities stalled the initial adoption of maglev transit systems, but maglev development is still strong in Japan, China, Germany, and the United States.

NASA is looking at how superconductors could be used for space. For example, the gyros that keep satellites oriented

Table 9.1 Electricity Overview, 1949–2002 (billion kilowatt hours)

	Net Generation											
	Electric Power Sector [a]											
Year	Electric Utilities	Independent Power Producers	Total	Commercial Sector[b]	Industrial Sector[c]	Total	Imports[d]	Exports[d]	Losses and Unaccounted for[e]	Retail Sales[f]	Direct Use[g]	Total
1949	291	NA	291	NA	5	R 296	2	(s)	43	255	NA	255
1950	329	NA	329	NA	5	R 334	2	(s)	44	291	NA	291
1951	371	NA	371	NA	5	R 375	2	(s)	47	330	NA	330
1952	399	NA	399	NA	5	R 404	3	(s)	50	356	NA	356
1953	443	NA	443	NA	4	R 447	2	(s)	53	396	NA	396
1954	472	NA	472	NA	5	R 476	3	(s)	54	424	NA	424
1955	547	NA	547	NA	3	R 550	5	(s)	58	497	NA	497
1956	601	NA	601	NA	3	R 604	5	1	62	546	NA	546
1957	632	NA	632	NA	3	R 635	5	1	62	576	NA	576
1958	645	NA	645	NA	3	R 648	4	1	64	588	NA	588
1959	710	NA	710	NA	3	R 713	4	1	70	647	NA	647
1960	756	NA	756	NA	4	R 759	5	1	76	688	NA	688
1961	794	NA	794	NA	3	R 797	3	1	77	722	NA	722

1962	855	NA	855	NA	3	R 858	2	2	81	778	NA	778
1963	917	NA	917	NA	3	R 920	2	2	88	833	NA	833
1964	984	NA	984	NA	3	R 987	6	4	93	896	NA	896
1965	1,055	NA	1,055	NA	3	R 1,058	4	4	104	954	NA	954
1966	1,144	NA	1,144	NA	3	R 1,148	4	3	113	1,035	NA	1,035
1967	1,214	NA	1,214	NA	3	R 1,218	4	4	118	1,099	NA	1,099
1968	1,329	NA	1,329	NA	3	R 1,333	4	4	129	1,203	NA	1,203
1969	1,442	NA	1,442	NA	3	R 1,445	5	4	133	1,314	NA	1,314
1970	1,532	NA	1,532	NA	3	R 1,535	6	4	145	1,392	NA	1,392
1971	1,613	NA	1,613	NA	3	R 1,616	7	4	150	1,470	NA	1,470
1972	1,750	NA	1,750	NA	3	R 1,753	10	3	166	1,595	NA	1,595
1973	1,861	NA	1,861	NA	3	R 1,864	17	3	165	1,713	NA	1,713
1974	1,867	NA	1,867	NA	3	R 1,870	15	3	177	1,706	NA	1,706
1975	1,918	NA	1,918	NA	3	R 1,921	11	5	180	1,747	NA	1,747
1976	2,038	NA	2,038	NA	3	R 2,041	11	2	194	1,855	NA	1,855
1977	2,124	NA	2,124	NA	3	R 2,127	20	3	197	1,948	NA	1,948
1978	2,206	NA	2,206	NA	3	R 2,209	21	1	211	2,018	NA	2,018

R = Revised; P = Preliminary; E = Estimate; NA = Not available; (s) = Less than 0.5 billion kilowatt hours.

Table 9.1 Electricity Overview, 1949–2002 (billion kilowatt hours) (continued)

	Net Generation											
	Electric Power Sector [a]											
Year	Electric Utilities	Independent Power Producers	Total	Commercial Sector[b]	Industrial Sector[c]	Total	Imports[d]	Exports[d]	Losses and Unaccounted for[e]	Retail Sales[f]	Direct Use[g]	Total
1979	2,247	NA	2,247	NA	3	R 2,251	23	2	200	2,071	NA	2,071
1980	2,286	NA	2,286	NA	3	R 2,290	25	4	216	2,094	NA	2,094
1981	2,295	NA	2,295	NA	3	R 2,298	36	3	184	2,147	NA	2,147
1982	2,241	NA	2,241	NA	3	R 2,244	33	4	187	2,086	NA	2,086
1983	2,310	NA	2,310	NA	3	R 2,313	39	3	198	2,151	NA	2,151
1984	2,416	NA	2,416	NA	3	R 2,419	42	3	173	2,286	NA	2,286
1985	2,470	NA	2,470	NA	3	R 2,473	46	5	190	2,324	NA	2,324
1986	2,487	NA	2,487	NA	3	R 2,490	41	5	158	2,369	NA	2,369
1987	2,572	NA	2,572	NA	3	R 2,575	52	6	164	2,457	NA	2,457
1988	2,704	NA	2,704	NA	3	R 2,707	39	7	161	2,578	NA	2,578
1989	2,784	P 62	P 2,847	P 4	P 115	RP 2,966	26	15	R 222	2,647	RP 108	R 2,755
1990	2,808	P 88	P 2,896	P 6	P 122	RP 3,024	18	16	R 199	2,713	RP 115	R 2,827
1991	2,825	P 108	P 2,934	P 6	P 132	RP 3,072	22	2	R 211	2,762	RP 118	R 2,880

1992	2,797	[P] 137	[P] 2,934	[P] 6	[P] 143	[RP] 3,084	28	3	224	2,763	[P] 122	[R] 2,886
1993	2,883	[P] 161	[P] 3,044	[P] 7	[P] 146	[P] 3,197	31	4	236	2,861	[RP] 128	[R] 2,989
1994	2,911	[P] 178	[P] 3,089	[P] 8	[P] 151	[RP] 3,248	47	2	[R] 224	2,935	[RP] 134	[R] 3,069
1995	2,995	[P] 200	[P] 3,194	[P] 8	[P] 151	[RP] 3,353	43	4	235	3,013	[RP] 144	[R] 3,157
1996	3,077	[P] 207	[P] 3,284	[P] 9	[P] 151	[RP] 3,444	43	3	237	3,101	[RP] 146	[R] 3,247
1997	3,123	[P] 207	[P] 3,329	[P] 9	[P] 154	[RP] 3,492	43	9	[R] 232	3,146	[RP] 148	[R] 3,294
1998	3,212	[P] 245	[P] 3,457	[P] 9	[P] 154	[RP] 3,620	40	13	[R] 221	3,264	[RP] 161	[R] 3,425
1999	3,174	[P] 356	[P] 3,530	[P] 9	[P] 156	[RP] 3,695	43	14	[R] 229	3,312	[RP] 183	[R] 3,495
2000	[R] 3,015	[P] 622	[P] 3,638	[P] 8	[P] 157	[RP] 3,802	[R] 49	15	[R] 231	[R] 3,421	[RP] 183	[R] 3,605
2001	2,630	[E] 932	[P] 3,562	[E] 7	[E] 150	[E] 3,719	38	18	138	3,397	[E] 205	3,602
2002	2,821	[E] 1,022	[P] 3,451	[E] 6	[E] 143	[E] 3,608	27	21	116	3,286	[E] 316	3,599

[a] The electric power sector (electric utilities and independent power producers) comprises electricity-only and combined-heat-and-power (CHP) plants whose primary business is to sell electricity, or electricity and heat, to the public—i.e., NAICS 22 plants. Due to the restructuring of the electric power sector, the sale of generation assets is resulting in a reclassification of plants from electric utilities to independent power producers.

[b] Commercial combined-heat-and-power (CHP) and commercial electricity-only plants. See Appendix G for commercial sector NAICS codes.

[c] Industrial combined-heat-and-power (CHP) and industrial electricity-only plants. Through 1988, includes industrial hydroelectric power only. See Appendix G for industrial sector NAICS codes.

[d] Electricity transmitted across U.S. borders with Canada and Mexico.

[e] Energy losses that occur between the point of generation and delivery to the customer, and data collection frame differences and nonsampling error. See Note 1 at end of section.

[f] Electricity retail sales to ultimate customers reported by electric utilities and other energy service providers.

[g] Commercial and industrial facility use of onsite net electricity generation; and electricity sales among adjacent or co-located facilities for which revenue information is not available.

Note: Totals may not equal sum of components due to independent rounding.

could use frictionless bearings made from superconducting magnets, improving the satellites' precision. Also, the electric motors aboard spacecraft could be a mere 1/4 to 1/6 the size of non-superconducting motors, saving precious volume and weight in the spacecraft's design.

Should any country ever establish a base on the moon, superconductors would be a natural choice for ultra-efficient power generation and transmission, since ambient temperatures plummet to 100K (–173°C, –280°F) during the long lunar night—just the right temperature for HTS to operate. And, during the months-long journey to Mars, a "table top" MRI machine (see Figure 9.1) [1] made possible by HTS wire would be a powerful diagnosis tool to help ensure the health of the crew. Worldwide, the current market for HTS wire is estimated to be $40 billion, according to industry analysts, and it is expected to grow rapidly.

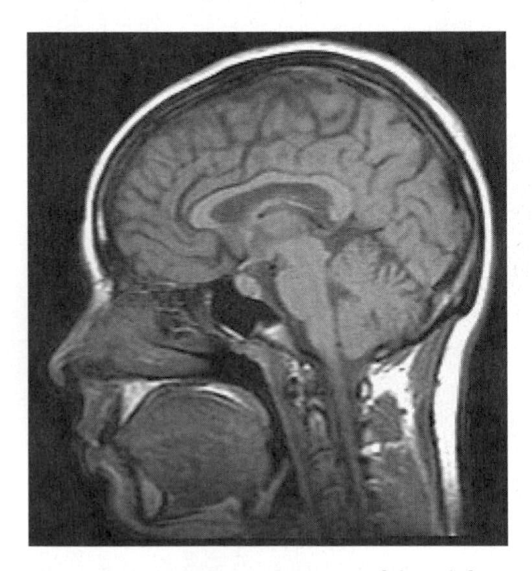

Figure 9.1 MRI scans, a powerful tool for medical diagnosis, use superconducting electromagnets to generate detailed images of body tissues. Most of today's MRI machines require expensive liquid helium to cool their low-temperature superconducting wire.

Behind the Scenes

The University of Houston has licensed this new wire-making technology to MetOx, a company founded in 1997. MetOx plans to begin full-scale production of this high-quality HTS wire in 2004.

Basically, the wire is made by growing a thin film of the superconductor only a few microns thick (thousandths of a millimeter) onto a flexible foundation. This well-known production method was improved upon in part through "Wake Shield" experiments flown on the space shuttle to learn about growing thin films in the hard vacuum of space, as shown in Figure 9.2.[1] Researchers learned how to grow higher-quality oxide thin films from the shuttle experiments and used that in the lab to improve the quality of their superconducting films.

Figure 9.2 The Wake Shield Facility being held out in space by the shuttle's robot arm. Image courtesy NASA.

Not surprisingly, the NASA group at Texas Center for Superconductivity at the University of Houston (TCSUH), can't reveal exactly how they make their HTS wire. The technologies springing from these NASA/industry research partnerships must be patented to achieve NASA's goal of using space to benefit American businesses.

But shouldn't this technology be shared with everyone? How many times have government and industry made excuses for why they can't reveal the inner workings of a new technological breakthrough under the guise of protecting U.S. businesses? The answer of course is many, many times, and there is nothing wrong with that as long as the new technology is made available immediately, not 40 or 45 years later like the video phone that was introduced at the World's Fair in 1960.

It makes one wonder whether these energy-saving technologies are a threat to big business—the threat being that they would have to re-tool (spending billions right away) their industrial operations in order to appease an awaiting public. The history of American big business has a track record of just the opposite in this area. Can you say "complacency"?

U.S. business has always operated on obtaining short-term profits, and making very, very long-term investments. This is why research and development departments within these companies have always received little funding, for fear that they might actually revolutionize that particular industry, which would cost trillions of dollars.

By the way, where are those flying cars that they promised us? Overall, it makes one wonder, and keeps conspiracy theories alive and well. Wait until you read Chapter 19 on free energy!

Nevertheless, in the years to come, it is hoped that the quality of HTS wire will translate into improvements in dozens of industries from power generation to medical care. Keep an eye on this one: The glamorous career of superconductors has only just begun.

Other HTS Applications

Finally, with the preceding applications in mind, let's take a detailed look at some of the other high-temperature superconductor applications:

- Superconductors to sustain Internet growth
- E-bomb
- Electrical power transmission for electrical storage devices
- High-speed quantum computers

Superconductors to Sustain Internet Growth

Irvine Sensors Corporation (Costa Mesa, CA) recently received an approximate $2 million research and development contract to demonstrate a superconducting digital router for high-speed communications. With the communications industry's ever-increasing need for speed, Irvine Sensors (under the contract) plans to exploit TRW's superconducting high-speed switching technology to develop a "super switch" that could evolve along with HTS technology. The initial demonstration goal of the R&D contract is a 256×256 switch that can operate at a minimum of 10 gigabits per second per channel and that can reconfigure in less than a nanosecond. This capability will be extended to a 4,096×4,096 switch with at least 40 gigabits per second per channel, if the full goals of the contract are met. Irvine Sensors is in the process of organizing a consortium of potential industry users to assist in defining potential product requirements and benchmarking the planned switch's performance. In order to support the ever-increasing need for faster and faster data transmission rates, conventional technology cannot handle the scalability requirements needed. TRW's superconducting technology addresses the power issues typically associated with high speed switches, while Irvine Sensors' chip-stacking enables the necessary electronics to be compressed into a space comparable to the size of an incoming fiberoptics data cable.

Recent studies estimate that data traffic on the Internet alone is now doubling every four months. Competing Silicon Germanium and Gallium Arsenide technologies are already having a hard time keeping up and are very size- and power-intensive. Stacked superconducting electronics may be a means to overcome these limitations and enable continued rapid growth of Internet and telecommunications traffic.

E-bomb

The next Pearl Harbor will not announce itself with a searing flash of nuclear light or with the plaintive wails of those dying of Ebola or its genetically engineered twin. You will hear a sharp crack in the distance. By the time you mistakenly identify this sound as an innocent clap of thunder, the civilized world will have become unhinged, for the "E-bomb," or electromagnetic pulse weapon, has come of age.

A fictional and dramatic preview of the E-bomb's potential damage was shown in Fox's *Dark Angel* TV show (cancelled): The year is 2020. The scene is Seattle, shortly after terrorists have set off an electromagnetic pulse bomb in the atmosphere that knocks out all satellites, toppling the economy and plunging the world into a 1930s-style depression in which politicians are for hire, cops are crooked, and the future looks drearier than Seattle in March.

The reality show followed in March of 2003, when the U.S. invaded Iraq. One of the first news reports went something like this: "The U.S. Air Force has hit Iraqi TV with an experimental electromagnetic pulse device called the E-Bomb in an attempt to knock it off the air and shut down Saddam Hussein's propaganda machine. Iraqi satellite TV, which broadcasts

24 hours a day outside Iraq, went off the air around 4:30 a.m. local time."

E-bombs can unleash in a flash as much electrical power (3 billion watts or more) as the Hoover Dam generates in 36 hours. And, although the Pentagon prefers not to use experimental weapons on the battlefield, the world intervenes from time to time.

America has remained at the forefront of electromagnetic pulse (EMP) device weapons development. It's believed that current efforts are based on using high-temperature superconductors to create intense magnetic fields, although much of this work is classified. It's an astoundingly simple weapon. It consists of an explosives-packed tube placed inside a slightly larger copper coil. The instant before the chemical explosive is detonated, the coil is energized by a bank of capacitors, creating a magnetic field. The explosive charge detonates from the rear forward. As the tube flares outward, it touches the edge of the coil, thereby creating a moving short circuit. The propagating short has the effect of compressing the magnetic field while reducing the inductance of the stator coil. The result is that two stage flux compression generators (FCGs) will produce a ramping current pulse, which breaks before the final disintegration of the device. Published results suggest ramp times of thousands of microseconds and peak currents of tens of millions of amps. The pulse that emerges makes a lightning bolt seem like a flashbulb by comparison.

Ultimately, the Army hopes to use E-bomb technology to explode artillery shells in midflight. The Navy wants to use the E-bomb's high-power microwave pulses to neutralize antiship missiles. And the Air Force plans to equip its bombers, strike fighters, cruise missiles, and unmanned aerial vehicles with E-bomb capabilities. When fielded, these will be among the most technologically sophisticated weapons the U.S. military establishment has ever built.

There is, however, another part to the E-bomb story, one that military planners are reluctant to discuss. While American versions of these weapons are based on advanced technologies, terrorists could use a less expensive, low-tech approach to create the same destructive power. Any nation with even a 1940s technology base could make them. The threat of E-bomb proliferation is very real. Scientists estimate that a basic weapon could be built for around $300.

Electrical Power Transmission for Electrical Storage Devices

Every appliance, from your toaster to your laptop computer, relies on a single aspect of subatomic physics: the negative charge of the electron. Charge is what makes electrical current flow through a maze of wires to do useful things, such as activating a heating element or encoding data. But another property of the electron, called spin, could greatly expand the particle's usefulness. Moving far beyond today's electronics, the emerging technologies of superconducting and spintronics may soon make it possible to store movies on a PalmPilot or build a radical new kind of computer.

The principle behind this trickery is deceptively simple. Ignoring for a moment the weirdness of the quantum world, the electron can be thought of as a tiny rotating bar magnet with two possible orientations: spin-up or spin-down. Engineers can distinguish between spin-up and spin-down electrons by the corresponding orientation of their magnetic fields, north-up or north-down. Conversely, a properly applied magnetic field can flip electrons from one state to the other. In this way, spin can be measured and manipulated to represent the zeros and ones of digital programming, analogous to the "current on" and "current off" states in a conventional silicon chip.

The first spin-related technology was the compass, a broadly defined piece of metal in which electron spins are mostly pointing in the same direction to generate a magnetic field. This field, in turn, attempts to align itself with Earth's magnetic pole. Physicists have used spin forever. Magnetism arises from the fact that electrons carry spin. But exploiting the magnetic properties of the electron doesn't really qualify as spintronics, according to physicists, until you start deliberately flipping the particle's spin back and forth and moving it from one material to another.

About a decade ago, when materials scientists set out to find ways to cram more data onto computer hard drives, the first major breakthroughs in full-fledged spintronics came at IBM's Almaden Research Center. A hard drive uses an electrical charge to place tiny patches of magnetic field in the recording material; it then reads back the encoded data by measuring which way the field points at different locations.

In 1988, the IBM project latched on to the work of two European scientific teams who had discovered a spin-related effect known as giant magnetoresistance. Starting with a magnetic material whose spins were all locked in one direction, the researchers had added a thin layer of metal and topped it off with another material in which the spins can flip. Current flowed easily from the top to the bottom of this composite if the spins were the same in both layers, but the current faced higher resistance if the spins were opposed. In theory, such a setup allowed a much more sensitive way to read back the data on a magnetic disk, but giant magnetoresistance seemed to occur only in expensive, pure crystals exposed to intense magnetic fields.

The Almaden team in 1991 found it could achieve the same effect in cheaper materials that responded to much weaker fields. The researchers eventually built a magnetic read head composed of one of these spintronic sandwiches. Transmitting digital data, magnetized patches on the spinning hard disk flip the spin state in the read head back and forth. A spintronic read head can detect much weaker magnetic fields than older devices can, so each bit of data can be much smaller. It's the world's most sensitive detector of magnetic fields at room temperature. Spintronics is why today's hard drives hold up to 100 gigabytes or more, compared to less than 1 gigabyte in 1997.

Now, researchers at Honeywell, Motorola, and the Naval Research Laboratory are trying to create spin-based computer memory. Based on the same principles, this is called magnetic random access memory, or M-RAM. A prototype design contains a series of tiny magnetic sandwiches placed on a silicon chip between crisscrossing arrays of wires. Electric current through the wires flips the spin, which stays put until it is changed again. Measuring the electrical resistance of a particular sandwich tells whether it represents a 1 or a 0.

Random access memory (information that is available only while the device is turned on) is refreshed 60 times a second by a surge of electricity in conventional desktop computers. M-RAM, in contrast, has almost no electrical demands. NASA is intrigued, because M-RAM could make it possible to build longer-lived spacecraft that perform more elaborate functions without requiring additional power. In more down-to-earth applications, M-RAM might lead to instant-on computers and cell phones with so much built-in memory that they could store entire conversations. You could do all sorts of things that you can't do today, like have video on your PDA. It is expected that IBM will be selling M-RAM by 2005.

Further ahead, spintronics could realize a long-sought, radical kind of data crunching known as quantum computing. According to the laws of quantum mechanics, an electron can be in both spin-up and spin-down states at the same time. That mixed state could form the base of a computer built around not binary bits, but the quantum bit, or qubit. It's not just a 1 or a 0, but any combination of a 1 and a 0. It's one of the first truly revolutionary concepts for computing that's come along in a long time. Feed a problem into a quantum computer, and instead of trying all possible results one at a time, it could calculate them all simultaneously. Barring any unforeseen breakthroughs, however, it will be at least 50 years before anybody builds a quantum computer. A discussion of how to build a quantum computer concludes this chapter.

The benefits of spintronics may spill over to other areas of electronics long before then. In 2002, scientists at the University of California at Santa Barbara and Pennsylvania State University demonstrated that they could drag a cloud of electrons from one semiconductor material to another without disrupting the spin state of the cloud. This achievement points the way toward spin-mediated versions of transistors, the on–off switches that form the building blocks of just about every device powered by a battery or plugged into a wall outlet.

Spintronics transistors might lead to faster, smaller, less-power-hungry versions of existing devices: New science enables new technologies. And the most exciting ones will be things you haven't even imagined yet.

High-Speed Quantum Computers

Sitting in a classroom building at Caltech, behold the world's most feeble computer network. It connects a grand total of two processors, crosses all of a basement corridor, and transmits a whopping single bit of information. Or it would if it were working, which it isn't. So, is it possible to build a High-Speed Quantum Computer?

"Generally large (complex) physical systems behave as classical objects," explains Physicist Dr. Mikhail Lukin (Harvard University). "In other words there is no chance to control its quantum mechanical properties. One of the challenges is to learn how to control quantum mechanical behavior of the complex systems. In other words, we would like to find out how to control quantum properties of large systems both in principle and in practice. And, that is the largest system that can be controlled this way." [4]

Given the pathetic specs, it might seem just a little surprising to hear that Caltech's network is widely considered to be one of the most challenging projects in all of computer science. That's because the one bit of data Caltech's network is designed to transmit won't be an ordinary one or zero of the sort that everyday networks traffic in. It will be a mixture of the two—a so-called quantum bit, or "qubit."

Physicists at Caltech are trying to build the world's first quantum computer network via superconducting technology. In a sense, they're getting a little ahead of themselves, since no one has yet come close to building a practical quantum computer—a computer, that is, that makes calculations on data in the weird multiple-reality state that is the hallmark of quantum mechanics. Still, the benefits of this stunningly radical approach to processing, promises to be so great that the young field of quantum computing has been steadily attracting researchers, not only from computer science, but also from physics, math, and chemistry. Just a few years ago most computer scientists doubted that a quantum computer could ever be built. Now, the tide of opinion seems to be turning, and the past year or two have seen some important advances. (See sidebar, "Lighting Up Quantum Computers.") For example, physicists at MIT have actually built a simple quantum computer. It can't do much. What it does is pick out one name from a list of four—but it does it faster than a conventional computer.

Lighting Up Quantum Computers

In 2003, according to physicists, they brought light to a complete halt for a fraction of a second and then sent it on its way, an achievement that could someday help scientists develop powerful new quantum computers. The research differs from work published in 2001 that was hailed at the time as having brought light to standstill. In that work, light pulses were technically "stored" briefly when individual particles of light, or photons, were taken up by atoms in a gas.

Harvard University researchers have now topped that feat by truly holding light and its energy in its tracks—if only for a few hundred-thousandths of a second. They have succeeded in holding a light pulse still without taking all the energy away from it.

Harnessing light particles to store and process data could aid the still-distant goal of building quantum computers, as well as methods for communicating information over long distances without risk of eavesdropping. The research may also have applications for improving conventional fiberoptic communications and data processing techniques that use light as an information carrier. The present research is just another step toward efforts to control light, but additional work is needed to determine if it can aid these applications.

According to Dr. Matthew Bigelow (University of Rochester), "The new research is an important scientific first. The new study is very clever. This is something that may ultimately spur the development of superior light-based quantum computers." [5]

So what's the big deal about quantum computing? Imagine you were in a large office building and you had to retrieve a briefcase left on a desk picked at random in one of hundreds of offices. In the same way that you'd have to walk through the building, opening doors one at a time to find the briefcase, an ordinary computer has to make its way serially through long strings of ones and zeros until it arrives at an answer. Of course, you could speed up the briefcase hunt by organizing a team, coordinating a floor-by-floor search, and then getting them all back together again to compare results. Likewise, ordinary computers can do this sort of thing by breaking up a task and running the components in parallel on several processors. That sort of extra coordinating and communicating, however, exacts a huge toll in overhead.

But what if instead of having to search by yourself or put together and manage a team, you could instantly create as many copies of yourself as there were rooms in the building? Could all the versions of yourself simultaneously peek in all the offices and then (best of all) every copy of yourself disappear except for the one that found the briefcase?

This is an example of how a quantum computer could work. Quantum computers would exploit the fact that under certain conditions the denizens of the atomic-scale world can exist in multiple realities—atoms and subatomic particles can be simultaneously here and there, fast and slow, pointing up and down. How? Not even physicists agree on that one, but countless experiments over the past seven decades have verified the bizarre phenomenon. By thinking of each of these different atomic states as representing different numbers or other types of data, a group of atoms with all their various combinations of potential states could be used to explore simultaneously all possible answers to a problem. And, with some clever jiggling, the combination representing the correct answer could be made to stand out.

Conventional computer chips are getting so jammed with ever tinier components that they may soon hit their physical limits in power and speed. Some researchers are hoping that quantum computers might break through those barriers. But, although a number of research teams are struggling mightily, even the most optimistic among them don't expect to do more than demonstrate some almost uselessly simple devices within the next three years or so.

Even then, the quantum future is not guaranteed. Any computer (quantum or otherwise) can't do much good unless it can be programmed to perform a practical task. And many researchers have been wondering whether quantum computers will be able to tackle real-world computing problems—or at least run them significantly faster than conventional computers can.

Most applications actually won't lend themselves to quantum computing. That's because the typical computer task, like calculating the orbit of a satellite or rotating a graphic image, requires computer logic that proceeds in serial fashion, each step depending on the results of the preceding one. Quantum computing can't speed up that sort of task. There isn't much advantage in having multiple selves, for example, if instead of looking for a briefcase in a single room you had to assemble a wristwatch out of parts scattered throughout all the rooms. Whether one person was to do the job or a thousand copies of that person, someone would still have to walk into each room, grab a component of the watch, and then add each piece, one at a time, in the correct order, to the wristwatch-in-progress. The desired result (in this case a completed wristwatch) requires that every searcher does part of the job. No one's contribution can be discarded.

In contrast, a suitable task for a quantum computer would be a problem in which one of the many possible combinations of quantum states can find and represent the answer all by itself. The other combinations, all chugging along toward wrong answers, must "collapse," as physicists put it, into the right answer. It's this selective collapsing that poses the challenge. After all, a large enough quantum computer could always be programmed to have its multiple-state atoms represent all possible answers. But what good would that do if there's no way of indicating which of the panoply of results is the right one?

To winnow out the desired result, physicists have come up with a general strategy. The approach is based on the ability of atoms to behave like waves rather than particles. Like two identical but opposite ocean waves colliding, atoms in multiple states can cancel each other out or reinforce each other, depending on how they're aligned.

Unfortunately, for a decade after the late physicist Richard Feynman first suggested the possibility of quantum computation in the early 1980s, no one could figure out a way to apply the phenomenon to a practical task. All that time, physicists were convinced that quantum computers would be good for only one thing: making calculations about quantum mechanics. It was as if, when computer chips were first developed, their designers had announced that the only thing the chips could be used for was learning more about the electrical properties of silicon.

Then, in 1994, AT&T researcher Peter Shor discovered the first practical chore a quantum computer might tackle. One of the more vexing problems in mathematics is the task of finding the prime-number factors of very large numbers.

NOTE

Prime numbers, such as 1, 3, 5, 7, and so forth, cannot themselves be broken into smaller whole-number factors.

Shor found that this problem could be reduced to the simpler one of determining when a certain complicated mathematical sequence starts repeating itself. Identifying a repeated sequence, Shor realized, was something a quantum computer could do. Roughly speaking, by encoding all the elements of the sequence onto the qubits, the states of the qubits that represent identical (and thus repeated) segments can be lined up to strengthen one another. After a while, these reinforced qubits wash everything else out, providing the answer. In theory, a quantum computer with 5,000 or so qubits could solve in about 30 seconds a prime-number problem that would take a conventional supercomputer 10 billion years.

It just so happens there is now an important application for this seemingly esoteric task. Computer data are protected from prying eyes by scrambling the characters of code that represent the data. The mathematical "key" to unscrambling the data is in the form of a very large number (typically 250 digits long) and its prime factors. Such encryption is considered unbreakable, because no conventional computer can figure out the prime factors of such large numbers in a reasonable amount of time.

But, in theory at least, a quantum computer could blow right through these prime-number encryption schemes. A quantum computer hacker would thus have clear access not only to credit card numbers and other personal information that routinely flies around computer networks (including the Internet), but also to government and military secrets. This explains why certain government agencies, operating on the assumption that it's better to lead than to follow, have been throwing millions of dollars at quantum computer research.

Quantum computer success wouldn't necessarily mean all of a physicist's data would become unsafe. Even if computer scientists were able to defy all predictions and build a working device in the near future, cryptographers would turn to schemes that aren't based on prime numbers. One such scheme already exists. It involves coding the data in the form of the shortest distance between two secret points in an abstract, multidimensional space. No one has yet shown that a quantum computer could solve this problem.

It seems that a quantum computer giveth as well as taketh away with regard to data security. That's because a quantum computer could (theoretically) be used to encrypt data in a multiple-state form that could be read properly only by other quantum computers specifically prepared by the sender to read data from the first one. According to physicists, you'd probably need only about a ten-qubit quantum computer to be useful for an encryption application.

Not only is it unlikely that quantum computers will destroy the integrity of the Internet, but they could also end up being a huge boon to it. In 1997, Bell Labs researcher Lov Grover discovered a way to apply quantum computers to a task that many of us engage in every day: searching out information hidden away somewhere in a vast repository of data. Finding information in a database is like the finding-the-briefcase problem. If different combinations of qubit states could each take a look at a different small segment of the database, then one of the combinations of states would come across the desired information.

Grover also figured out how to cause the combination of qubit states with the right answer to stand out. Again, speaking very roughly, the scheme depends on the fact that the qubit states representing "empty rooms" (that is, qubits that haven't found the desired data) are more similar to one another than they are to the qubit states with the answer, just as empty rooms more resemble one another than they do the room that holds the briefcase. Because of their similarity, the wrong qubit states can be combined in such a way as to cancel one another out. Eventually, the one set of qubit states representing the right answer remains.

The speedup offered by this sort of quantum search wouldn't be as dramatic because the difference between the "right" and "wrong" qubit states is more subtle than with the prime-number problem, thus slowing the cancellation process. For example, to search among 100 million addresses, a conventional computer would have to make about 50 million attempts before finding what it was looking for. A quantum computer would need some 10,000 tries. That's still a significant improvement, and it gets bigger with larger databases. What's more, database searching is such a fundamental computer task that any improvement is likely to have an impact on a large array of applications.

How would the Internet benefit? Right now, searching all publicly available web pages for certain key words takes a few seconds (assuming you have a good connection). But remember, the web is still in its infancy. In ten years, it might be thousands of times bigger, and growing, with far more people using it for far more chores. It's not hard to imagine all this activity bringing conventional computers to their silicon knees.

A number of theorists are struggling to come up with strategies for quantum software programs that will usefully solve still other sorts of problems. But, with many of these strategies, the process by which wrong answers are canceled out is so inefficient that they would provide only a modest improvement over conventional computers. There's a large class of problems for which quantum computers would be about twice as fast as classical computers. But physicists are after something sexier.

One possibility that physicists are taking a close look at is a quantum program that could determine whether two complex and very different-looking graphs that connect multiple points are in fact equivalent to each other. Such a program could prove invaluable to computer chip designers, for example, who often switch components around without knowing if they've actually changed anything. Another target is the "traveling salesman" problem, which essentially involves figuring out the shortest way to connect a large number of scattered points. This problem shows up in many forms, including the challenge airlines face in serving the most cities with the fewest possible planes. A nice bonus to solving either of these problems is that they're part of a large class of mathematical problems believed to be related, so cracking one of them could point the way to solving them all.

Few researchers are willing to predict whether quantum computing will ever go beyond a handful of applications. The overall trend has been encouraging, though, and despite the early suspicions of many, if not most physicists, the elusive nature of quantum mechanics would inevitably lead to the uncovering of subtle fundamental barriers to practical quantum computing. However, a deep and wide-ranging theoretical search has yet to turn up a single one.

Finally, what's the big rush? The history of computing suggests that hardware and software breakthroughs tend to occur ahead of the problems they end up solving. Maybe by the time you need to search databases so large that it takes months for an ordinary computer to get through them, quantum computers will be up and running. And that, of course, would free computer scientists to look for the next big thing.

Conclusion

These discoveries have generated intense interest among physicists, partly out of a desire to understand the basic mechanism of high-temperature superconductivity. It is known that the current carriers are Cooper pairs of holes and that conduction takes place along the copper-oxygen planes in these materials, but the mechanism that provides the attractive force within the Cooper pairs remains undecided.

The more general reason for interest in high-temperature superconductors is their possible applications. As previously explained, these include powerful magnets (for MRI, particle accelerators, etc.), lossless transmission lines for electrical power, and highly efficient motors and generators (where the resistive "copper losses" would be absent). However, all of the high-temperature superconductors are brittle ceramics; forming wires and coils from them has proved a formidable task.

Another remarkable property of superconductors is magnetic levitation: A magnet placed over a superconductor remains suspended above its surface because any vertical fall of the magnet induces circulating currents in the superconductor which oppose the motion. This has inspired designs for high-speed magnetically levitated trains, which are currently being tested.

Finally, high-temperature superconductors are currently being used to construct resonant cavities used in microwave communication equipment. Other possible applications rely on the Josephson effect: if two pieces of superconductor are separated by a thin (< 2 nm) oxide layer, Cooper pairs can tunnel through the oxide gap, even in the absence of an applied voltage. If a dc voltage V is applied across the junction, an alternating current flows whose frequency is $f = (2e/h)V = 483.6$ MHz where $V = 1$ microvolt. This allows small voltages to be measured reliably, since frequency can be measured with very high accuracy. A loop of superconductor which contains two Josephson junctions forms a SQUID that is sensitive to small magnetic fields, such as those produced in the heart or brain.

References

[1] "Astronomy Picture of the Day," [http://apod.gsfc.nasa.gov/apod/archivepix.html], NASA Headquarters, 300 E Street SW, Washington DC 20024-3210; Jet Propulsion Laboratory, 4800 Oak Grove Drive, Pasadena, CA 91109; NASA/Goddard Space Flight Center, Greenbelt, MD 20771, 2003.

[2] Energy Information Administration, Department of Energy, EI 30, 1000 Independence Avenue, SW, Washington, DC 20585.

[3] Dr. Steven Weinberg, physicist, Department of Physics, University of Texas at Austin, TX 78712. [This quote is part of an essay based on a commencement talk given by the author at the Science Convocation at McGill University in June 2003].

[4] Dr. Mikhail Lukin, Physicist, Physics Department, Harvard University, Cambridge MA 02138

[5] Dr. Matthew Bigelow, Scientist, Institute of Optics, University of Rochester, Rochester, NY 14627.

Biology and Paleontology

Biology: How the Basic Processes of Life Are Carried Out by DNA and Proteins

"Like most endeavors, life is seriously over-advertised and under-funded."
—Anonymous

Strands of genetic material called chromosomes are inside each of 100 trillion cells in the human body. All the information necessary to create your entire body (with all its differentiated cells) is contained in the nucleus, or core, of each cell. According to biologists, this chemically encoded information is enough to cover over eight thousand pages of written text. The information is stored in the tiny library known as your deoxyribonucleic acid (DNA). DNA makes up the genes and chromosomes of your physical structure.

Like beads on a thread, there exist nearly 100,000 genes arranged along the chromosomes. The fundamental unit of heredity is the gene. It instructs the body cells to make proteins that determine everything from hair color to your susceptibility to diseases. Specifically designed to carry out a single body function, each gene is actually composed of DNA.

DNA! What Is It?

According to biologists, DNA is a huge (macro) molecule encoding the information of who you are. This consists of not only everything about your physical body, but information about your emotional, mental, personal, and spiritual aspects as well. In addition, it consists of fine, hair-like strands in the form of a double helix. (See sidebar, "Strands of DNA.") Twisted so that the two legs, or strands, spiral around each other, DNA also looks like a long, wispy ladder with innumerable rungs that is coiled up again and again to fit in the tiniest of spaces. DNA would be over seven feet long, if all of it in any one cell of your body were pulled out of the nucleus and laid out end to end.

Strands of DNA

Spaghetti is a tangled mess in a pot of boiling water. But, let's say you want to look at a piece of spaghetti more closely. Simple. Just hook a strand around the handle of a wooden spoon, and it straightens out quite nicely. You can examine it easily.

At UC Berkeley, NASA-supported bioengineers want to do the same thing to DNA. Just like that pot of spaghetti, when researchers get a sample of DNA for analysis, it's often a tangled mess of coiled strands. They analyze it by chopping the strands into pieces, cloning the fragments, sequencing the pieces, sorting them according to size and, finally, fitting them back together again. In short, it's a complex and time-consuming process.

Thus, before deep space can be safely explored, protecting astronauts from deep-space radiation is a must. This is a still-unsolved problem, and one that must be successfully tackled. Most approaches to the problem rely on limiting astronaut exposure and shielding.

The UC Berkeley bioengineers' work suggests the possibility of finding astronauts who themselves are more resistant. Just as some versions of genes are more prone to the mutations that lead to problems like breast cancer, others may be more easily damaged by the intense radiation of deep space.

Because these methods are so straightforward, it might lend itself to a future technology: portable DNA analyzers. Astronauts on long space voyages could take one along and test their own genes for radiation damage. At the first sign of trouble, they could take precautions: spending the rest of the journey in heavily-shielded parts of the ship or avoiding spacewalks.

According to bioengineers, DNA biosentinels would benefit more people than just astronauts. The devices could be used to examine anyone's genes for, say, the tendency to react in a particular way to a medication or to develop a certain illness. Everyone (medical researchers, criminologists, pharmaceutical manufacturers) would like to have one of these devices. Of course, all of this is in the future for now. [1] [2]

DNA encodes biological molecules (such as proteins) that either control fundamental biological processes or make up basic structures of life. At about the same time Mendel and Darwin published their work, DNA was discovered as a major chemical of the nucleus. However, proteins were considered better candidates as molecules able to transmit large amounts of hereditary information from generation to generation during the early 1900s.

It seemed likely that the four chemical components of DNA were assembled in a monotonous pattern, and no specific cellular function had yet been found for DNA. On the other hand, proteins were important as enzymes and structural components of living cells. Proteins were also known to be polymers of numerous amino acids, which are called polypeptides. Most importantly, by carrying more structures than the four-letter alphabet of DNA (A to T and G to C), the 20-letter amino acid alphabet of proteins potentially could be configured into more unique information.

DNA is a polymer, or a very large molecule made by linking together a series of repeating units. The units are known as nucleotides in this case. A nucleotide is composed of a nitrogen-containing molecule called a base, a sugar molecule, and a phosphorous-containing group.

Thus, DNA is the primary genetic material of all living organisms—a molecule composed of two complementary strands that are wound around each other in a double helix formation. The strands are connected by base pairs that look like rungs of a ladder. Each base will pair with only one other: adenine (A) pairs with thymine (T), and guanine (G) pairs with cytosine (C). The sequence of each single strand can therefore be deduced by the identity of its partner.

Genes are sections of DNA that code for a defined biochemical function, usually the production of a protein. The DNA of an organism may contain anywhere from a dozen genes, as in a virus, to tens of thousands of genes in higher organisms like humans. The structure of a protein determines its function. The sequence of bases in a given gene determines the structure of a protein. Thus, the genetic code determines what proteins all organisms can make and what those proteins can do. Biologists estimate that only 1–3% of the DNA in your cells contain codes for genes; the rest may be used as a decoy to absorb mutations that could otherwise damage vital genes. [3]

As previously explained, the key to understanding how DNA works is to appreciate the fact that there are only four types of bases associated with DNA: adenine, cytosine, guanine, and thymine. Keep in mind that, in theory, there is no limit to the length of the DNA strand. Thus, a DNA strand can be composed of a long chain having millions of bases.

Coiled into a double helix, the DNA molecule is actually composed of two DNA strands. This can be thought of as resembling two wires twisted around one another. The researchers thus realized that the only way the bases on each strand could be properly aligned with one another in a double helix configuration was to place base A opposite T and G opposite C as they manipulated scale models of DNA strands.

The pairing of bases A to T and G to C was the only arrangement possible in the double-helix configuration according to biologists. This is a concept that has become known as base pairing. There are no restrictions on how the bases are to be sequenced on a DNA strand, although A-T and G-C pairs are always required. Thus, one can observe the sequences T-A-T-T or G-T-A-A or G-T-C-A. For example, according to biologists, when these sequences are joined together with their opposite number in a double-helix configuration, they pair as follows [3]:

T-A-T-T G-T-A-A G-T-C-A
A-T-A-A C-A-T-T C-A-G-A

On a DNA strand, any base can follow another. This means that the possible number of different sequence combinations is staggering! According to biologists, the average human chromosome has DNA containing 100 million base pairs. Taken together, all the human chromosomes contain about 3 billion base pairs. From these numbers, biologists can begin to appreciate the diversity of DNA and hence the diversity of living organisms. DNA is like a book of instructions. As previously explained, the alphabet used to create the book is simple enough: A, T, G, and C. The role and function of a DNA molecule defines the order in which these letters are arranged.

An Enzyme DNA Polymerase Chain Reaction (PCR)

It became apparent how DNA duplicated itself prior to cell division once the double-helix structure of DNA was discovered. The analogy of positive and negative photographic film is suggested by the concept of base pairing in DNA. The same information is contained in each strand of DNA in the double helix. One can make a positive print from a negative or from a positive, a negative. The unwinding of the DNA strands in the double helix starts the DNA replication process. Each strand is then exposed to a collection of free nucleotides. As dictated by the principle of base pairing (A with T and G with C) letter by letter, the double helix is recreated as the nucleotides are assembled in the proper order. Where before there was only one, the result is the emergence of two identical copies of DNA. When a cell divides, it can now pass on its genetic identity.

The phenomenon of DNA replication appeared to be only of academic interest to forensic scientists interested in DNA for identification purposes until recently. When researchers were able to perfect the technology of copying a DNA strand, all of this changed, however. This new laboratory technique is known as polymerase chain reaction (PCR). With the aid of a DNA polymerase, small quantities of DNA or broken pieces of DNA found in crime scene evidence can be copied. With as little as one-billionth of a gram of DNA, the PCR technique is capable of yielding useful information.

In order to separate the strands, PCR begins with heating the DNA. Primers (short DNA segments of known bases) are then added to combine with the strands. An enzyme, DNA polymerase, capable of synthesizing a specific region of DNA, is added, as well as a mixture of free nucleotides (G,A,T,C) to the separated strands. The enzyme directs the rebuilding of a double-stranded DNA molecule. By adding the appropriate bases one at a time, this in turn extends the primers, and results in the production of two complete pairs of double-stranded DNA segments. This completes the first cycle of the PCR technique, and the outcome is a doubling of the number of DNA strands—that is, from one to two. According to biologists, this cycle is then repeated 25 to 30 times, and over one million copies of the original DNA molecule is produced. For example, let's consider a segment of DNA that a biologist may want to duplicate by PCR. They pair as follows [4]:

-G-T-C-T-C-C-T-T-C-C-A-G-
-C-A-G-A-G-G-A-A-G-G-T-C-

Short sequences of DNA on each side of the region of interest must be identified in order to perform PCR on this DNA segment. Thus, short sequences are designated by boldface letters in the DNA segment as shown in the preceding example. If the PCR technique is going to work, these short DNA segments must be available in a pure form known as a primer.

According to biologists, the first step in the PCR process is to heat the DNA strands to about 94°C (201°F). The double-stranded DNA molecules separate completely at this temperature.

The second step is to add the primers to the separated strands. According to biologists, this allows the primers to combine or hybridize with the strands by lowering the test tube temperature. They pair as follows [4]:

-G-T-C-T-C-C-T-T-C-C-A-G-
C-A-G-A
C-C-A-G
-C-A-G-A-G-G-A-A-G-G-T'C-

The third step, according to biologists, is to add the DNA polymerase and a mixture of free nucleotides (G, A, T, C) to the separated strands. The polymerase enzyme directs the rebuilding of a double-stranded DNA molecule when the test tube is heated up to 72 °C (161°F). This results in the production of two complete pairs of double-stranded DNA segments, thus extending the primers by adding the appropriate bases one at a time [4]:

-G-T-C-T-C-C-T-T-C-C-A-G-
C-A-G-A-G-G-A-A-G-G-T-C-
-G.T-C-T-C-C-T-T-C-C-A-G
-C-A-G-A-G-G-A-A-G-G-T-C-

The first cycle of the PCR technique is now completed, and the outcome is a doubling of the number of DNA strands—that is, from one to two. The cycle of heating, cooling, and strand rebuilding is then repeated, resulting again in a doubling of the DNA strands. Thus, four double-stranded DNA molecules will have been created from the original double-stranded DNA sample upon completion of the second cycle. According to biologists, typically 25 to 30 cycles are carried out to yield over one million copies of the original DNA molecule. Each cycle takes less than two minutes to perform. [4]

Understanding DNA Typing

The knowledge that, within the world's population, there are numerous possibilities for the number of times a particular sequence of base letters can repeat themselves on a DNA strand is the key to understanding DNA typing. The possibilities become even greater when one deals with two chromosomes, each containing different lengths of repeating sequences. The combination of PCR with Short Tandem Repeat (STR) analysis is the most widely used technique for DNA typing.

Locations (loci) on the chromosome that contain short sequence elements that repeat themselves within the DNA molecule are known as STRs. Because they are found in great abundance throughout the human genome, they serve as helpful markers for identification. According to biologists, what is important to appreciate is that the repeating sequence is relatively short in length, three to seven bases, and that the entire strand of an STR is also very short—that is, fewer than 500 bases in length. This means that STRs are much less susceptible to degradation and may often be recovered from bodies or stains that have been subject to extreme decomposition.

In order to understand the utility of STRs in forensic science, let's look at one commonly used STR known as HUMTH01. The DNA segment contains the repeating sequence AATG. There have been seven HUMTH01 variants identified in the human genome. These variants contain five through eleven repeats of AATG. [5]

Thus, HUMTH01 is extracted out of biological materials and amplified by PCR during a forensic examination. The ability to copy an STR allows extremely small amounts of the molecule to be detected and analyzed. Once the STRs have been copied or amplified, they are separated on an electrophoretic gel. By examining the distance the STR has migrated on the electophoretic plate, one can determine the number of AATG repeats that exist in the STR. Every person has two STR types for HUMTH01, each inherited from one parent. Thus, for example, one may find in a semen stain HUMTH01 with six repeats and eight repeats. This combination of HUMTH01 is only found in approximately 3.5% of the population. [5]

There are hundreds of different types of STRs found in human genes. This is what makes STRs so attractive to forensic scientists. The smaller the percentage of the population from which these STRs can emanate, the more STRs one can characterize. This gives rise to the concept of multiplexing. Using the technology of PCR, one can simultaneously extract and amplify a combination of different STRs. For example, one system on the commercial market is the STR Blue kit. According to biologists, this kit provides the necessary materials for the coamplification and detection of three STRs—D3S1358, vWA, and FGA. This is called triplexing. The design of the system ensures that the size of the STRs does not overlap, thereby allowing each marker to be viewed clearly on an electrophoretic gel. [5]

Known as the Combined DNA Index System (CODIS), the forensic science community in the United States has standardized thirteen STRs for entry into a national database. The probability of identity is the probability that two individuals selected at random will have an identical STR type. The smaller the value of this probability, the more discriminating will be the STR. A higher degree of discrimination and even individualization can be attained by analyzing a combination of STRs (multiplexing). Because STRs occur independently of each other, the probability of biological evidence having a particular combination of STR types is determined by the product of their frequency of occurrence in a population. Thus, the more impressive the frequency of occurrence of the analyzed sample in the general population, the greater the number of STRs characterized. [5]

In order to provide analysts with one additional piece of useful information along with STR types (i.e., the gender of the DNA contributor), manufactures of commercial STR kits typically used by crime laboratories have made some provisions. Located on both the X and Y chromosomes, the focus of attention here is the amelogenin gene. This gene, which is actually the gene for tooth pulp, has an interesting characteristic in that it is shorter by six bases in the X chromosome as compared to the Y chromosome. Hence, when the amelogenin gene is amplified by PCR and separated by electrophoresis, a male with an X and a Y chromosome will show two bands. A female having X and X chromosomes will have just one band. Typically, these results are obtained in conjunction with STR types.

Nuclear and Mitochondrial DNA

Normally, it's assumed that the subject of attention is the DNA found within the nucleus of a cell when one describes

DNA in the context of a criminal investigation. Actually, a human cell contains two types of DNA—nuclear and mitochondrial. According to biologists, the first constitutes the 23 pairs of chromosomes contained within the nuclei of your cells. Each parent contributes to the genetic makeup of these chromosomes. Nevertheless, mitochondrial DNA (mtDNA) is inherited solely from the mother and is found outside the nucleus. [6]

Cell structures found in all of your cells are known as *mitochondria*. According to biologists, they are the power plants of your body, providing about 90% of the energy that your body needs to function. What's important from the forensic science perspective is that a single mitochondrion contains several loops of DNA, all of which are involved in energy generation. Further, since each cell in your body contains hundreds to thousands of mitochondria, this effectively means that there are hundreds to thousands of mtDNA copies in a human cell. This compares to just two copies of nuclear DNA located in that same cell. Thus, forensic scientists are offered enhanced sensitivity and the opportunity to characterize mtDNA in situations where nuclear DNA is significantly degraded, such as in charred remains, or may be present in small quantity such as a hair shaft. Interestingly, in situations where authorities cannot obtain a reference sample from an individual who may be long deceased or missing, a mtDNA reference sample can be obtained from any maternally related relative. Nevertheless, through the use of mtDNA analysis, all individuals of the same maternal lineage will be indistinguishable. [6]

It must be noted that forensic analysis of mtDNA is more rigorous, time consuming, and costly when compared to nuclear DNA profiling. At the same time, mtDNA analysis is significantly more sensitive than nuclear DNA profiling. (See sidebar, "Bringing DNA Profiling to the Masses.") So, only a handful of public and private forensic laboratories receive evidence for this type of determination at this time.

Bringing DNA Profiling to the Masses

Researchers are developing technologies that could someday deliver genetic self-knowledge to the masses. From the string of three billion "letters" in your DNA, each of you might learn which diseases you are susceptible to in time to take preventive action. If you got sick, person-to-person variations in your genes would guide doctors in choosing exactly the right allergy medication, antidepressant, or painkiller for you. Eventually, personal genomes could help athletes tailor their training, dieters draw up meal plans, and addicts map out recovery programs. In a couple of decades, you might even carry digital cards encoded with your personal gene maps right alongside your medical insurance cards.

Such information won't do much good until scientists learn far more about how subtle DNA variations affect health and how those predispositions can be overcome. Personal genomics will also require vastly faster and cheaper DNA-analyzing technology than exists today, and the race to develop it is already on.

Genetic tests are nothing new. Patients, prospective parents, and even fetuses can be tested for some gene variants, such as those linked to cystic fibrosis, metabolic problems, and certain cancer risks. But those tests are mostly slow and expensive, focusing on one or a handful of genes at a time. True personal genomics means examining the tens of thousands of genes strung out along the 6-foot length of human DNA. Full genome sequencing, although much faster than it was in 1997, still takes a laboratory full of machines, each analyzing a small part of the DNA, together with computers for stitching the results together.

Now scientists have been given a portrait of the human genome. They can begin cataloging the person-to-person variations that help make each of you distinct, and they can set to work on miniaturized labs-on-a-chip that can rapidly scan an individual's genome for those variations. But other researchers point out major challenges, including developing an optical reader able to reliably detect the tiny fluorescent markers as they flash past, along with software capable of making sense of the data.

Researchers across the country are working on an even more ambitious scheme to sequence individual genomes letter by letter, as the DNA molecule snakes through an infinitesimally small hole in a membrane. Each chemical unit passing through the hole would block it by an amount that depends on the unit's shape, thus reducing an electric current passing through the hole. The current's ups and downs would be translated into a complete DNA sequence. Although the technology looks promising in theory, it faces a serious stumbling block: Researchers have not yet figured out how to make a hole that is reliably small enough (about 2 nanometers across, one tenth the size of a virus) to hold a DNA strand snugly.

Concerns about privacy, in-utero testing and abortion, insurance discrimination, and even employment discrimination may hamstring the personal genomics revolution before it gets started. But if those issues can be resolved and the scientific hurdles can be overcome, the genetic revolution may eventually give you personal health profiles that will help you lead a better and healthier life. [7]

As previously mentioned, nuclear DNA is composed of a continuous linear strand of nucleotides (A, T, G, C). On the other hand, mtDNA is constructed in a circular or loop configuration. According to biologists, each loop contains a sufficient number (approximately 16,569) of As, Ts, Gs, and Cs to comprise 37 genes involved in mitochondrial energy generation. In the human population, two regions of mtDNA have been found to be highly variable. These two regions have been designated Hypervariable Region I (HV1) and Hypervariable Region II (HV2). The process for analyzing HV1 and HV2 is tedious: It involves generating many copies of these DNA regions by PCR and then determining the order of the ATGC bases constituting the hypervariable regions. This process is known as sequencing. The FBI laboratory, the Armed Forces DNA Identification Laboratory, and other laboratories have collaborated to compile a mtDNA population database containing the base sequences from Hypervariable Regions I and II. [6]

Most laboratories will simply report the number of times these sequences appear in the mtDNA database, once the sequences of the hypervariable regions from a case sample are obtained. Currently, however, the mtDNA database is too small to allow for reasonable estimates of the frequency of occurrence of mtDNA sequences in the population. Nevertheless, mtDNA typing will not approach STR analysis in its discrimination power even under the best of circumstances. Thus, mtDNA analysis is best reserved for samples for which nuclear DNA typing is simply not possible.

Nuclear Hair DNA Typing

An examination of morphological characteristics is still the current approach for the examination of hair specimens. However, this technology has been extended to the individualization of human hair by recent major breakthroughs in nuclear DNA typing. Forensic hair examiners can link human hair to a particular individual by characterizing the nuclear DNA present in the hair root or in follicular tissue adhering to the root. When pulled from the head, many hairs will be found with a follicular tag—a translucent piece of tissue surrounding the hair's shaft near the root. This has proven to be the richest source of DNA associated with hair.

The FBI initiated a program to compare human head and pubic hairs through mitochondrial DNA in 1996. As compared to nuclear DNA, there are many more copies of mitochondrial DNA located in your cells. For this reason, the success rate of finding and typing mitochondrial DNA is much greater from samples, such as hair, which have limited quantities of nuclear DNA.

Preservation and Collection of Biological Evidence

Only eclipsed by the fingerprint, the advent of DNA profiling has vaulted biological crime scene evidence to a stature of importance. In fact, the high sensitivity of DNA determinations has even changed the way police investigators define biological evidence. In the past, crime laboratories were usually able to extract some useful information from a blood or semen stain, or from a hair deposited at the crime scene. Today, according to biologists, the sensitivity of PCR means that 1 nanogram (one billionth of a gram) or less of DNA can yield sufficient information to individualize evidence. With this technology in hand, the horizon of the criminal investigator extends beyond the traditional dried blood or semen stain. This includes a bedsheet containing dead skin cells, or stamps and envelopes moistened with saliva, or a cup that has come in contact with a person's lips. [8]

With a minimum amount of personal contact, the evidence collector must handle all body fluids and biologically stained materials. All body fluids must be assumed to be infectious. Hence, wearing disposable latex gloves while handling the evidence is required. The possibility that the evidence collector will contaminate the evidence can be reduced significantly by the wearing of Latex gloves. During the evidence collection phase of the investigation, these gloves should be changed frequently. The wearing of face masks, shoe covers, and possibly coveralls, constitute safety considerations and the avoidance of contamination.

When a transfer of blood between a victim and suspect can be demonstrated, blood has great evidential value. For this reason, all clothing from both victim and suspect should be collected and sent to the laboratory for examination. This procedure must be followed even when the presence of blood on a garment does not appear obvious to the investigator. Laboratory search procedures are far more revealing and sensitive than any that can be conducted at the crime scene. In addition, blood should also be searched for in the less than obvious places. For example, the criminal may have wiped his or her hands on materials not readily apparent to the investigator. Efforts must be made to find towels, handkerchiefs, or rags that may have been used and then hidden. Also, the examination of floor cracks or other crevices that may have trapped a quantity of blood should be given full attention.

Because the accumulation of residual moisture could contribute to the growth of DNA-destroying bacteria and fungi, the packaging of biological evidence in plastic or airtight containers must always be avoided. Each stained article should be packaged separately in a paper bag or in a well-ventilated box. If feasible, the entire stained article should be packaged and submitted for examination. If this is not possible, the dried blood is removed from a surface with the aid of a sterile cotton-tipped swab lightly moistened with distilled water from a dropper bottle. A portion of the unstained surface material near the recovered stain must likewise be removed and placed in a separate package. This is known as a substrate control. The forensic examiner might use the substrate swab as a control to confirm that the results of the tests performed were brought about by the stain and not by the material on which it was deposited. However, this practice is normally not necessary when DNA determinations are carried out in the laboratory. Also, the collected swabs must not be packaged in a wet state. This is a very critical point.

Until delivery to the laboratory, all packages containing biological evidence should be refrigerated or stored in a cool location out of direct sunlight, but blood mixed with soil is one common exception. Microbes present in soil will rapidly degrade DNA. Therefore, blood in soil must be stored in a clean glass or plastic container and immediately frozen.

When an analyst can compare each of its DNA types to known DNA samples collected from victims and suspects, only then will biological evidence attain its full forensic value. According to forensic scientists, for this purpose, at least 7 cc of whole blood should be drawn from individuals by a qualified medical person. The blood sample should be collected in a sterile vacuum tube containing the anticoagulant ethylenediamine tetraacetic acid (EDTA). In addition to serving as an anticoagulant, EDTA inhibits the activity of enzymes that act to degrade DNA. Prior to delivery to the laboratory, the tubes must be kept refrigerated (not frozen) while awaiting transportation to the laboratory. Besides blood, there are other options for obtaining control DNA specimens. The least intrusive DNA control and one that can readily be used by non-medical personnel is the buccal swab. Here, cotton swabs are placed in the subject's mouth and the inside of the cheek is vigorously swabbed, resulting in the transfer of buccal cells onto the swab. There are some interesting alternatives available to evidence collectors that include: a toothbrush, combs and hair brushes, a razor, and earplugs, if an individual is not available to give a DNA control sample. [8]

Contamination Issues

On the TV shows *CSI*, *CSI: Miami*, *NCIS*, and *Crossing Jordan*, you have seen forensic scientists capable of detecting extremely small quantities of DNA from biological evidence. With increased sensitivity comes a greater chance that accidental contamination can be detected in crime scene evidence. Introducing foreign DNA into a stain while collecting it can cause contamination. It can also occur when items of evidence are in contact with each other during a transfer of DNA.

The fortunate thing here is that the presence of contamination during an examination of DNA band patterns in the laboratory can be readily revealed. For example, with an STR, one will expect to see a two band pattern. If one observes more than two bands, it becomes apparent that one could be dealing with a mixture of DNA from more than one source. In order to minimize the possible occurrence of contamination of biological evidence, there are some relatively simple steps that crime scene investigators can take:

- Always collect a substrate control for possible subsequent laboratory examination.
- Always package each item of evidence in its own well-ventilated container.
- Always wear disposable latex gloves when collecting biological evidence.
- Pick up small items of evidence such as cigarette butts and stamps with clean forceps. Disposable forceps are to be used so that they can be discarded after a single evidence collection. [9]

Now that you know what DNA is, let's briefly look at why DNA and protein could not be produced by random chance.

DNA and Proteins

One of the most important discoveries of the 20th century was the discovery of the DNA molecule. It has had a powerful effect on biological research. It has also brought quandary and confusion to evolutionary scientists. If they cared to admit the full implications of DNA, it would bring total destruction to their theory.

This part of the chapter will primarily and briefly discuss the DNA code and the components of protein—and will show that each are so utterly complicated as to defy any possibility that they could have been produced by chance events. Yet random actions are the only kind of occurrences that evolutionists tell us have ever been used to accomplish the work of evolution.

The significance of all this is immense. Because of the barrier of the multi-billion DNA code, not only was it impossible for life to form by accident, it could never thereafter evolve into new and different species! Each successive specification change would require a totally new and different, but highly exacting, code to be in place on its very first day of its existence as a unique new species.

So, according to biologists, with over 100,000 different proteins to manufacture, how the heck does your body get it right? (See sidebar, "What Really Are Proteins?"). When one thinks of the amount of information the body needs to keep track of—eye, hair, and skin color; protein sequence; toenail size; etc.—it would seem a task for a supercomputer to record all of the necessary information. In essence it is. But not a supercomputer made of silicon wafers and TV screens. Rather one made of an intricate biomolecule called DNA. [10]

What Really Are Proteins?

Proteins are polymers of amino acids. While there are hundreds of thousands of different proteins that exist in nature, they are all made up of different combinations of just 20 amino acids. Proteins are large molecules that may consist of hundreds or even thousands of amino acids.

Peptide Bonds

In proteins, many peptide bonds form between amino acids to create long chains (thus proteins are also called polypeptides because they contain many peptide bonds). Proteins serve many purposes in the body. Structural proteins such as keratin and collagen are the main ingredients in your hair, muscles, tendons, and skin and help give structure to the body. In addition to adding structure, other proteins perform a wide range of functions in the human body. Amylase is a protein that helps your body digest starch; the protein hemoglobin is responsible for transporting oxygen in the blood stream; insulin helps regulate the storage of glucose in the body; and the list goes on and on. There are an estimated 100,000 different proteins in the human body alone. Each has a different structure and performs a different function in the body.

Because proteins perform such specific tasks in the body, each protein has to be manufactured in the body to suit the individual's specific needs. When an animal eats protein, that protein is broken down in the digestive tract into its individual amino acids. These amino acids are then recombined in the body in the specific sequence needed to form whichever protein the animal needs at that point in time. Thus, protein in food is just a source of amino acids. Some good food sources of protein include beans, milk and cheese, and fish and meats. [10]

As previously explained, DNA is in the family of molecules referred to as nucleic acids. One strand of DNA has a backbone consisting of a polymer of the simple sugar deoxyribose bonded to something called a phosphate unit. According to biologists, the backbone of a strand of DNA resembles the following [10]:

sugar-phosphate-sugar-phosphate-sugar-phosphate-sugar-phosphate-...

What is impressive about DNA is that each sugar molecule in the strand also binds to one of four different nucleotide bases. These bases, Adenine (A), Guanine (G), Cytosine (C), and Thymine (T), are the beginnings of what you will soon see is a molecular alphabet. Each sugar molecule in the DNA strand will bind to one nucleotide base. Thus, you'll see that a single strand of the molecule looks more like the following as a biologist's description of DNA unfolds [10]:

C T G A ...
sugar-phosphate-sugar-phosphate-sugar-phosphate-sugar-phosphate-...

Each strand of DNA contains millions or even billions (in the case of human DNA) of nucleotide bases. According to your genetic ancestry, these bases are arranged in a specific order. The order of these base units makes up the code for specific characteristics in the body, such as eye color or nose-hair length. Your body's DNA uses 4 letters (the 4 nucleotide bases) to code for millions of different characteristics, just as you use 26 letters in various sequences to code for the words you are now reading. [10]

Two strands of DNA cross-linked together actually make up each molecule of DNA. Each nucleotide base in the DNA strand will cross-link (via hydrogen bonds) with a nucleotide base in a second strand of DNA forming a structure that resembles a ladder. These bases cross-link in a very specific order: A will only link with T (and vice-versa), and C will only link with G (and vice-versa). Thus, the following looks like a biologist's picture of DNA [10]:

sugar-phosphate-sugar-phosphate-sugar-phosphate-sugar-phosphate-...
G A C T ...
C T G A ...
sugar-phosphate-sugar-phosphate-sugar-phosphate-sugar-phosphate-...

The discovery that the structure of DNA is actually a double helix was made by James Watson, Francis Crick, and Rosalind Franklin in 1953. In other words, like the cord of a telephone, the DNA ladder previously described coils around itself.

When more genetic material is needed (such as during reproduction, to pass on characteristics from parent to offspring), the specific base-pairing of DNA aids in the reproduction of the double helix. When DNA reproduces, the 2 strands unzip from each other and enzymes add new bases to each, thus forming two new strands.[10]

All the information needed to produce everything in the human body lies within this coil of DNA. A strand of DNA may be millions, or billions, of base-pairs long. Different segments of the DNA molecule hold code for different characteristics in the body. A gene is a relatively small segment of DNA that codes for the synthesis of a specific protein. This protein then will play a structural or functional role in the body. A larger collection of DNA that contains many genes and the support proteins needed to control these genes is known as a chromosome.[10]

Protein Synthesis

How does a gene code for a protein? According to biologists, protein synthesis is a two-part process that involves a second type of nucleic acid along with DNA. This second type of nucleic acid is ribonucleic acid (RNA). RNA differs from DNA in two respects: The sugar units in RNA are ribose as compared to DNA's deoxyribose. RNA does not bind to the nucleotide base Thymine because of this difference. Instead, RNA contains the nucleotide base Uracil (U) in place of T (RNA also contains the other three bases: A, C, and G).[11]

Transcription

The two DNA strands in a gene that codes for a protein that unzip from each other, is the first step of protein synthesis. Similar to the way DNA replicates itself, a single strand of messenger RNA (mRNA) is then made by pairing up mRNA bases with the exposed DNA nucleotide bases. Remember that U is paired with each of DNA's A bases, since mRNA does not contain the base Thymine.[11]

Translation

The mRNA leaves the cell nucleus and travels to a cellular organelle called the ribosome (it is beyond the scope of this chapter to conduct a full discussion of the cell, nucleus, and ribosome) after it is manufactured. In the ribosome, the mRNA code is translated into a transfer RNA (tRNA) code which, in turn, is transferred into a protein sequence. In this process, each set of three mRNA bases (the mRNA base triplet is called a codon) will pair with a complimentary tRNA base triplet (called an anticodon). Each tRNA is specific to an amino acid. Therefore, amino acids are linked together by *peptide bonds* as tRNA's are added to the sequence. This will eventually form a protein that is later released by the tRNA.

Biologists are left with a protein that consists of the following chain after the processes of transcription and translation are complete:

Aspartic Acid-Leucine

Although this protein is only two amino acids in length, proteins normally consist of hundreds or thousands of amino acids. [11]

So, what is the challenge here? In other words, how are the basic processes of life carried out by DNA and proteins? The great problem here is to determine what all the proteins do and how one can reproduce life from this knowledge. The answer to the problem lies somewhere in the genome project.

The Challenge

Sometime between now and the year 2033, the DNA sequences of many life forms including humans will be determined. This is known as the genome project: the greatest biological development in science history.

The Genome Project

Scientists have cracked the code! This is the longest, tiniest imaginable, most important, oldest code: the code of human life—the DNA sequence of humanity. As previously explained, the numerics are staggering. Written in just a four-letter alphabet (A, T, C, G), the human genome is around three billion letters long, or about one billion words in length since each word (a codon) is three letters long. And there are around 600 billion-trillion copies of the human genome on Earth (6 billion people times 100 trillion cells per person). [12]

According to biologists, it took about three billion years to create (the age of life on Earth) and only 15 years to decipher if one starts at the beginning of the Human Genome Project. Alternatively, it might be argued that it has taken several hundred thousand years (the age of Homo sapiens) for humans to look inside themselves and figure out their vital essence.

The most important accomplishment in the biological research field to date has been the human genome project. This is the crown jewel of 20 years of biological research. The human genome project is a massive project built on the scientific endeavors of decades of dedicated investigators, on a scale unmatched in the history of biology.

The genome encodes the proteins that regulate numerous biological processes that form the structural elements of life. Genes provide the characteristics that distinguish one individual from another and allow these features to be passed from one generation to the next through reproduction, thereby providing the microscopic mechanism for evolution. The genome is often called the blueprint for life for these reasons. In short, the sequence of the human genome and similar sequences for other organisms, comprise the Books of Life—the bible of biology, so to speak.

The genome is composed of chromosomes. According to biologists, there are 24 different types of chromosomes in humans, which are labeled chromosome 1 through chromosome 22, and chromosome X and Y. Thus, the Great Code is contained in 24 volumes. Humans, like other higher forms of life, are diploid (that is, their chromosomes are duplicated in the nucleus of a cell). There are 23 pairs, 22 of which are matched: There are two copies of chromosomes 1 through 22, and then either an XX pair for females or an XY pair for males. Each chromosome consists of a long DNA molecule wrapped into a compact form around proteins known as histones—roughly the way thread is wound about a bobbin. DNA is comprised of two long chains of nucleotides bound and twisted about each other to form a helix. As previously explained, the nucleotides are of four types: adenine (A), guanine (G), cytosine (C), and thymine (T). Specifying the nucleotide sequence as a series of "biological letters," such as CTATGAT, determines the DNA molecule. [12]

According to biologists, the remarkable scientific accomplishment that has been achieved is to provide nearly complete DNA sequences for the 24 human chromosomes. These sequences will be precisely known within a relatively short period of time. Eventually, the genomes of almost every living creature on Earth will be part of the scientific data bank, the sum of which constitutes the Library of Life.

Certain sections of the DNA that code proteins are known as genes. Messenger ribonucleic acid, abbreviated mRNA, transports the information in the DNA to the protein-producing machinery of a cell. In a given cell, certain genes are turned on, meaning that they are allowed to generate the proteins that they code, while other genes are switched off. In other words, the function of a cell is determined by the genes that are turned on.

The genetic code determines the amino acid sequence of a protein coded by a gene. Without this "biological Rosetta stone," Nature's four-letter texts would be as incomprehensible as a message from an alien civilization. According to biologists, less than 1.5% of the genome encodes proteins; the rest consists of non-coding sequences, a sizeable fraction of which is junk, meaning that it appears to have no present biological purpose. In fact, the human genome is a genetic jungle full of sequences of "freeloaders," "parasites," "hitchhikers," "ancient viral invaders," and "evolutionary fossils" that are all competing for space on the DNA molecule. The "hitchhikers," scientifically known as transposons, have copied themselves and jumped from place to place. It appears that some stretches of sequences date back to the days of unicellular life in the Pre-Cambrian, more than 700 million years ago. Bacteria carry much less excess baggage. With a minimum amount of junk DNA, the bacteria coding regions appear one after another. [12]

According to biologists, the male-defining Y chromosome is a junk yard, full of repetitive, non-functional nucleotide sequences. Furthermore, in the Y chromosome, there are many copies of sperm-production genes. In other words, it is as though males are afraid of sterility or are trying to defend themselves against female invasion. What is worse is that evolution has reduced it to a little stump in comparison with the other chromosomes, and it will be stuck with these features for a long time. Because the Y chromosome does not recombine (that is, it does not undergo sequence shuffling during reproduction), it is slow to evolve. On the other hand, this renders it useful in molecular anthropology, which uses DNA to deduce various relations among *Homo sapiens* during the past 200,000 years.

The DNA sequences will eventually provide a comprehensive list of all the proteins that the body makes, even though many proteins have been studied in detail. Responsible for many human diseases are defects in the proteins, which are caused by sequence errors in the genes.

Individuals are 99.9% identical from the viewpoint of the human genome. Yet, according to biologists, the residual 0.1% leads to several million spelling differences, with some such variations leading to dramatically higher risks of certain cancers and other diseases. These differences are known as polymorphisms, of which the most important type are single nucleotide polymorphisms, or SNPs (pronounced "snips"). Thus, a main source of genetic variation are SNPs. So what have you already learned from the Human Genome Project? [12]

Surprise! According to biologists, the human genome appears to contain only a third as many genes as had been previously estimated. Scientists had expected to find as many as 100,000 genes. But the latest results suggest somewhere between 26,000 to 40,000 genes, with 30,000 being the favored figure. However, the old rule "one gene—one protein" appears to be wrong. A single gene may be able to initiate the manufacturing of several proteins, depending on the circumstances. This means that the number of distinct proteins in a human body probably numbers around 100,000. This is because a gene consists of exons separated by introns. All the coding for the protein is in the exons. The introns are removed and the exons are spliced together when a protein is made, but there are often several ways in which the splicing can be done, as it turns out.[12]

It also appears that in biological sophistication, the total number of genes is not a leading factor. The roundworm, for example, has 19,000 genes, while the fruit fly possesses 13,600. These organisms are relatively simple invertebrates: The roundworm has only 960 cells, whereas a human has 100 trillion, and the 100 billion brain cells in a human should be compared to the 300 neurons of the roundworm.[12]

Of course, there are other results: A larger amount of junk DNA has been indicated in the initial findings. For a long time, biologists have known that much of the genome consists of repeating elements that have copied and inserted themselves into the sequence and whose only purpose appears to be to reproduce themselves. Although most junk DNA seems to serve no extant biological purpose, it might play a role in evolution. It should be somewhat humbling that it makes up more that 98.5% of the genome. In other words, less than 1.5% of the genome is used for coding proteins. So, before the sequencing projects were done, this small percentage was found to be half of what was thought to be the number. [12]

Another interesting result is that whole blocks of genes are copied from one chromosome to another. As a protective mechanism, this might have occurred in evolution tens of millions of years ago. According to biologists, Chromosome 19 is the biggest culprit, sharing genetic blocks with 16 other chromosomes. It also appears to be the one most densely packed. Large-scale block transfers have also been seen in the genome of the mouse. These duplicated fragments of DNA that have gotten inserted back into the chromosomes have shaped the size and architecture of the genome of these mammals. [12]

Vast regions of repeating sequences are also contained in the human genome. Scientists estimate that almost 50% of the genome consists of these repeaters. In human chromosomes, two freeloaders, called LINE1 and Alu, make up, respectively, about 17% and 10% of the DNA. [12]

NOTE

The percentage of thymine (T) in the genome is the same as adenine (A) because these two nucleotides appear in complementary positions in the two strands that make up DNA. Likewise, the percentage of guanine (G) is the same as cytosine (C).

A lot of progress has been made with the annotation of the human genome. So, for genetic biologists, what is the next great challenge? [12]

Unfortunately, biologists do not presently know how to combine a specific set of proteins to provide a cell with a particular function. Nature miraculously does this automatically. It is like throwing all the small parts that you have constructed from the blueprint manuals into adjacent piles and having a car amazingly emerge. So, the next great goal in understanding life is to figure out how proteins collectively interact to carry out cellular processes. At the genetic level, biologists must learn to deduce the biological consequences of having a whole ensemble of genes turned on.

So, how important is it to understand the human genome? To understand the human genome is to understand life at a fundamental level. [12]

Life at the elementary level is well understood in general terms. Its processes are metabolism, transcription of DNA into RNA, translation of RNA into proteins, and DNA replication. A more detailed understanding of life will be achieved as biologists learn more about genes and proteins.

The Importance of Understanding the Human Genome

Putting in order more than three billion chemical units that encode the instructions on how to build and operate a human being is essentially sequencing the human genome. But those instructions are written in a language biology does not fully understand. Indeed, some have described the genome as a parts list minus information on how the parts connect or what they do, and leading scientists are quick to point out that just knowing the raw data set that makes up the genome is not an end in itself.

Only after scientists have figured out how the parts go about making the machine that is the human body will the usefulness of the genome emerge. This is what the biology of the new millennium is all about. It is an accelerated science based as much on bits and semiconductor chips as on microscope slides and test tubes. Compressing what would have been a decade or more of research into a day or two of database queries, scientists can now see the genetic correlation with disease and seek out starting points in the genes of humans and other creatures.

Actually, an industry known as bioinformatics has grown up around the idea that biology will increasingly depend on sorting and manipulating huge amounts of data. Industry analysts forecast that the market for genomics information and the technology to use it will reach an annual $7 billion by 2006.

The genetic code cracking, or sequencing, is being done mainly by two groups of scientists: the Human Genome Project and Celera Genomics of Rockville, MD. The Human Genome Project, an international public consortium with funding over 15 years of about $3 billion, downloads all its sequencing data every day to freely accessible databases run by the National Center for Biotechnology Information, the European Bioinformatics Institute, and the DNA Data Bank of Japan. Meanwhile, Celera has sold its genome database and related information to pharmaceutical and biotechnology firms. Celera's and the Human Genome Project's efforts to sequence the genome have often been described as a race—a metaphor that irks most scientists in the public project. [13]

The Next Step in the Discovery Process

The next step in the discovery process is annotating the genome by attaching function and meaning to its genes. But first, the genes must be located, separating them from the 95% and more of the genome that appears uninvolved in the workings of human cells. Spread throughout the billions of bases that make up the genome, there are an estimated 25,000–150,000 genes, and little is known about the majority of them. [13]

The sequencing of the genomes of other species is an important help to the annotation process. Owing to decades of study, many genes and their functions have already been identified in fruit flies and mice, and estimates of the similarity between the genes of mice and men run high. So, by matching up known genes in mice or flies with sequences in the human genome, scientists get a clue to the function of human genes. Celera worked on the fruit fly commonly used in genetics experiments and sequenced the 120-million base pairs of its genome as a test before tackling the human genome. Both that company and a public–private group, the Mouse Sequencing Consortium, have now moved on to mice, which have about the same number of base pairs—about three billion—in their genome as humans. A rough draft of the mouse genome was recently completed by Celera.

A key outcome of sequencing the genome is the measurement of the genetic variation among humans. That variation in the sequence of bases is thought to be very slight. Perhaps only one base in 1,000 will differ from person to person, making us 99.9% identical.

Currently, a consortium of drug companies, computer firms, and public institutions are cataloging the places in the genome where a gene is spelled differently from person to person. These genetic locations, SNPs, could be the keys to why some people are prone or resistant to certain diseases. At last count, the consortium has identified 630,334 SNPs. Celera (whose business depends on such value-added information as SNPs) claims to have found at least 4.8 million more.

In order to make use of the raw data coming out of the project daily, biologists, computer scientists, and engineers have a great deal of work ahead of them. This is a very rich and complicated text that has had glosses on its meaning added over the 3.5 billion years since life began on earth, and it's very unlikely that biologists are going to be able to extract its full meaning in the course of the next month, year, or frankly, century.

Now, let's examine what all this newly found genome knowledge will bring. The answer is a revolution, the genetic revolution. The human nucleotide sequences mark the beginning of a whole new approach to biology. [13]

How This Knowledge Will Affect Your Life

Wondering what happened to all that knowledge that scientists got from mapping the human genome? It launched a new race to identify the genes that give you diseases like cancer, high blood pressure, diabetes, and Alzheimer's. The winner gets to make remarkable new medicines and conduct new medical procedures that will extend your descendents' life expectancy to an average of 123 years by 2150, 150 years by the year 2175, and perhaps to 300 years or more by the 23rd century.

However, if sequencing (analyzing) the results of the human genome project are accelerated by the use of a biochemical supercomputer (see sidebar, "The Biochemical Supercomputer"), your average life expectancy could be extended to 115 years by the year 2047 through advances in genetic medicine, organ regeneration, and cloning. In other words, if you are in your mid 50s now, you would have a pretty good chance of reaching 100 years of age with the help of technological breakthroughs in computer storage and data manipulations. So, if biologists learn to quickly manipulate the DNA sequences at will, then they will be able to quickly change and create life at will.

The Biochemical Supercomputer

The element silicon is so closely identified with computers that most people would be more likely to associate it more readily with California's high-tech valley than with the periodic table. But such thinking may soon have to be radically revised, as high-speed computation moves beyond chips and machines to include the tools of biochemistry and genetics: test tubes, slides, solutions, even DNA.

DNA is present in every living organism, and the appeal of the molecule as a super-computer mechanism lies in its demonstrated ability to store a vast amount of information—indeed, all of the instructions for replicating life. Although the chemistry set won't be replacing your PC anytime soon, scientists have demonstrated how these information-laden molecules might perform calculations in future computers.

Instead of using zeroes and ones to encode information using electrical current, the "memory" in a DNA computer takes the form of thousands of DNA strands that are synthesized in a lab. Each strand contains a different sequence of the chemical bases—symbolized by the letters A, C, T, and G—that make up all DNA molecules. To sift through all these strands, scientists subject the DNA memory to various enzymes that eliminate certain strands of DNA, leaving only the strands of bases that represent correct answers.

Recently, scientists at the University of Wisconsin have reported that they had found a way to perform a simple calculation using strands of DNA that had been attached to a gold-plated surface. Previous experiments with DNA computing had allowed the DNA to float freely in a test tube, but this method will allow the wet chemical steps required for a calculation to be automated. It's a route to scaling up DNA computing to larger problems.

Another group at Princeton University recently reported on a way to use RNA (a chemical cousin of DNA) to perform a similar calculation. To demonstrate that their technique works, the team calculated the answer to a simple version of a classical chess dilemma called the "knight problem." The computer must determine in which positions a chess player can place the knights on the board so that none can attack another. The scientists encoded each strand of RNA to represent a possible configuration of knights. Then they performed a series of steps in a test tube with chemicals designed to eliminate RNA strands representing wrong answers, and then they analyzed the remaining RNA strands to see if they all corresponded to correct answers. Almost 98% of the supposedly "correct" strands did in fact correspond to correct chess configurations—a surprisingly high success rate for a preliminary experiment.

At the moment, it is still much faster to use a PC to perform such calculations but silicon-based computers perform their magic simply by running through every possible answer one by one at the speed of electrical current. Because of DNA's power to store information (a few grams of the material could store almost all the data known to exist in the world), scientists believe that such biochemicals will eventually be the most efficient medium of storing and manipulating information. But its real advantage over a conventional computer is that rather than analyzing each possible answer in sequence, the DNA computer would act on the entire library of molecules (or answers) simultaneously.

Although scientists are optimistic about the ability of the technique to find the right answers with 100% accuracy in the future, other researchers are quick to point out that the field is in its infancy, as compared with conventional computing methods, and that for many applications, silicon-based microchips will always be better. Silicon computing won't go away, and the applications that it's used for won't go away.

What's really needed, according to most researchers in the field, is a "killer" application particularly suited for the way DNA computing solves problems. Such real-world problems might involve the encryption of large amounts of military information, or they might involve some combination of silicon and DNA computing. Scientists will just have to see how far they can push the technology, and see how far they can take it. [14]

The technological benefits of this knowledge will be staggering. Let's take a look at some of them.

Breaking Cancer's Genetic Activity

With the aid of a gene chip and the mapping of the human genetic code, cancer fighting strategies are shifting away from a tumor's shape and size toward a tumor's genetic and molecular profile. Researchers at the University of California–San Diego (UCSD), as well as other scientific institutes are part of a booming nationwide effort to gather the molecular profiles of all cancers. Already, investigators have deciphered profiles of melanomas, leukemias, and breast cancers, among many others, and have found intriguing results. Cancers that look the same under the microscope may actually be completely different diseases because they have distinct genetic characteristics, and therefore they require different treatments. Indeed, a new molecular perspective is expected soon to enable doctors to look beyond traditional cancer classification techniques (the size, shape, and location of tumors) and make critical decisions for patients based on a sound understanding of their tumor's biology.

Eventually, the entire diagnostic vocabulary will require an overhaul. In the future, clinicians will identify cancer not by the site where it arises, but rather by the kinds of molecular defects it has. In other words, a typical patient would be diagnosed with a certain kind of cancer that just happens to be in the breast or lung. Then, the treatment would match the patient's tumor type, not simply its location.

It's not that scientists had an epiphany about cancer. It's just that they now have the tools to effectively investigate its genetic complexity. Until recently, cancer researchers laboriously studied one gene at a time, but advances in molecular biology and computer science are combining to accelerate cancer exploration. In particular, a new tool called the gene chip, invented by Stanford biochemist Patrick Brown, is now allowing researchers to sort through thousands of genes at a time, doing the work of years in just a few days.

The gene chip, known more technically as a DNA microarray, is a thumbnail-sized glass wafer embedded with thousands of genes. Despite its name, the chip has no relation to a microprocessor and involves just a few simple steps to produce and use. In fact, many labs are building their own machines to make the chips. Each gene chip analysis gives a read-out of the distinct patterns of genes switched on or off in a cell, effectively letting the researcher peer inside and get a comprehensive snapshot of the cellular dynamics at work. [15]

The new science is still being feverishly worked out in the lab, but already there are signs of its clinical potential. Indeed, normally cautious scientists are using superlatives they generally avoid to describe how molecular profiling will change the ways doctors detect, diagnose, and treat the diseases. [15]

Spotting Cancer Early

Oncology's first rule has always been to spot cancer early, when it's most vulnerable. That's why a lot of effort has gone into refining early-detection instruments such as mammography, X-rays, and CT scans. These technologies, however, are far from perfect. They can miss critical tumors and raise alarms over benign ones. A growing number of researchers now believe the future of early detection lies in discovering cancer's molecular signposts deep within the body, even before the disease becomes visible or symptomatic.

As cancer silently moves about the body, it leaves an incriminating trail of genes and proteins—a residue fundamentally different from that shed by normal tissue. Several teams of scientists are working on blood, saliva, and urine tests to catch those hallmark gene and protein patterns. One group recently reported promising results for a potential ovarian cancer detection test, which they hope to replicate for prostate, breast, lung, and pancreatic cancers. Right now there is no recommended screening test for the general population, and the symptoms of this "silent" disease are so nonspecific that they are often confused with something else, like indigestion or bloating. [15]

According to researchers, more than 80% of ovarian cancer cases are advanced by the time of detection, and only about a third of those patients survive five years or more. Determined to take luck out of the equation, the National Cancer Institute (NCI) has been working with other government scientists on a test that would spot in one drop of blood the distinctive pattern of proteins associated with ovarian cancer. The test gives results in 30 minutes, though it is still very much at the experimental stage. [15]

To devise the simple test, the NCI's team of researchers took blood samples from cancer patients and healthy patients and ran them through a device that sorts proteins by size and electrical charge. The device spewed out a chart for each sample's protein pattern resembling a bar code you would get in a grocery store. Those barcodes were then fed into an artificial intelligence system which "learned" to sift out those with cancer from those without by using a distinctive pattern of only a handful of proteins. To check the "trained" computer's reliability, the scientists blindly fed it a set of unknown samples. It identified all 50 cancer patients correctly (including 18 that were early stage and thus highly curable) and it picked out 63 of 66 of the noncancerous samples. Three healthy women would have gotten a false positive. [15]

The NCI calls the study "potent proof" that molecular techniques in principle can be used for early detection but cautions against premature conclusions. The test still needs validation by other researchers using many more patients, and it needs to be piloted in the real world to see if the benefits of screening healthy people outweigh the harms. If it can pass these tests, a cancer that whispers may finally be heard.

The Name Game

Once cancer is detected, one of the most challenging problems is figuring out what kind of beast it is. Will it grow aggressively, or hibernate for decades? Will it respond to treatment, or is it a feisty breed? Many scientists consider today's methods for diagnosing cancer (looking at the size, shape, location, and microscopic appearance of a tumor) as hopelessly primitive. A molecular portrait would dramatically sharpen these blunt diagnostic instruments. [15]

Consider tumor size, for example. Clinicians have long believed that all small tumors are basically the same with an equal likelihood of becoming more aggressive, but a recent molecular analysis has proved this isn't so. Some tumors are apparently predestined, even from their minuscule beginnings, to be malignant, while others are not. [15]

Using gene chips, the researchers sorted through some 25,000 genes in several dozen small breast tumors of young women. The patients had already undergone surgery, and their fates were known. The final data showed that indeed the molecular program of the cell—not tumor size—dictated outcome. It was a cluster of merely 70 genes that predicted whether a tumor would spread in the future. [15]

The results could radically change breast cancer treatment if they are validated in clinical trials. Right now, nearly all women get backup chemotherapy after surgery to try to prevent the cancer from spreading. But studies have previously estimated that about 75% of patients would do fine without it. [15]

Molecular profiling may also define a disease more accurately than the traditional microscopic analysis. For example, the most common form of non-Hodgkin's lymphoma is now diagnosed by its appearance under a microscope; the cells are large and diffuse. But specialists have wondered why it is that only 40% of patients with the disease respond to chemotherapy. A team of researchers from the National Institutes of Health had a sneaking suspicion that they had always been dealing with two different cancers masquerading as one, simply because they were indistinguishable under the microscope.[15]

Using gene chips, they were able to sort through about 10,000 genes in each patient. After the data crunching, they could actually subdivide the patients into two groups based upon the unique genetic profiles of their tumors. One group had a great response to treatment, while the other did poorly. The groups differed in about 1,000 genes. Indeed, the scientists found two distinct types of cancer, the medical equivalent of discovering a new kind of biological species.[15]

Targeting Tumors

Molecular profiling's potential goes far beyond giving good news or bad. It provides scientists with a detailed blueprint of the abnormal circuitry of cancer cells so that they have a better idea of how to design therapies to target tumors at their most vulnerable points.[15]

Indeed, scientists envision future cancer treatments as radically different from the broad slash-and-burn of chemotherapy and radiation—more akin to antibiotics. Infectious disease experts ask: What is the organism causing it? What treatments can we target at it? Molecular profiling tells scientists what the DNA changes are that are causing the disease, and then they can pick drugs to target those abnormalities.[15]

In the lymphoma study, for example, the scientists searched for a molecular soft spot in the genetic profiles of patients who didn't respond to chemotherapy. They found that those patients had a unique defect in a set of genes—a defect that prevented cells from dying off in the usual way. A clinical trial is now going on for the chemotherapy-resistant lymphoma patients, targeting those defective genes.

Some cancers have been considered simply too complex to decipher at the genetic level, and scientists have chosen not to spend their limited time on them. But molecular profiling might put even those tumors on the radar screen. One example is multiple myeloma, a cancer of the blood that afflicts 15,000 new people a year. Now, using gene chips, scientists have already begun to make sense of the genetic chaos of multiple myeloma. Indeed, scientists now believe myeloma is really an umbrella term that describes at least four distinct types of disease. And, in a recent study, scientists have identified several new possible drug targets for these variations. Of course, years of hard work lie ahead to further explore these novel targets and design therapies to strike them, but a huge step has been taken forward. [15]

Targeted therapies already have shown promise. Herceptin, for example, kills certain kinds of breast tumors by hooking onto a key molecule on the surface of a cell. But Herceptin and other similar drugs were developed through painstaking analysis, one gene at a time. With the gene chip, the discovery of new Herceptins will take place much faster. [15]

Clearing the Hurdles

The quest to identify molecular signatures of cancer has picked up speed quickly, in large part because Patrick Brown (inventor of the gene chip) and his team have shared the secrets of the gene chip from the start. But there are significant hurdles, too. First, biologists have never had to reckon with such vast amounts of data.

With terabytes and terabytes of data, it can be tricky to extract useful bits of information from all the noise. The entire culture of biology needs to change, to incorporate statisticians as collaborators or rigorously train biologists in those methods.

Another major stepping stone is standardizing the technology. Since many labs make their own gene chips, they decide which genes go on a particular chip. It's also important for all of the unprocessed data to be available on a public central database so scientists can validate and improve on one another's work.

No one doubts that this new technology will have a major impact. For some cancers, it may dramatically change the way they are detected, diagnosed, and treated. For others, it may not make much difference at all. But scientists say a new, comprehensive molecular understanding of cancer, brought to light by this technology and its future spin-offs, will someday transform cancer from a death sentence into a chronic, but manageable, disease.

Now, let's look at how to grow your own organs. Bioengineers foresee a time when this will be possible due to the technology fallout from the human genome project.

Growing Your Own Spare Parts

The human body is an infuriating mix of fallibility and intolerance. The immune system stubbornly fights attempts to replace vital organs when they wear out and break down. More than 90,000 Americans are waiting for transplants. Those lucky enough to receive a donor organ will still face a daunting fight for survival, forced to swallow a daily dose of immunosuppressive drugs. But, in years to come, patients may be able to bypass the waiting lines and overcome problems of organ rejection by getting spare body parts made from their own cells. Bioengineers envision a time when all replaceable organs will be grown in the lab.

The process of growing an organ resembles the construction of a building. Bioengineers begin with a blueprint of the organ, and from it build a frame, or scaffold. Then comes the addition of cells (the walls, floor, and ceiling of the organ), and finally the installation of an energy source or a blood supply. So far, researchers have grown skin, cartilage, a few nerves, some fingers, and bladders. Their eventual aim is to custom-tailor a complete organ using the patient's own cells as raw materials, artificial polymers as scaffolding, and enzymatic enticements to persuade the blood system to supply nourishment.

Bioengineers at MIT have turned organ blueprinting into an advanced art. A human liver consists of dense tissue filled with minuscule nooks and crannies. It is composed of at least six types of cells and performs multiple functions, from storing vitamins to filtering toxins from the blood. Using computer technology developed to design airplane parts, bioengineers have captured some of this complexity in a three-dimensional blueprint. [16]

An organ requires a three-dimensional support structure once it exists on a computer screen. Bioengineers at Harvard Medical School forge plastic organ scaffolds that are porous and inviting to human cells. The tiny holes riddling the polymer provide homes for cells. Bioengineers can trap the cells physically inside the polymer, or they can trap them chemically. As the cells grow, the polymers dissolve. The structure that remains consists almost entirely of human cells, and hence, it stands a better chance of evading the body's immune system. [16]

With the ultimate goal of growing a human artery (a body part in great demand for bypass surgery), bioengineers at Duke University have also been using cell-growing techniques in pig experiments. Heart surgeons typically reroute blood around a clogged artery with a vein removed from the leg, but within five years, one out of three patients ends up with blocked vessels again. Bioengineers start with a tubelike scaffold made of the polymer polyglycolic acid (PGA) and coat the inside with smooth muscle cells isolated from a pig's carotid arteries. These cells maintain the elasticity and strength of artery walls. Rocked by gentle pulsing waves (like a fetus in a womb) the crayon-size tube is bathed in a nutritive soup for the next eight weeks.

By the end of that period, the PGA scaffolding has broken down and the cells have begun secreting collagen fibers, the substance that gives healthy organs their structure. Finally, bioengineers add a layer of slippery endothelial cells, which prevent the blood from clotting, on top of the smooth muscle cells, and then put the fabricated arteries back into the pig. In a small trial on four pigs, the arteries worked for nearly a month. Clearly, bioengineers are a long way from growing a viable human artery, but the pig experiments are a promising start. [16]

Bioengineers at the University of Michigan have taken on an even more elaborate engineering challenge: developing a bioartificial kidney. As blood courses through a kidney, it filters out urea, an ammonia-based compound that is excreted in urine. With the urea, a healthy kidney removes valuable sugars, salts, and water, substances that it then returns to the bloodstream. Dialysis separates out the same constituents, but doesn't replace those that aren't toxic. So, bioengineers have also constructed an external blood-filtering device (similar to a dialysis machine), but one that uses cellular parts. This device has only been tested on dogs. It first separates the urea and other substances from the blood, and then routes the filtrate through dozens of tubes as fine as fishing wire and lined with kidney cells. Those cells reabsorb sugars and salts and pass them back through the tubing's porous walls into the dog's blood. The cells also manufacture cytokines, proteins that signal the immune system and help avert deadly bacterial infections, which are the primary cause of death among people who experience acute kidney failure. Bioengineers are eagerly awaiting authorization from the Food and Drug Administration so that they can begin testing the device on humans. Eventually, bioengineers hope to use the same biotechnology to build an implantable kidney. [16]

Another challenge is keeping an implanted organ nourished. Bioengineers at Carolina Medical Center in Charlotte, North Carolina, design and grow breast tissue which may one day replace saline implants. Developed with funding from a company called Reprogenesis, the tissue grows from cells taken from the patient's own fat. [16]

The scaffold of the breast tissue consists of sodium alginate. This is a spongy seaweed-derived substance mostly used as an ice-cream thickener. To the alginate, bioengineers attache a peptide that will signal the vascular system that the tissue is blood-vessel friendly. Then they seed the alginate with cells, which multiply and break down the scaffold. Once implanted, the tissue will be infiltrated by blood vessels carrying nutrients. Bioengineers have successfully tested the technique only in rats, but have high hopes for its success in humans.

Other organs are in the pipeline. Bladders built by bioengineers have worked in dogs in the Harvard lab, where they also develop replacement kidneys and windpipes. In addition to the bioengineers' scaffolds, MIT has begun crafting a polymer that conducts electricity. The polymer provides a trellis for nerve tissue to grow along and can carry the neuron's connective electrical impulses when placed between two ends of a severed nerve.

Corneas (the transparent surface of the eye) have been painstakingly constructed from layers of human cells by tissue engineers of Laval University in Quebec City, Canada. These may be closest to successful implantation because the engineers bypassed an artificial scaffold by cajoling the cells into producing their own collagen, which is what normal human cells do. This strategy may allow doctors to avoid any inflammation triggered by artificial scaffolds.

Most bioengineers offer only cautious time lines when asked when their organ of choice will be ready. The estimates of 10 or 20 years are discouraging for those waiting on transplant lists. The scientists are doing their best to coax Mother Nature along in the mean time.

So, if bioengineers can grow organs, why not grow a complete human? Or better yet, why not turn the clones into cures?

Cures from Clones

The notion that cloning could provide life-giving new cells for sick people is more than a fantasy. Recently, scientists have tested some of that promise in animals (see sidebar, "Cloning That Woolly Mammoth"). They've taken body cells and cloned them to generate embryos, then showed that stem cells from those embryos can grow new tissue and cure diseases in the animals that provided the original cells. Among their promising preliminary results are:

- Embryonic stem cells injected directly into sheep became disks of bone and perhaps also replacement liver and brain cells.
- In sheep, replacement neurons have cured spina bifida.
- Old cows have been given new young immune system cells.
- Patches of heart muscle, skeletal muscle, skin, and cartilage have been grown from cloned embryos and transferred back into the parent cows.
- Tiny kidneys grown from cloned cells (the first organs to be made this way) appear to have functioned properly for three months in cows. [17]

Cloning That Woolly Mammoth

According to scientists, resurrecting dinosaurs is practically impossible. But what about cloning a woolly mammoth, that creature with long, curved tusks and small ears emblematic of the last ice age?

Resuscitating this hairy cousin of the elephant that went extinct 4,000 to 10,000 years ago is equally improbable to some. Not to some bioengineers. They are out in the Siberian Arctic unearthing what they say is one of the best-preserved woolly mammoths ever found, a specimen that may provide perfect DNA. That string of genetic code could be seeded in an elephant's egg, which could give birth to a pure woolly mammoth—as well as to a huge debate about what bioengineers can do with, and to, Nature.

This 10-foot-tall mammal, weighing more than two tons, has remained buried under 4.5 feet of permafrost in the steppes of northern Russia for the past 20,380 years. Bioengineers believe they have a 100% chance of finding parts of the mammoth still intact. The mammoth died at 47 years old.

The mammoth's head and trunk were closest to the surface. They have deteriorated because they were exposed to climate changes, but the rest of the body was not exposed. Its woolly skin is in good condition. Bioengineers think its internal organs and possibly even its stomach contents are as well.

Finding intact parts, however, does not necessarily mean finding clonable DNA, if the animal is well preserved, there should be little problem in getting DNA.

DNA Degradation

According to bioengineers, it's been successfully done before. However, that DNA has generally been low in quality and concentration, so one can expect it to be highly fragmented and low-yield. Also, over time, DNA tends to accumulate damage that cannot be repaired. You need the entire genome to be completely intact for cloning to work.

What bioengineers would do is something similar to the process that created the famous sheep Dolly: extracting the nucleus of one adult mammoth cell and inserting it into an empty egg cell. The embryo would then be implanted in the uterus of an Asian elephant. This is the mammoth's closest living relative. It would be used as a surrogate mother that would gestate it as its own, but without transferring to the baby any of the elephant's genes.

Chill Out

It is very critical that bioengineers keep the mammoth from thawing. Bacteria and fungi begin colonizing an animal's tissues from the moment of death and consuming it until all is buried by snow, even though it might be frozen for centuries. In order to find DNA good enough for cloning, the amount of time elapsed between death and freezing is therefore the critical factor.

A crane finally raised the animal within a chunk of its frozen tomb, and a Russian helicopter took it a few miles east to the caves of Khatanga, a locale chosen to keep everyone protected from the Siberian blizzards while the DNA is recovered. The mammoth was found in 1997 by a Siberian family, who claimed only the animal's 9.5-foot-long curved tusks. They were probably used to dig into the snow in search of plants to eat.

Not That Long Ago

New findings suggest that woolly mammoths existed as recently as 4,000 years ago on Wrangell Island, although the general consensus is that woolly mammoths became extinct 10,000 years ago. Paleo-Indians may have hunted them for food.

Exactly why they became extinct is still a mystery. One theory is that they died of heat during a global warm spell. Other theories suggest that these massive losses were due to disease transmitted during the first contact between ice age mammals and Paleo-Indians. Just as natives were affected by European-borne diseases during the conquest of the Americas, mammoths and other immunologically naive creatures could have been wiped out in a very short period of time by disease associated with humans. If scientists in Siberia succeed in retrieving pristine DNA from the woolly mammoth, the next step is tackling both the science and ethics of recalling a creature from extinction.[18]

Plans are now being drafted to determine whether cloned embryonic stem cells can be turned directly into insulin-producing cells in mice and primates, and whether cloned cells could cure lupus, stroke, and arthritis. Scientists have demonstrated other stem cell cures using cells taken from ordinary embryos. Such cells can generally survive only in "immune-deficient" animals that are incapable of rejecting them, but they hint at the promise of the perfectly matched cells that clones could provide. In one such study, a paralyzed rat regained movement thanks to stem cells implanted in the spinal cord.

Still, it's a long way from cures in animals to human therapies. Scientists estimate that treatments for the average person are still at least 10 years away. But what are the ramifications for cloning in general?[17]

Ramifications

Scientists see a revolution in medicine that will render many of today's drugs and treatments obsolete. Essentially, cells yielded from human research cloning are the same stem cells that the Bush administration decided are promising enough to fund, only better. Unlike existing stem cell lines, stem cells created through cloning would provide a patient with a fresh supply of cells with his or her own genetic code. Gone would be transplant failures and the need for immune-suppressing drugs. In the same way that antibiotics and vaccines rid the world of infectious plagues a half century ago, these cells could for the first time eradicate the chronic, degenerative diseases of our day, such as cancer, Alzheimer's, and heart disease.

Because body cells are rejuvenated by an egg's proteins, therapeutic cloning would also tackle aging itself, replenishing the body with younger, more vigorous cells than even the most healthy cells already in place. And, because DNA removed from a body cell can be tinkered with before it is placed into an egg, scientists hope someday to add factors—genes for immune cells, for example, that would make a patient resistant to AIDS.

The dream of cloning to cure disease has its detractors, and not just among politicians. Some biologists are set on avoiding moral issues by trying to coax adult stem cells back to an embryonic state. For example, they may be able to tinker with the genetics of generic stem cells to make them less likely to be rejected, thus making DNA-specific matches unnecessary.

But, in the absence of a proven better alternative, to halt the work now would be like taking penicillin away from doctors in the last century. Indeed, many medical developments have been at least temporarily halted because of ethical qualms. Religious leaders found vaccines objectionable because they interfered with God's plan for who should get sick, and in vitro fertilization was condemned in the 1970s by many of the same conservative ethicists who today oppose therapeutic cloning. Organ transplants were once seen as objectionable, and recombinant DNA technology (the ability to create synthetic genes) was banned from top universities like Harvard and MIT for years, for fear that horrible and dangerous creatures would be produced. However, much of the opposition melted when the technology was used to create a synthetic form of insulin to treat diabetics, and today recombinant DNA is used in virtually every research lab in the world.

It's still too early to say whether the United States will accept or reject therapeutic cloning. Scientists still have mountains of work ahead of them. It takes not just an embryo, but the nurturing of stem cells and the ability to transform those stem cells into specialized types before any clinical applications can be used in humans. If history is any guide, there is evidence that the public attitudes will warm toward this new and intimidating medical technology.

Next, let's look at why scientists are tinkering with the natural limits on life span. It's because they can, due to the benefits of mapping the human genome. Is that such a good idea? This writer thinks it is!

Immortality

For much of her life, Jeanne Calment's big claim to fame was that she sold pencils to Vincent van Gogh in her youth. But by the time she died in August of 1997, the charming Frenchwoman had made her way into the *Guinness Book of World Records* for a grander accomplishment. She had become the longest-lived human on record, dwelling 122 years, 5 months, and 14 days on Earth.

That remarkable record is almost certain not to last long. As medical research teases apart the body's natural aging processes, the promise for extending the average life span beyond Calment's years is growing rapidly. Over the next 50 years you'll be completely reshaped by biology. Bioengineers will double human life span, but this will be a small part of all that will happen. Scientists, theologians, and bioethicists are preparing for the time when people will live much longer than is the norm today. Some see great benefits, others point out ethical quagmires, and all acknowledge that an unprecedented leap in life span will have ramifications as monumental as any event in human history.

Over the past century, vaccines, antibiotics, and good sanitation have upped the average American's life expectancy by decades, to today's 75 years for men and 80 for women. Now, new understandings of the genes and chemicals involved in aging may not only help humans live far past that age, but more importantly, may also help people retain health and strength in those latter years. The discovery of biological clocks ticking away in each of our cells, and a knowledge of how to reset those clocks, opens the possibility that a human would never die—at least not from old age.

Two basic sources of aging affect all of us, and science is finding ways to outwit them both. In the first category are mishaps—assaults from outside the body and mistakes from within—that weaken the system over time until organs simply can't function any longer. Just the act of living contributes to this slow decay: Whenever you take a bite of food, for example, the body makes free radicals, unstable and highly active molecules that tear destructively through cells. Although the body tries to repair the damage, it's never quite as good as new. Using the newest techniques of genetic medicine, scientists are finding ways to prevent this wear and tear. At the University of Colorado–Boulder, for example, geneticists have found that by tweaking a certain gene in roundworms to create a "super antioxidant gene," they can double the worms' life span. The altered gene produces an abundance of chemicals that find and destroy free radicals before they ever wreak their havoc.

Turning the Clock Back

Research on the second natural aging process points to even more dramatic possibilities. Genetic codes dictate how long any given organism should keep functioning, and these codes vary from species to species; that's why some insects live only a day, cats live about 20 years (some have lived to be 37 years), and a type of holly plant in Tasmania lives more than 50,000 years. Evolutionary biologists at the University of California–Irvine have created fruit flies with special "longevity genes" simply by breeding flies at older and older ages. Nature doesn't give a damn about you after you reproduce. After 40, you're toast—you don't have to be fit any longer, and you decay. But by finding flies whose genes enabled them to reproduce later and later in life, biologists have created a strain of flies that live twice as long as normal fruit flies. Finding longevity genes in humans could yield similar results. The beauty of longevity genes is that they not only allow creatures to live longer, but actually delay the onset of senescence. It's as if you reach the age of 20 and stay put there for decades. Then, at age 40 or 50 or 60, the clock starts ticking again and you are 21.

But even if everyone were to receive special longevity genes, each cell in the body nonetheless carries the plans for its eventual death right in its DNA. At the end of each chromosome is a long strand of "nonsense" DNA called a telomere. Each time a cell divides, the strand gets a bit shorter, like a burning candlewick. When it runs out, the cell can no longer divide and is left to age and die, but scientists have found that the enzyme telomerase can rebuild the strand over and over again. With the help of this substance, human cells have been enjoying immortality in a petri dish at places like Geron Corp. for more than 200 normal lifetimes now. [19]

No one has yet figured out how to make an immortal human being from a bunch of immortal human cells, but some researchers are coming close to at least creating human parts that might not wear out. For example, at Advanced Cell Technology in Worcester, MA, scientists have found that by merging a body cell, such as a lung cell, with a cow egg stripped of its cow DNA, they can "reprogram" the cell to rebuild its telomere strand, and also grow a whole new organ. Recently, scientists created a new bladder with this method and successfully transplanted it into a dog. The company is now trying to develop heart "patches" that can strengthen ailing hearts in humans. Scientists predict that in the next ten years, it will be routine for today's 40-year-olds to trade in old parts for new and improved ones by the time they reach age 50. [19] That is, if science proceeds unchecked by those with ethical objections.

When human DNA is merged with an egg, it creates "pre-embryonic" cells that in theory might be able to develop into a human being. Many cells that other researchers use are gathered from embryos discarded by in-vitro fertilization labs. Congress will continue to debate whether to allow such cells to be used in federally funded labs, and it may ban them altogether in the future. That would be a grave mistake. Scientists are not recklessly searching for a fountain of youth. They're trying to cure disease and prevent suffering. Aging research can battle almost every known disease, from diabetes to cancer.

The Good, the Bad, or the Evil

Even if scientists could avoid the controversial cells now needed to advance research, they would still hit opposition from those who find the very idea of dramatically extending life to be immoral. Scientists can make it socially despicable. Just like nuclear testing, they can decide that they don't want it. Ethicists use equally strong words about the selfishness of the endeavor. It is evil to focus energy on trying to live longer than 80 years when many poor people now don't live past 40. Others foresee a dangerous future where overpopulation leads to draconian rules about childbearing, and where only a favored few are allowed to live a second century.

Finally, apart from social justice concerns, many theologians feel that attempts to live indefinitely are a slap in the face to God. God created death for a good reason, they say, and to destroy death would be to destroy our own humanity. But others, scientists and theologians alike, celebrate their coming longevity.

Conclusion

In the early 20th century, physicists uncovered the dynamics of the atom, which is known as quantum mechanics. That discovery led to the electronics revolution and the technology that you so much enjoy today. Now biologists will lead the way. Coming is the biotechnological revolution. It will last for decades, perhaps even several centuries.

Biologists are already entering the age of genetic-based medicine. The new knowledge of the human nucleotide sequences will accelerate the development of therapeutic drugs that function at the molecular level. More accurate medical diagnoses will be available. Doctors will be able to address the fundamental causes of countless human disorders and will have a better chance of predicting the side effects of drugs. On the horizon are cures for cancer and heart disease.

Eventually, scientists will be able to identify all of the genes contributing to a given disease. Individuals will know for which sicknesses they are most at risk, giving them the possibility of making health-driven lifestyle changes or of taking preventative medical steps. Doctors will be able to tailor treatment to individuals.

The day will come when a medical checkup consists of a DNA readout, and genetic flaws will be corrected soon after or even before birth. Scientists will tell you how your physical abilities, intelligence, external characteristics, and personality are affected by the variations in SNPs. Genetic manipulation will provide ways to overcome the limits imposed by your evolutionary past.

The human genome sequence is a powerful tool for gaining insight into your genetic heritage and where you stand in the evolutionary scheme of things. The evolutionary tree can be determined by comparing the genomes of Earth's species.

Eventually, you will be able to take control of your own biological destiny when scientists learn to manipulate the human genome at will. No longer will you be at the mercy of the forces of natural selection. You will be able to modify in part your vital essence. This will not be the intention in the beginning. Initially, the goal will be to correct defective genes. But gradually, genetic manipulation will expand to allow couples to select features of their offspring. "Pro-choice" will take on a new meaning. At some point, scientists will have almost complete mastery of the genome. Moreover, genetic manipulation will not be only confined to humans. Long before it is used on mankind, it will applied to animals and plants.

One can imagine the genetics-dominated world of the late 21st century: There will be fruits, vegetables, and meats that are genetically modified for higher nutritional value. Sheep, mink, pigs, cows, and other livestock will have their genes adjusted to yield higher output. Zoos will house unusual animals that differ notably from the animals from which they were derived. In place of refineries will be vast vats of swamp-like liquids containing bacteria, who, like domesticated farm animals, will produce high-tech, genetically designed products that will provide a wide range of humanity's needs: food, energy, chemicals, and medicines.

The completion of the human genome project does not represent an end for genomics. The human sequence still contains gaps and probably will not be complete for at least another two to three years. Furthermore, there are lots of genomes of other important species to be sequenced. Bioengineers have recently finished sequencing the mouse genome, and other groups will eventually turn to farm animals and possibly crop plants. The various large genome centers, no longer bonded by a common grant or cause, are likely to disperse intellectually and pursue separate lines of activity. Biotech labs around the nation also plan to build on their automation capability and become resource centers for DNA-microarray analysis, performing highly multiplexed routine genetic analysis as part of large studies to track the effects of genetic variation within human populations.

The sequencing biochemical supercomputers will evolve with them. Various biochemical companies are preparing to launch a series of capillary electrophoresis machines designed not for production, but for rapid analyses taking less than 40 minutes. Sequencing will become resequencing or comparative sequencing—the search for genetic differences between individuals.

Finally, manipulating the genes of humans and living creatures will allow mankind to do what has been traditionally attributed to God. Indeed, President Clinton described the human genome as "the language in which God created Man."

References

[1] NASA Headquarters, 300 E Street SW, Washington DC 20024-3210; Jet Propulsion Laboratory, 4800 Oak Grove Drive, Pasadena, CA 91109, 2003.

[2] Gil U Lee, Linda A. Chrisey, and Richard J. Colton. "Direct Measurement of the Forces between Complementary Strands of DNA," Chemistry Division, Code 6177, Naval Research Laboratory, Washington, DC, 20375-5342 and the Center for Bio/Molecular Science and Engineering, Code 6900, Naval Research Laboratory, Washington, DC 20375-5348, USA, 1994.

[3] "What Is DNA Repair?," National Institutes of Health, 37 CONVENT DR MSC 4255, Bethesda, MD 20892-4255, 1996.

[4] "20-mer Polymerase Chain Reaction Procedure (for MJ Research Thermal Cycler)," U.S. Department of Agriculture (USDA), Washington, DC, 2003.

[5] "How Does Forensic Identification Work?" Oak Ridge National Laboratory (ORNL), 1060 Commerce Park, MS 6480, Oak Ridge, TN 37830, 2003.

[6] Max Ingman, "Mitochondrial DNA Clarifies Human Evolution," American Institute of Biological Sciences, 1444 I St. NW, Suite 200, Washington, DC 20005 USA, 1998.

[7] B. Alex Merrick, "The Human Proteome Organization (HUPO) and Environmental Health," Environmental Health Perspectives, c/o Brogan & Partners, 1001 Winstead Drive, Suite 355, Cary, NC 27513 USA, 2001.

[8] "Collection of Biological Material in Gassing Cases," California Criminalistics Institute, 4949 Broadway Room A104, Sacramento, CA 95820, 2003.

[9] "Contamination in Sequence Databases," National Center for Biotechnology Information, National Library of Medicine, Building 38A, Bethesda, MD 20894, 2000.

[10] Dr. Ron Rusay, Dr.Veikko Keränen, Tomi Laakso, Erik Jensen, and Thomas Dunham, "Music from: DNA/Proteins/Math," Science & Technology Education Program, Lawrence Livermore National Laboratory, 7000 East Ave., L-428, Livermore, CA 94550, 1999.

[11] "Protein Synthesis," Eureka! Science, Corp., PO Box 42615, Indianapolis, IN, 46242, 2003.

[12] "The Human Genome Project Information," Oak Ridge National Laboratory (ORNL), 1060 Commerce Park, MS 6480, Oak Ridge, TN 37830, 2003.

[13] "The Science behind the Human Genome Project: Basic Genetics, Genome Draft Sequence, and Post-Genome Science," Oak Ridge National Laboratory (ORNL), 1060 Commerce Park, MS 6480, Oak Ridge, TN 37830, 2004.

[14] Hratch G. Semerjian, William F. Koch, James R. Whetstone, and W. Mickey Haynes, "Chemical Science and Technology Laboratory," NIST, 100 Bureau Drive, Stop 3460, Gaithersburg, MD, 20899-3460, 2004.

[15] "Cancer Survivorship Research and Support," NWHIC (a service of the U.S. Department of Health and Human Services), Office on Women's Health, 8550 Arlington Blvd., Suite 300, Fairfax, VA 22031, 1999.

[16] Sylvia Pagan Westpha, "Growing Human Organs on the Farm," *New Scientist*, 147–151 Wardour Street, London, W1V 4BN, UK, 2003.

[17] Senator Chuck Grassley, "Human Cloning: Must We Sacrifice Medical Research in the Name of a Total Ban?" Statement of Senator Chuck Grassley of Iowa, Hearing of the Committee on the Judiciary, 135 Hart Senate Bldg., Washington, DC 20510, 2002.

[18] Jim Wilson, "Cloning a Mammoth," *Popular Mechanics*, 810 Seventh Avenue, 6th Floor, New York, NY 10019, 2004.

[19] Richard Haigh and Mirko Bagaric, "Immortality and Sentencing Law," Volume 2, May, *The Journal of Philosophy, Science & Law*, School of Public Policy, 685 Cherry Street, Atlanta, GA 30332-0345, 2002.

Biology: Protein Folding

Viewed from the summit of reason, all life looks like a malignant disease and the world like a madhouse.
—Johann Wolfgang von Goethe (1749–1832)

Mad Cow disease. Alzheimer's disease. Cystic fibrosis. An inherited form of emphysema. Many cancers. Recent discoveries show that all these apparently unrelated diseases result from protein folding gone wrong. Aside from these diseases, there is something amiss when proteins fold, as evidenced by the many unexpected difficulties biotechnology companies encounter when trying to produce human proteins in bacteria.

What exactly is this phenomenon? What is the protein folding problem? (See sidebar, "The Protein Folding Problem in a Nutshell.")

The Protein Folding Problem in a Nutshell

The elucidation of protein structure plays a quintessential role in the development of the biologist's understanding of evolution and the function of biological processes. Currently, protein structures are determined experimentally by X-ray crystallography and nuclear molecular resonance (NMR) spectroscopy. These methods, although accurate, are time consuming and suffer from inherent drawbacks. There is a need for the development of theoretical methods which allow for the ab-initio calculation of protein structure.

It is important that the elucidation of protein structure can be performed at the same rate as its sequencing of genomes. The phenotype (product) of a gene is a protein, and the pressure of natural selection is on the phenotype, and not on the DNA. In order to be able to understand the meaning of large amounts of genomic data, biologists need to understand how proteins (enzymes, etc.) create specific functional structures.

The Protein Folding Problem refers to the combinatorial problems involved in enumerating the conformations of a given protein molecule. Cyrus Levinthal (1968) outlined this in a simple paradox: Let each amino-acid residue in a 100-residue protein have 6 possible conformations, leading to 6^100 possible conformations available for this protein. This calculation does not include sidechain conformations, which will increase the number of degrees of freedom further. The question is now: How does the protein fold, given this large number of possible conformations? These simple calculations urge the development of new, efficient, and accurate search methods.

This project attempts to address the two main problems involved in the protein folding problem. The first problem is the understanding of the energetics involved in protein folding. This is addressed here from a thermodynamic viewpoint, developing empirical force fields parameterized using experimentally determined protein structures. The second problem deals with the search problem outlined in the preceding Levinthal's Paradox. The developed search methods and force fields are applied to simulations of fragments of proteins (see Figure 11.1) [1], which are known to contain structure, independently of the rest of the structure, and to small peptide chains that have been shown by NMR to have a structure in the solution (see Figure 11.2). [1]

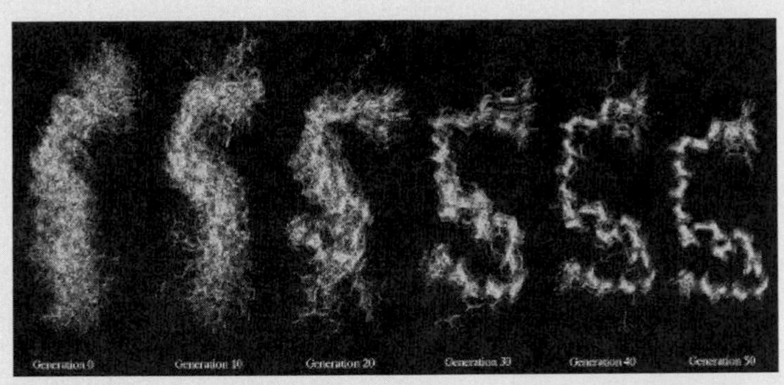

Figure 11.1 An illustration of Genetic Algorithm (GA) simulation of a small 22 Amino Acid peptide corresponding to the membrane binding domain of Blood Coagulation Factor VIII. The simulation attempts to select the most likely conformation of the peptide using an objective energy function that describes the mechanics of the peptide chain. The Genetic Algorithm emulates the process of natural selection in order to search the energy space efficiently. This is illustrated in the figure by a set of snapshots taken during a 50-generation simulation. It is seen how, gradually, one conformation is selected above the others through crossovers and recombination of fragments within the population of structures. The GA search method is able to exploit parallel computing resources to the full. The simulation here is performed using the 19-node IBM SP2 at NIST.

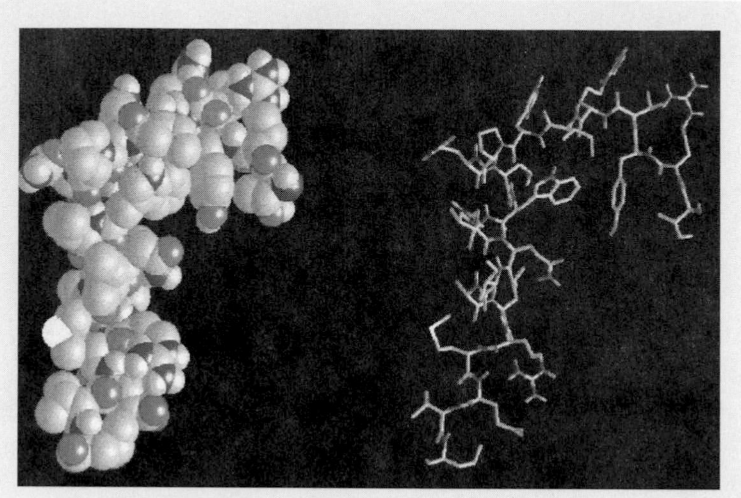

Figure 11.2 Structure of Blood Coagulation Factor VIII membrane binding domain. The structure is the best obtained in a series of Genetic Algorithm simulations. The structure shown is similar to that obtained through NMR experiments. The peptide is shown in two representations: (A) Corey Pauling Koltun (CPK), illustrating the compact nature of the molecule, and (B) Stick rendering, showing the chemical composition of the peptide.

Scientists have learned that proteins are fundamental components of all living cells: your own, the bacteria that infect you, and the plants and animals you eat. Furthermore, all of the following are also proteins: the hemoglobin that carries oxygen to your tissues; the insulin that signals your body to store excess sugar; the antibodies that fight infection; the actin and myosin that allow your muscles to contract; and the collagen that makes up your tendons and ligaments (and even much of your bones).

In addition, machines known as ribosomes string together amino acids into long, linear chains in order to make proteins. Like shoelaces, these chains loop about each other in a variety of ways—they fold. As with a shoelace, only one of these many ways allows the protein to function properly, yet lack of function is not always the worst scenario. For example, too much of a misfolded protein could be worse than too little of a normally folded one. Why? Because a misfolded protein can actually poison the cells around it.

The Importance of Protein Folding

For many years, the importance of protein folding has been recognized. Almost a half-century ago, Linus Pauling discovered two quite simple, regular arrangements of amino acids (the a-helix and the b-sheet) that are found in almost every protein. And in the early 1960s, Christian Anfinsen showed that the proteins actually tie themselves: No shaper or folder is needed, because if proteins become unfolded, they fold back into proper shape of their own accord.

These discoveries ended up being very important for understanding Alzheimer's disease and cystic fibrosis. Of course, neither Pauling nor Anfinsen nor the committees that awarded them their respective Nobel prizes knew at the time that the discoveries would be so important. And when Pauling, at least, was doing his breakthrough studies, he could hardly have imagined the enormity of today's biotechnology industry. What scientists did know is that any process that was so fundamental to life as protein folding would have to be of the utmost practical importance.

Research did not stop with Pauling and Anfinsen. Indeed, scientists now know that Anfinsen's conclusions needed expansion: Sometimes a protein will fold into a wrong shape and some proteins, aptly named chaperones, keep their target proteins from getting off the right folding path. These two small but important additions to Anfinsen's theory hold the keys to protein-folding diseases. [2]

Protein folding can go wrong. Biochemists have known this since antiquity (but didn't know they knew). When an egg is boiled, the proteins in the white unfold. But when the egg cools, the proteins don't return to their original shapes. Instead, they form a solid, insoluble (but tasty) mass. This is misfolding. Similarly, biochemists have always cursed the tendency of some proteins to form the insoluble lumps in the bottom of their test tubes. Biochemists now know that these, too, were proteins folded into the wrong shapes.

Biochemists lacked the tools to study these insoluble lumps until recently and did not expect such masses to be particularly interesting. The prevailing view at the time was that the lumps were just hopelessly tangled and completely amorphous masses of protein fibers (aggregation). Researchers eventually discovered that these aggregates of incorrect folding could be highly structured, but before this crucial insight and proper investigative tools were developed, biochemists simply threw their fouled test tubes away.

Protein Deposits in Tissues

Physicians have noticed that certain diseases are characterized by extensive protein deposits in certain tissues as far back as the start of 20th century. Most of these diseases are rare, but Alzheimer's is not. It was Alois Alzheimer himself who noted the presence of "neurofibrillary tangles and neuritic plaque" in certain regions of his patient's brain. Tangles are more or less common in diseases that feature extensive nerve cell death; plaque, however, is specific to Alzheimer's. The major question, which has only recently been answered, is whether plaque causes Alzheimer's or, like tangles, is a consequence of it. [3]

It has been shown that neuritic plaque (unrelated to the plaque that clogs atherosclerotic blood vessels and causes heart attacks) is composed almost entirely of a single protein after further investigation. A key to understanding the disorder was eventually provided by deposits of large amounts of a single, insoluble protein around the degenerating nerve cells of Alzheimer's disease.

It was development of the biotechnology industry that unexpectedly spurred interest in insoluble protein gunk. This industry can produce proteins (often otherwise difficult-to-obtain human proteins) quickly and economically in bacteria. To their surprise, however, scientists who worked for biotech companies often found two things: Protein that was supposed to be soluble instead precipitated as insoluble inclusion bodies within the bacteria, and proteins that were supposed to be secreted into the surrounding medium instead got stuck at the bacterial cell wall. This puzzling activity led scientists to seriously study just what goes wrong during protein folding.

Continued Research

The National Institutes of Health and the National Science Foundation continued to finance research in several laboratories in the decades after Anfinsen's work. Working in relative obscurity, these protein biochemists tried to discover how a completely unfolded protein, with hundreds of millions of potential folded states to choose from, consistently found the correct one—and did so within seconds to minutes.

Could there be specific, critical intermediates (partially folded chains) in the folding process? This turned out to be a difficult question to answer. Partially folded chains don't stay that way very long; they become fully folded chains in a fraction of a second. Nevertheless, researchers had not only found clear evidence for the existence of partially folded proteins by the early 1980s, but also realized the key role these played in the folding process. [4]

One study involved the difficulty in getting bovine growth hormone to fold properly. Although the unfolded proteins were not sticky, and the fully folded proteins were not sticky, the partially folded molecules stuck to each other—a first clue as to the origins of misfolded lumps (at least for purified proteins in test tubes). It still remained unclear why misfolding occurred in cells under certain circumstances but not under others.

Temperature-Sensitive Mutations

One of the first serious investigations of protein misfolding occurred in the early 1980s. These studies focused on temperature-sensitive mutations (mutations allowing growth at 75°F but not at 100°F) in the tailspike protein of bacteriophage P22. Neither bacteriophage P22, a virus that infects certain bacteria, nor its tailspike protein has any practical importance in themselves. Faced with thorny problems, however, scientists often look for experimental systems that will allow them to get a foothold or find a way around them. In this case, they thought that a large protein whose folding passes through multiple stages would be a good system for looking at folding pathways within cells. Many temperature-sensitive mutations were never examined for their effect on folding, but had already been isolated in bacteriophages. [4]

Their hopes were realized: Despite having only one amino acid altered, the majority of the temperature-sensitive mutations they found caused the tailspike protein to end up as insoluble gunk at high temperatures. It was now possible to analyze what went wrong in a protein's folding process, since these folding failures were occurring in bacterial cells growing in the laboratory.

The obvious guess at the time was that the mutant proteins were less stable. After all, the temperature scale is fundamentally defined by how much atomic-scale shaking or motion is going on. In other words, the more shaking there is, the higher the temperature. This implies that a less stable protein is more likely to fall apart at elevated temperatures, and it might therefore be more likely to end up (like cooked rhubarb) as insoluble gunk.

With this in mind, all of this turned out not to be the case. If the mutant chains were allowed to fold up at a low temperature, and were then heated, they were as stable as a wild-type enzyme. It turned out to be a partially folded intermediate, on the route from the random shoelace to the correctly folded protein, that was sensitive to temperature. Thus, these intermediates would stick to themselves and be unable to reach the properly folded state at higher temperatures.

Again, this turned out to be a general problem in the folding of many proteins. In other words, they have to pass through partially folded states in which they are delicately poised between folding all the way to the correct state or becoming seriously stuck as a result of premature entanglement with other molecules. Recognizing that it was the intermediates, and not the fully folded proteins, that were in trouble opened the way to understanding some aspect of a range of diseases.

Hereditary Diseases

Collaborators have conducted similar studies in connection with a human disease over the past several years. The minor differences between their results and others are very revealing.

Familial Amyloidotic Polyneuropathy

Peripheral nerves and other organs are damaged by deposits of amyloid-type protein in the hereditary disease familial amyloidotic polyneuropathy (FAP). Although the disorder is quite rare, extensive genetic studies have shown that the disease results from mutations in the protein transthyretin. As with the P22 tailspike protein, transthyretin contains large amounts of b-sheet structure and normally consists of several identical amino acid chains (four in this case) associated into a single, three-dimensional structure. [5]

FAP results from any of more than 50 distinct mutations within the transthyretin protein, each altering a single amino acid. After studying several of these, scientists found that their four-chain structure is less stable under mildly acid conditions than is the wild-type structure. This contrasts with the P22 tailspike mutations, which fold slowly but are stable once folded. It also appears that transthyretin aggregation takes place from a monomeric unfolding intermediate, rather than the folding intermediate involved in P22 tailspike aggregation (the pathway may or may not be the same in both directions). [6]

In both cases, however, the single-chain intermediates have structures that nature has designed for association with other chains of the same type. Therefore, in order to alter their normal linkage with two or three other chains into an endless series of linkages that creates insoluble gunk, it apparently takes only a very small change in the shape of these intermediates.

There is yet another contrast between the P22 tailspike mutations and those in transthyretin: From the P22 virus's view, the problem with the tailspike mutations is that not enough normal protein is made. People with transthyretin mutations, on the other hand, have all the normal transthyretin they need to carry out its usual function (transporting the thyroid hormone). The problem is that, as the protein is being broken down, it forms insoluble gunk, and the insoluble gunk poisons the tissues where it is deposited. [6]

Alzheimer's

FAP is a rare disease; not so Alzheimer's, which afflicts 10% of those over 65 years old and perhaps half of those over 85. Every year Alzheimer's not only kills over 100,000 Americans, but also costs society $132.2 billion to care for its victims.

In 1991, several different research groups found that individuals with specific mutations in their amyloid precursor protein developed Alzheimer's disease as early as age 40. The body processes amyloid precursor protein into a soluble peptide (small protein) known as Ab. Under certain circumstances, Ab then aggregates into long filaments that cannot be cleared by the body's usual scavenger mechanisms. These aggregates then form the b-amyloid, which make up the neuritic plaque in Alzheimer patients. So, the consistent association of amyloid precursor protein mutations with early-onset Alzheimer's has finally answered a long-debated question: The deposit of neuritic plaque is part of the pathway leading to the disease, not a late consequence of it. [7]

Researchers chemically synthesized fragments of the 40-amino-acid-long peptide in order to help understand the Ab aggregation process. By using these fragments, they showed that the key step is getting started. Specifically, the precursor fragments have to form a specific nucleus, which then grows into the amyloid process. Possibly the slowness of this first step is why Alzheimer's disease is almost entirely limited to older people, and it could be that the mutations in amyloid precursor protein that lead to early-onset Alzheimer's are the ones that make it progress more quickly and easily. [7]

Even so, Ab remains soluble in most people. Most individuals who develop Alzheimer's disease have the normal form of amyloid precursor protein, indistinguishable from that in people who never acquire the disorder. Why the same form of Ab aggregates in some people's brain but not in others, remains a mystery, although a recent discovery has suggested an intriguing possibility.

Biochemists know that people with different genetic variants of the protein apolipoprotein E (apoE) have quite different risks of developing Alzheimer's disease. Compared to those with the most common variant, known as apoE3, those with the apoE4 variant are significantly more likely to develop the disease. Although other studies disagree, some studies suggest that those with the apoE2 variant may be at lower risk.

Because apoE is best known as part of the complex that transports cholesterol and other fatty materials in the bloodstream, these findings are particularly surprising. What could a fat-transporting protein have to do with Alzheimer's disease? It may be significant that small amounts of this protein are associated with neuritic plaque and that apoE binds to Ab in the test tube. However, the results of this finding are in dispute. [7]

Researchers report that adding apoE to a test-tube solution of soluble Ab causes rapid formation of plaque-type b-amyloid fibers—and that apoE4 does so more rapidly than apoE3. Others, however, have obtained opposite results: apoE prevents fibril formation. Thus, whereas some suggest that apoE acts as a pathological chaperone, one that actually promotes misfolding, other researchers believe that it exerts a normal chaperone's protective effect. In either case, apoE's influence on the folding of Ab may play a major role in development of Alzheimer's disease.

Mad Cow and Creutzfeldt-Jacob Diseases

Mad Cow disease and its human equivalent, Creutzfeldt-Jacob disease, is perhaps the most interesting example of a protein-folding disorder. These diseases, along with the sheep version known as scrapie, have had the scientific community in an uproar for years. They are infectious diseases transmitted by prions, or protein particles. Prions seem to be pure protein; they contain neither DNA nor RNA. Yet an infectious agent is necessarily self-replicating. "How," scientists asked themselves, "could a pure protein replicate itself?" The answer now starting to emerge may be viewed as a variation on the concept of the pathological chaperone, only in this case the protein serves as its own chaperone. [8]

Constantly being produced by the body is the protein whose aggregation damages nerve cells in Mad Cow disease. Normally, though, it folds properly, remains soluble, and is disposed of without problem. But, suppose that somehow a small amount misfolds in a particular way so as to become a scrapie prion. If this scrapie prion bumps into a normal-folding intermediate, it shifts the folding process in the scrapie direction and the protein, despite its perfectly normal amino acid sequence, ends up as more scrapie prion. And the process continues: So long as the body keeps producing the normal protein, a little bit of scrapie prion can keep on creating more and then more. The prion then is replicating itself without needing any nucleic acid of its own, in effect.

What old-school scientists find even more strange is that the process resembles something akin to genetics. Different strains of these diseases, with somewhat different clinical symptoms, "breed true" as they are transmitted from one animal or human to another. Moreover, these strain differences are associated with slight differences in the protein deposits that apparently cause the disease.

NOTE

Scientists have recently used these strain differences to show that a few Britons truly have Mad Cow disease, the form seen in cattle, rather than the usual human form of Creutzfeldt-Jacob disease.

Other experiments have shown how genetics is possible without nucleic acids, just as replication can occur without DNA or RNA. Thus, when researchers mix seed quantities of two different scrapie prion strains in separate test tubes with large amounts of normal protein, each test tube produces more of the specific scrapie prion strain that was added. That is, each strain induces the normal protein to fold in exactly the same way as the original seed. The strain breeds true in the test tube, just as it does in the body. Genetics without nucleic acid is truly possible in the world of protein folding, odd as it may seem.

Miniscule Protein

Many human diseases arise from protein misfolding, leaving too little of the normal protein to do its job properly, unlike the examples of FAP, Alzheimer's disease, and Mad Cow disease, in which the problem derives from accumulation of toxic, insoluble gunk. Cystic fibrosis is the most common hereditary disease of this type.

Recent research has clearly shown that the many previously mysterious symptoms of this disorder all derive from lack of a protein that regulates the transport of the chloride ion across the cell membrane. More recently, scientists have shown that by far the most common mutation underlying cystic fibrosis hinders the dissociation of the transport-regulator protein from one of its chaperones. Thus, the final steps in normal folding cannot occur. So, in retrospect, the normal amounts of active protein are not produced.

An even greater example of the mutations studies in P22 tailspike protein is the hereditary form of emphysema. Investigators have found that one of the most common mutations producing this disorder greatly slows the normal folding process, just as the P22 temperature-sensitive mutations do. As with the tailspike mutations, the resulting buildup of a crucial folding intermediate leads to aggregation, which deprives affected individuals of enough circulating alantitrypsin to protect their lungs. The result is emphysema.

There is a far more common instance of misfolding, as intriguing as these examples may be, which leaves too little normal protein to do its job. The protein's job is to block cancer development in this case.

Scientists have learned over the past couple of decades that most cancers result from mutation in the genes that regulate cell growth and cell division. The most common of these genes, involved in roughly 40% of all human cancers, is p53. Thus, to prevent cells with damaged DNA from dividing before the damage is repaired (or to induce them to destroy themselves if the damage cannot be fixed) is the sole function of the p53 protein. In other words, p53 exists to prevent cells from becoming cancerous. [9]

The p53 mutations associated with cancer fall into two classes. The first keeps the protein from binding to DNA; the other makes the folded form of the protein less stable. In the second group, there is simply never enough properly folded protein around to block the division of DNA-damaged cells. It will be interesting to see how many of the p53 mutants fall into this second class and whether some way can be found to stabilize them.

Protein Misfolding Treatment

Finding ways to treat human disease is the whole purpose of studying protein folding. The story of protein folding has not yet led to treatments for the diseases involved, but this could happen within the next decade.

The key is to find a drug that can either stabilize the normally folded structure or disrupt the pathway that leads to a misfolded protein or a small molecule. Although many molecular biologists and protein chemists believe this will be quite difficult, others are more optimistic.

However, it is difficult to pinpoint where the search for treatment currently stands. One scientist notes that the bulk of that work is tied up in the patent stage: Companies are pursuing it, but have published little on the subject. Nevertheless, one research group has shown that both thyroid hormone and the related compound 2, 4, 6-triiodophenol (TIP) can stabilize transthyretin. Since TIP neither blocks the action of thyroid hormone nor exerts any hormone-like effects of its own, it appears to be a promising treatment for FAP. Developing small-molecule therapies is quite straightforward for proteins like transthyretin that naturally bind small molecules, but these therapies are more difficult to apply to proteins that do not have a small-molecule binding site. [9]

One of the few other groups currently publishing their research on small-molecule structure stabilizers is working to stabilize p53 (an acknowledged difficult target). In fact, by using two different approaches, one laboratory has obtained encouraging results. [9]

On the way are treatments based on biochemists' growing knowledge and continued research of protein folding. The saga that began with Pauling's fundamental studies of protein structure and Anfinsen's investigation of what some call the second genetic code will reach its practical fruition when these treatments arrive.

Form and function are powerfully related in proteins. One of the keys to understanding how proteins work lies in learning how they fold into particular shapes. With that in mind, let's now look at how over the last ten years biophysicists have been able to simulate the protein folding process from theory by using computers.

Understanding the Mechanism of Computational Protein Folding

Biophysicists are studying the mechanisms of protein folding by using computers to generate protein molecules from scratch. However, they don't start from a pile of atoms, and they don't have to simulate the bonding process. Because biophysicists are interested in non-bonding interactions, they start with a sequence of the amino acid building blocks and watch as the forces between atoms and among the amino acids change the structure of the molecules.

The researchers study the amino acid alanine as one of their model systems. Taking what they learn from studying this small molecule, they help write the rules that allow them to see inside larger biomolecules, such as genes and enzymes. Their work leads to better tools for biomolecular imaging and rational drug design. [10]

Twisting Proteins

Proteins are large molecules made up of chains of amino acids folded into complex shapes or conformations. Once on their own in solution, these chains twist and shimmy with energy. Each atom is constantly reacting to the positions of its neighbors. However, all is not chaos. The chains tend to fold into recognizable types of conformations.

The protein chain settles down as it folds. It seeks a shape with low energy, one in which the atoms are not repelling each other. Most likely to be the protein's native conformation is the structure with the lowest energy—for example, a coiling alpha helix when you build a chain of the amino acid alanine (polyalanine).

Biophysicists don't have a method yet for watching proteins fold under a microscope. Instead, researchers approach the question from two directions. They either settle for using X-ray crystallography to examine protein crystals, or they use computers to simulate the folding process. Biophysicists today focus on building protein chains in this virtual environment. [10]

Fundamental Biophysics

Based on their knowledge of fundamental physics, biophysicists are both theorists and experimentalists who study the way a protein should fold. They allow a structure to grow and change in a computer model and calculate the potential energy of the molecule, repeating the process and looking for the best among an endless landscape of solutions.

Biophysicists are solving the multiple-minima problem, which is much like searching for a needle in a haystack or a flake of gold on a beach full of sand. Remember, there isn't one answer to the problem. Instead, biophysicists find the best among many answers, or the global minimum. It becomes impossible even to calculate all possible answers, once you begin to deal with interesting biomolecules. Instead of guessing their way toward the best answer, they use random methods intelligently. If biophysicists' equations are right, then the solution and the global minimum will be the structure most likely to occur in nature (the native conformation). This is one of the fundamental measures of their success. [10]

The power of biophysicists' approach to modeling protein folding rests in its reliance on math, not estimation. As you move from a single amino acid to a chain, or polypeptide to a full-blown protein molecule, the number of calculations needed to find the global minimum solution grows exponentially with the addition of each link in the chain. Even supercomputers cannot yet handle whole enzyme molecules using this method.

Working within the constraints of biophysicists' resources (usually the best in the world) and their model system (a well-known amino acid), they are able to refine the fundamental understanding of the physics of protein folding. They are now refining the measures and tools used by others studying larger systems and designing new molecules. [10]

Programming Potentials

Protein modelers start out with a chain of amino acids that is already bonded together. They focus on non-bonded interactions between the atoms and groups of atoms in the chain.

Based on the chain's potential energy, it gives directions to fold into the most relaxed state possible. Thus, a measure of the energy stored in the molecule is the potential energy.

The Van der Waals force is the most powerful force in these interactions. By becoming repellent in close quarters, it is a stabilizing force that pulls distant atoms towards each other but prevents them from colliding. This is the dominant force at work inside a lively molecule that contributes to the calculation of its potential energy. Modelers also incorporate the interactions between atoms and groups of differing electrical charges. [10]

By building up from the behaviors of individual atoms to the folding of molecules, biophysicists have also devised their own formula for calculating the potential energy of a molecule. Researchers refer to this method as a potential function or simply as a potential. Biophysicists are constantly refining their version of the potential function. As one of the test molecules, they use the amino acid alanine, which is well understood from laboratory studies.

Molecule Modeling

Today, much of biophysicists' work focuses on proteins, which are large molecules composed of amino acids. They use a supercomputer to simulate the folding behavior of polyalanine, a small polypeptide (a molecule with many peptide bonds). By adding molecules of the amino acid alanine one at a time to a chain, biophysicists form polyalanine. They seek out the best possible shape or conformation for the new polypeptide each time a piece is added to the chain. Because biophysicists' methods take into consideration the interactions between each possible pair of links, the problem gets more difficult each time you add a link to the chain.

The researchers chose alanine as their model molecule because its structure and behavior are well understood through empirical study and because it is relatively small. Based on observation, this information gives them a basis for comparison with the results of their theoretical simulations. However, polyalanines are more difficult to build in a test tube than in a computer! So, when they are testing and tuning their methods, the biophysicists also simulate other proteins.

Often referred to as the second half of genetics, is understanding the mechanism of protein folding. Computational approaches have been instrumental in the efforts. Simplified models have been applied to understand the physical principles governing the folding processes and will continue to play important roles in the endeavor. Encouraging results have been obtained from all-atom molecular dynamics simulations of protein folding. A recent microsecond-length molecular dynamics simulation on a small protein, villin headpiece subdomain, with an explicit atomic-level representation of both protein and solvent, has marked the beginning of direct and realistic simulations of the folding processes. With growing computer power and increasingly accurate representations together with the advancement of experimental methods, such approaches will help biophysicists achieve a detailed understanding of protein folding mechanisms. [10]

Computational Protein Folding

Determined primarily by their structures, proteins support life by carrying out important biological functions. Subjected to evolutionary pressure, only those proteins that are helpful to the survival of living beings have been retained. Though their folding time may not be a subject of active refinement of evolution, proteins are required to be able to adapt well-defined structures soon after being synthesized and transported to their designated locations within cells to perform their functions. Such a requirement sets the upper limit for their folding time and is one of the important aspects of proteins that sets them apart from other polymers, including other nonprotein polypeptides. The astronomically large number of possible conformations suggests that proteins use some sort of "directed" mechanisms to fold. An elucidation of protein folding mechanisms must address how proteins fold into their well-defined three-dimensional structures within a limited time. This part of the chapter reviews briefly the history of computational protein folding studies, discusses the recent developments in more detail, and presents a perspective of the future.

Proteins can fold into and subsequently maintain well-defined structures under the right physiological conditions. This is determined by sequences through delicate balances of enthalpy and entropy; weak interactions, including van der Waals, electrostatic, and hydrogen-bonding forces; and a balance between protein intramolecular interactions and the interactions with solvents that also play major roles in protein folding. A major motivation for the mechanistic studies has been the need to understand the roles of these interactions in determining protein structures, since such an understanding can help to improve the accuracy of protein structure prediction. Because of the close association between protein structures and their functions, understanding how protein sequences determine their structures has often been referred to as the second half of genetics. With the explosive growth of genomic sequence data, the need for reliable structural prediction methods that can complement the existing experimental approaches such as X-ray crystallography and NMR spectroscopy is compelling. In this regard, an appealing aspect of the physically based modeling is its generality. These models use the physical interaction energies as the primary criteria to analyze protein structures. The same set of physical principles that drives protein folding also dictates substrate and ligand binding, as well as the induced conformational changes that are often associated with protein functions and are important for a detailed understanding of biochemical processes. Understanding protein folding would inevitably aid in the understanding of these processes. The relatively recent discovery of folding-related diseases reinforces such a need. Despite great progress made using a variety of approaches, it is still difficult to establish detailed descriptions of the protein folding processes, and such descriptions are the necessary steps toward the comprehensive understanding of the mechanisms of folding. [10]

Lattice Modeling

Protein folding has come of age through computational studies. Among the early successes was an Ising model simulation on the unfolding and hydrogen exchange of proteins in which a two-state transition was observed, which was not surprising given the three-dimensional nature of the model. Biophysicists have studied the folding of myoglobin without using a computer by representing the protein at the secondary structure level and treating each helix as a uniform rigid body cylinder. Using this highly simplified representation, they concluded that the folding was a nucleation process, similar to that of crystal growth. Also, in the study of the folding of a four-helix bundle, a similar representation has been applied recently in combination with a Brownlan dynamics approach. [10]

In the late 1970s, a more detailed representation also appeared. Using a combination of Langevin dynamics and energy minimization, biophysicists studied folding of bovine pancreatic trypsan inhibitor (BPTI) and Carp Myogen. In these studies, biophysicists represented each amino acid by two particles. They observed highly complex folding processes in which secondary structures were seen both forming and breaking, and challenged the notion that folding was preceded by forming stable secondary structures first. This pioneering work marked the beginning of physically based models in the studies of protein folding, albeit at a somewhat crude level. The level of approximation, both in the representation and the parameter, naturally implied a certain level of uncertainty and sometimes even significant error.

Biophysicists also noted that hydrogen bonds seem to slow down the folding process, a finding that has yet to be clarified by further studies. Nevertheless, biophysicists should recognize the pioneering nature of the work, which helped to topple the then-popular view that stable secondary structure always forms first in the folding process. The fact that most current structure prediction methods use a similar representation to that of biophysicists is a strong testament to the power of such an approach. Fifteen years later, using a residue-level lattice model, biophysicists have successfully simulated the folding of some small proteins. Interestingly, the parameters were obtained by analyzing protein structures deposited in the Protein Data Bank (PDB).

The advantages of lattice models are clear. The highly simplified models allow efficient sampling of conformational space. This was particularly important at the time when the speed of the most powerful computer was many orders of magnitude slower than a current personal computer. When designed properly, the model can give a well-defined global energy minimum that can be calculated analytically. In fact, one can enumerate all energy states and calculate the corresponding free energies in such models. One can also control other features of the energetic surface. When carefully parameterized, lattice models can be applied to structure prediction and can give encouraging results. The lattice model also allows Monte Carlo simulations that give ensemble averages. This was a critical advantage as well, because at the time, all experiments were conducted macroscopically and could only give ensemble-averaged results. Single-molecule studies were much later developments. This type of model has contributed a great deal to biophysicists' understanding of protein-folding mechanisms and has enjoyed widespread application.

Aimed at two distinct objectives, there are two types of lattice model simulations. One simulation pioneered by a group of biophysicists, was designed to understand the basic physics governing the protein folding process. A key feature of this type of lattice model is its simplicity (the size can range from 3^2 to 5^3 lattice points). A good example of such an approach has been shown by other biophysicists, who, through lattice model simulations, postulated that proteins have a funnel-like energy landscape with a minimally frustrated character that guides proteins toward their native states. The postulate deviated markedly from the old pathway doctrine and elevated biophysicists' understanding at the conceptual level.

The second useful simulation example was done by other biophysicists who emphasized the importance of hydrophobic interactions. Some of this work has been reviewed previously. Recently, this type of approach has been extended to residue-level off-lattice models. Similar to the approaches of other biophysicists, this approach assured the foldability of the model by systematically biasing the energetic surface toward the native state of that particular protein under study in a process consistent with the diffusion-collision model. Because this type of model has not been designed for real proteins, tests on these models have been limited to the studies of general features of protein folding. Nevertheless, a good deal can be learned from these studies. For example, a prediction has been confirmed recently by experiments that a small set of amino acids (hydrophobic and hydrophilic) can be combined to produce foldable protein-like peptides.

Other lattice models by biophysicists belong to the second category. These models are geared toward realistic folding of real proteins and are therefore parameterized using real proteins as templates by statistical sampling of the available structures and are often referred to as statistical potentials (or knowledge-based potentials). Along the same line was the approach by other biophysicists who developed a residue-based off-lattice model. Because the residue-level representations are applied to real proteins that have large numbers of energy minima, in contrast to the simplified lattice models previously described, their energetic surface can no longer be described exactly, even though exhaustive sampling can be conducted for short sequences (shorter than 100 amino acids). More importantly, the pair-wise discrete neighboring "energy" for the interactions between the nearest neighbors allows only a small number of possible conformations. The lattice coordinates also impose restrictions to the representation, though a high-coordination lattice model has been developed as well. The successes in predicting protein structure are indeed very encouraging, given their highly simplified approaches.[10]

Residue-Level Models

The need to develop methods that can reliably differentiate native states from the non-native ones is a constant driving force in the computational study of protein folding. The most widely used approaches in protein structure prediction have been based on residue-level models (either lattice or off-lattice models) with typically statistical potentials obtained from the structural database or PDB. A growing trend in the community has been the development of atomic-level statistical potentials in attempts to improve the accuracy. The application of all-atom representations with physical potentials in structural prediction, on the other hand, has been limited. A typical application would be at the final stage—a minor refinement of the structures using limited energy minimization designed to eliminate the bad contacts. It has been pointed out that the gas phase energy calculated by all-atom molecular mechanics is a poor descriptor of the quality of the structures. This is not surprising given the critical role that solvent plays in determining protein structures, and in fact is reassuring, because gas-phase energy alone should not be able to discriminate good structures from the bad ones. The accuracy was dramatically improved with the inclusion of the solvent effect as expected.

Through a combination of an all-atom representation of protein and a continuum model of solvent, an improved level of accuracy has been obtained. Biophysicists have studied the folding of Alanine-based peptides and noted interesting features from the simulations, including the role of electrostatic interactions between the successive amides, which favored extended conformations and built energy barriers to helix folding and intermediate states. Other biophysicists have adopted a similar approach and have applied it to the studies of the folding free energy of chymotrypsin inhibitor 2 and that of G-peptide, using unfolding simulations, and tested this approach on a set of proteins and on two peptides. Continuing along this line, was the work of still other biophysicists who proposed the use of the mean solvation force to represent solvent, and tested this method on alanine-dipeptide. This type of model tries to strike a balance between the accuracy of the representation and the computational cost. Application of the continuum solvent model can significantly reduce the number of particles included in the calculation and, hence, the computational cost, even after considering the overhead due to the added complexity of the continuum solvent model. The new development reflects considerable improvement in the level of sophistication, in addition to the improvement due to the differences between all-atom and residue-level models of protein. Compared to the ad hoc approach of biophysicists in parameter generation, the present parameters were based on quantum mechanical calculations and refined against experiments. The solvent model has also been improved substantially from initially simple solvation-free energy approaches to today's solvent model based on macroscopic electrostatics. As pointed out by many researchers in the field, a common deficiency of the continuum solvent models is that the simulated events can occur at time scales much smaller than those found in experiments, which, in many cases, can be corrected by taking into account the viscosity of the solvent. A more serious problem can arise when solvent plays a structural role. This becomes an important issue in protein folding, since proteins can have substantial solvent molecules in the interior in some important states, such as molten globule states. Furthermore, prior to reaching the native state, studies have suggested that solvent plays a role as a lubricant. In addition, ejection of a solvent molecule from the interior may contribute a nontrivial portion to the free-energy barriers.[10]

All-Atom Representation of Both Solvent and Protein

The all-atom representation of both solvent and protein is at an even higher level of sophistication. A hallmark of models of this type is that their parameters are obtained through high-level quantum mechanical calculations on short peptide fragments. Such an approach has several advantages. It assures the generality and allows further refinement upon the availability of more accurate quantum mechanical methods and upon the need for such an improvement. Such models also allow further extension. For instance, active efforts have been undertaken to parameterize polarization energy that can be integrated seamlessly into present simulation methods. Some of the earlier developments have been reviewed. Because the detailed models require both a large number of particles, typically more than 10,000, and a small time step of one to two femto-seconds (10^{15} seconds), direct simulation of the folding processes, which take place on a microsecond or larger time scale, has been difficult. Therefore, such models have been applied to study the unfolding processes of small proteins that can be accelerated substantially by raising the simulation temperature, by changing solvent condition, by applying external forces, and by applying pressure. The detailed representation has allowed direct comparisons with experiments, and encouraging results have been obtained. Limited refolding simulations were also attempted, starting from partially unfolded structures generated from the unfolding simulations, and considerable fluctuations were observed. These short-time refolding simulations have also identified the transition states in the vicinity of the native state. Care must be taken, though, because the short-time refolding simulations can only sample the conformational space in the vicinity of the unfolding trajectories. Equilibration of water in this type of short-time refolding simulation is needed to avoid simulating a trivial collapse process of water equilibration when the system is brought to room temperature and to restore faithfully the room-temperature solvent condition that has been distorted significantly due to the entropy-enthalpy imbalance at high temperature. Furthermore, this imbalance, inherent in the typical unfolding simulations may also be reduced by conducting the unfolding simulations at moderate unfolding temperatures; such that both temperature and pressure can be maintained at experimentally relevant conditions.[10]

The attempt to reconstruct the free-energy landscape is a powerful extension of unfolding simulations. Using the weighted histogram method, biophysicists have calculated the free-energy landscapes of folding a three-helix bundle, the segment B1 of streptococcal protein G, and the Betanova from restrained unfolding simulations. They demonstrated funnel-shaped free-energy landscapes and the existence of multiple folding pathways, and showed that the shapes of the funnels are also dependent on the type of proteins. Biophysicists have also observed that ejection of water from the interior of the intermediate state contributes to the free-energy barrier of folding, suggesting the role that water may play in the folding process in addition to its role as solvent. Such an observation is only possible with the explicit inclusion of solvent in the simulation. It is noteworthy that the application of the restraint functions is an integral part of the methodology, because it ensures a sufficient number of transitions between neighboring states and hence ensures the reversibility that is absent in the unrestrained unfolding simulations. Nevertheless, the weighted histogram method has also been applied in the analysis of unrestrained unfolding trajectories. Free energy profiles (or probability profiles) have also been generated directly from the unfolding trajectories, but it is unclear at what temperature the profiles were generated.

So, how do proteins reach their native state? This is a central question concerning the elucidation of the protein folding mechanism: Therefore, direct simulation of protein folding using an all-atom model has been termed the "holy grail." Encouraging developments have been made in the simulations of the folding processes of small peptide fragments with explicit representations of both solvent and peptides. Biophysicists also studied the formation of ß-turn in aqueous solution. Case and coworkers studied the transitions between two conformations of a ß-turn motif and most of the simulated distances agree with NMR data. Biophysicists have also studied the folding and unfolding processes of a short ß-peptide in methanol at temperatures below, around, and above T_m of the peptide. They observed reversible formation of secondary structure in a simple two-state manner within 50 nanoseconds. The estimated folding free energies, based on the population of the states, are in qualitative agreement with experimental observations. Other biophysicists have studied the formation of undecamer peptides at the water-hexane interface. Biophysicists also recently developed a method called Self-Guided Molecular Dynamics (SGMD) and applied it to studying the folding of a 16-residue peptide. These studies provided detailed atomic-level descriptions of the formation of isolated secondary structure motifs in their respective solvent environments. [10]

Direct Simulations of the Early Stages of the Folding Process of Small Proteins

The application of such models in the direct simulations of the early stages of the folding process of small proteins is an exciting development. This includes a 36-residue villin headpiece subdomain (HP-36) and a zinc-finger-like protein BBA1, on the microsecond time scale. Such "top-down" simulations were perceived as impossible for the study of protein folding in the foreseeable future. Even though both the villin headpiece subdomain and BBA1 are small, they both share the features common to other proteins. The villin headpiece subdomain is a helical protein with well-defined tertiary and secondary structures. Its three helices form a unique type of fold with Helix 1 aligned perpendicular to the plane formed by Helices 2 and 3. The three helices are held together by a tightly packed hydrophobic core. Its melting temperature T_m is about 70°C (158°F), and the estimated folding time is about 10 microseconds, making it one of the fastest folding small stable proteins. BBA1 was designed by biophysicists using the zinc finger as the template. By forming its secondary structures first, unfolding simulations suggested that BBA1 might fold. [10]

In a simple downhill process, the simulation on HP-36 indicated that its initial collapse phase was accompanied by partial formation of the native helices and reduction of hydrophobic surface. The observed time scale of helix formation, 60 nanoseconds, agrees qualitatively with experimental observations on other proteins. The importance of the initial collapse phase has been understated somewhat in the past and has often been termed as the dead time burst phase, perhaps due to the fact that experimental studies of these ultrafast processes have been difficult until recently. Because early-stage species can often form in the burst phase and they may lead to the subsequent formation of other species and intermediates, the characteristics of these early stage species may affect the folding kinetics, whether or not they themselves are productive intermediates. The observed concomitant formations of both helical domains and hydrophobic clusters in the simulation suggest a way to lower the entropy cost in the subsequent folding processes by reducing the protein internal entropy in the early stages. The reduced protein entropy can be partly compensated by the entropy gained due to releasing water from the surface, as shown in the simulation by the strong correlation between the radius of gyration and the solvation-free energy (SFE), and by the large decrease of the SFE during the collapse process. These early stage nascent domains may (1) dissipate, (2) aggregate, or (3) grow. Some of these domains may well be the "nuclei" for the later-stage intermediate structures. Formation of native-like domains in the early stage helps the formation of the later-stage intermediate structures and perhaps the formation of the native structure. Conversely, the non-native domains will eventually dissipate, and those that are retained in the later-stage intermediate species will be difficult to dissipate and contribute to the free energy barriers. Correlation calculations indicated that the collapse was driven both by a lowering of the internal energy of the protein and burial of hydrophobic surface. [10]

Also observed in the simulation were considerable fluctuations between compact and extended states. This suggested a shallow free-energy landscape in the vicinity of extended conformations. The residence times of the compact conformations are much longer than those of the extended conformations, suggesting that the compact conformations are energetically more favorable than the extended conformations, thus indicating that the free-energy landscape is rugged and the existence of intermediate states is likely.

By going through many intermediate states of varying degrees of stability, proteins can fold from fully unfolded states to the native state. In fact, since these intermediate states can be referred to as the "landmarks" of the protein folding free-energy landscape, studying the mechanism of protein folding, in a sense, is to study these intermediate states, the relationship between them, and the relationship between them and the native state. One microsecond, even though it was more than two orders of magnitude longer than the longest simulation conducted up to 1998 on proteins in water, is still an order of magnitude shorter than the shortest estimated folding time for a protein; thus, it is unrealistic to expect the protein to reach the native state during such a simulation. However, this time scale appears to be sufficient to observe some marginally stable intermediates, if they exist. This was indeed the case in the simulation. A marginally stable intermediate was observed in the simulation and lasted for about 150 nanoseconds. The main-chain structure of the intermediate was remarkably similar to the native structure, including partial formation of Helices 2 and 3 and a closely packed hydrophobic cluster. The solvation free energy, calculated using the method and parameters developed by biophysicists, reached a level comparable to that of the native structure. Further analyses using the molecular mechanics–Poisson Boltzmann/surface area (MM-PB/SA) method indicated that the free energy of the intermediate state is the lowest among all the states sampled during the simulation, and both were significantly higher in free energy than the simulation starting from the native structure. This suggests that the reason the simulation did not reach the native state was because of kinetics and not force-field artifacts. Both the solvation free-energy and MM-PB/SA free-energy calculations were independent from the simulation, yet both of their results were consistent with the simulations. By being consistent with the long residence time of the state, both calculations indicated that the intermediate state was the most favorable one sampled during the simulation. [10]

It is generally perceived (perhaps even among most specialists in the field) that molecular dynamics simulation is deterministic in a way that is different from stochastic algorithms (such as Monte Carlo), in which the built-in randomness ensures that the simulation will asymptotically approach the ergodic limit. Because of that, it has been argued, molecular dynamics simulation methods are nonergodic. This holds true, however, only to a limited extent. Recent studies by two groups of biophysicists have demonstrated that molecular dynamics simulations on complex systems are inherently chaotic. The biophysicists demonstrated that near-identical simulations differing by a root-mean-square deviation (RMSD) of 0.02 Å (angstrom) resulted in two very different trajectories within 1 nanosecond, and the resulting structures can differ by an RMSD of as much as 5.0 Å. Yet the RMSD can be reduced substantially to within 2 Å after rigid-body alignment, clearly indicating that the trajectories sampled the same conformational free-energy basin. The biophysicists further demonstrated that despite short-time (<1 picosecond) chaos which is due to physical interaction (van der Waals forces), and not the algorithmic instability, all trajectories remained close to each other with an RMSD of less than 2 Å, similar to the typical thermal fluctuation. [10]

It is not surprising that earlier tests on simple harmonic systems failed to reveal such an important aspect, because the source of chaotic behavior is the nonlinearity of the physical interactions. This has important implications to biophysicists' simulations of protein folding. Because of the presence of chaos in the system, which is a source of randomness, the simulations have a stochastic character in the long-time scale, and hence can asymptotically approach the ergodicity limit, as well as having deterministic behavior in the short time scale. Simulations of similar conditions are expected to sample the areas that are close to each other in the phase space and produce similar trajectories differing in detail. Due to the randomness, which is a source of uncertainty, one should focus on the qualitative behavior in the analyses of individual trajectories, such as the time scale of the events and general trend. The randomness also contributes to the level of fluctuations exhibited in the folding simulations. Thus, multiple simulations are needed for statistically meaningful results, when one wants to focus on the detail, such as the role of individual hydrogen bonds and contacts.

Finally, biophysicists have therefore also conducted two additional simulations on HP-36, each to 0.5 microseconds starting from different states. In both simulations, HP-36 reached compact states within about 50 nanoseconds with the radius of gyrations comparable to that of the native state. More importantly, substantial secondary structures were formed during the initial collapse phase. Both the time scale and the concomitant occurrence of the hydrophobic collapse and the secondary structures were consistent with the biophysicists' earlier simulations and with experiments. One of these two trajectories also reached a state with structure resembling that of the intermediate state found earlier, including similarities of both topology and formation of a tightly packed hydrophobic cluster. [10]

Conclusion

The understanding of the mechanisms of protein folding has been quite an evolving field. Increasingly detailed studies, both at structural level and time scale, have highlighted recent developments. For instance, earlier studies would try to address questions like cooperativity, high rate of folding, ability to converge on the native state from so many starting conformations in the unfolded state, and secondary structures. Many current studies try to address questions on transition states, the relationship between protein structures and the folding processes, and characterization of intermediate and molten globule states. Many of these developments have been catalyzed by the advancements of experimental methods, which have been summarized recently. Hydrogen exchange experiments have been applied to study the dynamics of proteins. An atomic force microscope (AFM) can probe the stability of proteins that can be compared directly with simulations. Solution structure techniques can be used to study some of the intermediate states, including molten globule, partially folded, and unfolded states. Ultrafast kinetic experiments can provide increasingly detailed information on the early-stage folding processes. Single-molecule techniques have made it possible to monitor the dynamics and folding and unfolding processes of individual molecules.

Encouraging progress has been made recently, so in order to study the folding process, biophysicists are now in an era of active application of molecular dynamics simulations. Because of the vital importance of water in protein folding and in the cell, the explicit representation of both solvent and proteins is expected to play an increasingly important role in biophysicists' understanding of protein folding mechanisms. The goal of understanding the folding of small proteins should be achievable in the future, along with the promises of even more accurate models and considerably faster computer speed (together with the advancement of experimental approaches).

A major goal is the prediction of structures from sequences, besides the understanding of protein folding mechanisms. It has become increasingly clear that structural prediction methods have made considerable progress, along with the development of protein design, despite biophysicists' lack of comprehensive understanding of the folding mechanisms. The quality of knowledge-based structure prediction methods depends on the available experimental structures. They can reach reasonable accuracy for proteins with sequence homologies to proteins with known structures at above 50%, but the quality decreases substantially for lower-sequence homologies. Other related issues include the need to predict the binding affinities of small molecules to proteins when the flexibility of either ligands or proteins plays a role and the need to understand conformational changes using simulation methods. These reinforce the need for further understanding of protein folding mechanisms. [10]

One of the most challenging areas in biophysical chemistry is the subject of protein folding. Given the complexity of protein structures, diverse folding processes should be expected, including the possible role of certain folding-assisting domains within large proteins. Furthermore, the complexity of in vivo folding processes is, as yet, another challenge. Therefore, full understanding of protein folding mechanisms is indeed a daunting task. Despite this, recent developments have made us believe that an eventual solution may lie ahead. From a theoretical perspective, an immediate objective is to accurately replicate the complete folding process of small fast-folding proteins on computers, including atomic details. Such simulation results would provide the data for developing abstract models at a conceptual level that describe general and unambiguous features of protein folding mechanisms. The success of such simulations would itself be a strong testament to the accuracy of the method and parameters. The diversity of protein structures and the complexity of the in vivo folding process can, in principle, be dealt with by a combination of experiments and further simulations. Biophysicists may also be able to answer questions such as whether or not a particular part of a protein is designed to assist the folding of the rest of the protein, as found in "intramolecular chaperons," and how this assistance occurs. The mechanism of chaperon-assisted folding processes can also be better understood. But the understanding of more complex folding processes will be more difficult to achieve without the basic understanding of the folding process of single-domain, small-, and fast-folding proteins. [10]

References

[1] John Moult and Jan T. Pedersen, "Parallel Cooperative Algorithms for Protein Folding," NIST, U.S. Department of Commerce, 1401 Constitution Avenue, NW, Washington, DC 20230, 2003.

[2] Xinhai Ni and Howard K. Schachman, "In Vivo Assembly of Aspartate Transcarbamoylase from Fragmented and Circularly Permuted Catalytic Polypeptide Chains," The Protein Society, 9650 Rockville Pike Bethesda, MD 20814-3998, 2001.

[3] Stephen H. McLaughlin, Shanti N. Conn, and Neil J. Bulleid, "Folding and Assembly of Type X Collagen Mutants That Cause Metaphyseal Chondrodysplasia-type Schmid," School of Biological Sciences, 2.205 Stopford Building, University of Manchester, Manchester, M13 9PT, United Kingdom, 1999.

[4] Jose C. Martinez, M. Teresa Pisabarro, and Luis Serrano, "Obligatory Steps in Protein Folding and the Conformational Diversity of the Transition State," EMBL, Meyerhofstrasse 1, 69117-Heidelberg, Germany. Departamento de Quimica Fisica, Facultad de Ciencias, Universidad de Granada, 18071-Granada, Spain, 1998.

[5] Alfred Goldberg, "Introduction to Protein Folding and Disease," *Nature Reviews Drug Discovery*, Porters South, 4 Crinan Street, London N1 9XW, 2004.

[6] Arthur L. Horwich and Jonathan S. Weissman, "Deadly Conformations: Protein Misfolding in Prion Disease," Cyberdyne Systems, LLC, one of Eugene Oregon's, 1997.

[7] "Chemists 'Put the Twist' on Protein Building Block," Protein Design, © 2001-2003 by David C. Yee.

[8] Francis C. Assisi, Mathematical Model for Mad Cow Disease, INDOlink.com, Inc., 2440 Camino Ramon, Suite 225, Bishop Ranch 6, San Ramon, CA 94583, 2004.

[9] "Protein Misfolding, Not Mutant Gene, Key to Lethal Sleep Disorder," University Of Chicago Medical Center, ScienceDaily LLC, 1999.

[10] "Researchers Use Computers to Redesign Protein Folding," Howard Hughes Medical Institute, 4000 Jones Bridge Road, Chevy Chase, MD 20815-6789, 2001.

Paleontology: How Present-Day Microbiological Information Can Be Used to Reconstruct "The Ancient Tree of Life"

I feel again a spark of that ancient flame.
—Virgil (70–19 BC)

Scientists can trace the "ancient tree of life" back to the origin of life's three great domains, two of microbes and one of more complex creatures as shown in Figure 12.1. But where did those lineages come from billions of years ago? One idea is that there is no single root. Instead of one lucky organism giving rise to others, imagine a welter of simple "cellular entities" freely swapping genes. If so, the first modern cells probably annealed out of a kind of universal genetic pool.

"Ecologist John Muir (1838–1914) once said 'that if you try to pick one thing out of the universe, you'll find it attached to everything else.' Today," explains Dr. Peter Koehler (Geologist instructor for Northern Arizona State), "microbiologists and geneticists are proving that statement correct. We are finding out that we have a strong kinship with all other living inhabitants of planet Earth. This makes sense as all inhabitants of planet Earth, for the most part, deal with relatively close envi-

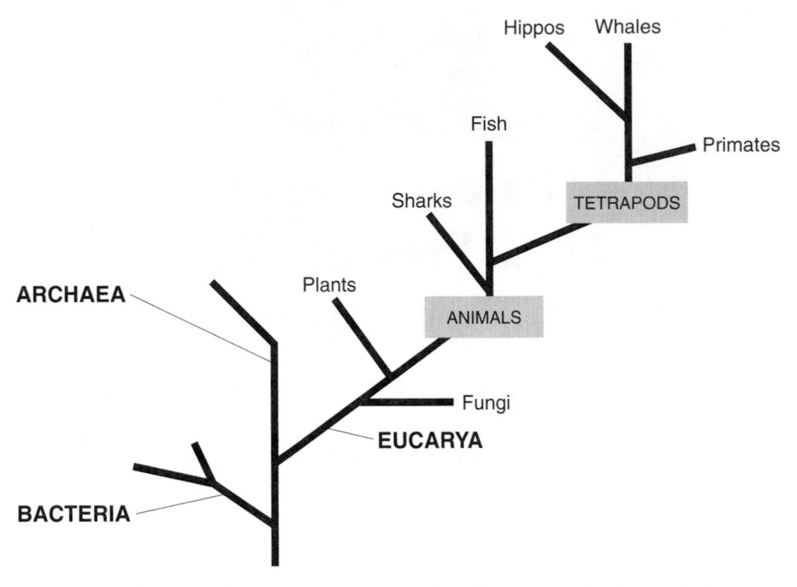

Figure 12.1 Based on comparisons of molecules and physical features, scientists are reconstructing an ancient family tree of living things. This illustration shows the key branching points along one of the major limbs or domains of life.

ronmental conditions. As organisms come and go, we should remember that species respond individualistically to environmental pressures, further branching (growing) the tree of life. I think the tree of life analogy is generally incorrect (unless you consider the roots of course), as it more resembles a web of life, with early (simple) organisms at the center and the most evolutionary advanced (species living today) radiating towards the edges, such that we, humans, are more akin to Fungi than to grasses occupying to the opposite (furthest) end of the web.

"With the extraplanetary endeavors of today, one may wonder that if remnants of past life are found on any other of our neighboring planetary bodies, what genetic similarities might they posses to some Earth dwelling inhabitants. As we share similar environmental pressures within our solar system, the island biogeography theory (MacAuthur and Wilson) suggests greater differences further from the genetic continent. Should we expect to find genetic linking, as we are finding out on our own planet, supporting a universal cosmic continuum?" [1]

"This is a very good question," responds Dr. R.L. Folk (UT of Austin world-renowned expert in sedimentary rocks and fossils of bacteria), "and we do not know the answer to this question yet. My feeling is, in the beginning, you had to start with very small genetic change, very small DNA change, and minimal organisms. I think these minimal organisms exist today, and some people call them 'nanobacteria,' though some people do not believe the very existence of nanobacteria." [2]

The Three Domains of the Ancient Tree of Life

When Charles Darwin published *The Origin of Species* in 1859, the ancient tree of life was the only illustration he used. It was a rough sketch—a statement of principle, more than anything else—showing organisms branching outward as their descendants evolved into new species. Aided by a flood of genetic data and powerful computers, biologists can finally contemplate finishing Darwin's sketch.

It may take years or decades of collecting specimens and analyzing data, but even a partial ancient tree of life will be a powerful tool. Virtually everything scientists do in biology is comparative. You can't understand an organism without comparing it to others along the ancient tree of life. It will also aid medical scientists, who will pick over the ancient tree of life for clues to how diseases emerged and where cures might be found. For the rest of the scientific community, the ancient tree of life already offers lessons in humility, such as humans' close kinship to the fishes and the surprising status of mushrooms, closer to us than they are to potatoes.

The naming of animals and classifying nature seems to be a basic human instinct. These days, the effort is called phylogeny, and it is not just an exercise in naming. Its goal is to trace evolutionary relationships, by cataloging and comparing the characters—DNA sequences, the shapes of ankle bones or seed pods—that hint at which organisms are close relatives and which have long been separated. It's like a planet-wide game of "one of these things is not like the others": You put the monkeys and rabbits on one branch, the bananas and carrots on another, and pretty soon you've got the rudiments of a phylogenetic tree, or ancient tree of life, as discussed in the sidebar, "The Ancient Tree of Life."

The Ancient Tree of Life

One surprise in the ancient tree of life tree was the appearance of bacteria that thrive in high temperatures (hyperthermophiles), near or at the root of the tree. The ancient tree of life, and supporting research from other fields, led to the speculation that life originated in very hot environments, perhaps in the hydrothermal vent systems found deep underwater around the globe. (See sidebar, "Quantifying the Changing Salinity and Carbonate Budgets.") The resulting speculation that life arose in a hot environment has become widely accepted among researchers in the original tree of life, and it has taken on the status of fact in textbooks that explain the origin of life.

When micropaleontologists look at the RNA tree, they see that all of the deep branches are organisms that grow at temperatures above 80°C (176°F). A lot of micropaleontologists have taken that to mean life must have originated at high temperatures. But, for several reasons, it's more likely that life originated at low temperatures. If micropaleontologists could find a low-temperature organism that's even more primitive, that would prove the point.

"All the deep branches of the RNA tree of life contain hyperthermophiles," explains Dr. Julie K. Bartley, Assistant Professor for the Department of Geosciences. "This leaves three possibilities: 1) The origin of life occurred at high temperatures; 2) the origin of life occurred at low temperatures, but some event caused all the low-temperature forms to become extinct; and 3) we're missing some low-temperature critters near the base. Interestingly, the third possibility doesn't exclude the possibility of a high-temperature origin of life, because the low-temperature preference in those hypothetical basal organisms could have originated independently." [4]

"We have 'nanobacteria' in all sorts of low-temperature situations," explains Dr. Folk, "including inside of our body. It has already been found. It's time for the biologists to wake up and realize what's under their noses." [2]

Long Branch Attraction

Micropaleontologists approached reconstruction of the tree of life by paring down the parts of the ribosomal Ribonucleic acid (rRNA) gene (a class of RNA found in the ribosomes of cells) to include in the analysis. They chose parts of the gene that changed the least over time. This decision sidesteps some of the errors that can affect less conservative approaches.

For example, long-branch attraction arises because the assumptions in computer models used to construct evolutionary trees underestimate the rate of DNA change. If the part of a gene under study evolves more quickly than the average, the problem is even worse. Choosing slowly evolving locations helps match the data to the assumptions. The approach reduces systematic errors. [5]

It's not generally appreciated that molecular sequence analysis is a field in its infancy. Thus, it is an inexact science, in which there are few analytical tools that are based on general mathematical principles. As a result, many—perhaps most—phylogenetic trees reconstructed from molecular sequences are incorrect, because they make mathematical assumptions that are not met by the data being analyzed. Frequently, these incorrect assumptions lead to long-branch attraction.

Long-branch attraction can be caused by one or more of three pitfalls of sequence analysis. For all three, the effects are the same. In trees artificially produced by long-branch attraction, rapidly evolving sequences (represented by long branches on unrooted phylogenetic trees) will be placed with other rapidly evolving sequences, even if the sequences are only distantly related. In comparison with most problems in molecular biology, which can be solved by requiring more data, long branch attractions are diabolical. When long branch attractions are present, if longer sequences are used, the incorrect solution will be even more strongly supported.

The downside of using less data is that the approach increases random error. This is why micropaleontologists plan to use more genes in the future.

The ancient tree of life that today's micropaleontologists came up with differs in some striking ways from the one that has become standard. First, the hyperthermophilic bacteria do not appear near the trunk of the tree. Instead, the reanalysis shows them up among the leaves, part of a large, bushy crown group that includes most of the bacteria.

In a way, this result does not come as a surprise to micropaleontologists. There are a lot of adaptations required to live at high temperatures, designed basically to keep the molecules together. They tend to fall apart at high temperatures. So micropaleontologists have always felt more comfortable with the idea that it was a secondary set of adaptations some time after life had originated that allowed these organisms to radiate into high-temperature environments.

One could have guessed from the discovery of hyperthermophilic organisms that they are specialized, and thus very evolved. The statistics for finding hyperthermophiles near the trunk of the tree are weak.

Other research supports the doubt cast on hyperthermophiles as common ancestors. Research from French and American researchers suggests that the subunit, or nucleotide, makeup of rRNA in the earliest ancestral organisms is incompatible with life at high temperatures.

Furthermore, a key enzyme that helps protect bacterial DNA in high-temperature environments was acquired by horizontal gene transfer from Archaea. When the genome of the hyperthermophilic bacterium Thermotoga maritima was completed, 24 percent of its genes showed more similarity to archaeal genes than to bacterial genes. Similar studies on other hyperthermophilic bacteria produced similar results. [5]

But the real surprise is the organism that did get placed close to the trunk of the tree of life in micropaleontologists' reanalysis: an unusual bacterial group called planctomycetes, which thrive only in moderate temperatures. Planctomycetes are intriguing because they combine features found in all three domains of life: Archaea, Bacteria, and Eukarya. The distinction among the three domains of life goes like this: Eukaryotic cells all have their DNA packed into a nucleus with a double membrane; bacterial and archaeal cells have no nuclear membranes. Many biochemical differences separate Archaea and Bacteria, including a key component of their cell walls.

Planctomycetes play differently. Instead of dividing in two like most bacteria, they reproduce by budding, a little like yeast. They lack that key cell-wall chemical found in all other bacteria, but not found in archaeans and eukaryotes. The punch line, though, is the presence in planctomycetes of a nuclear membrane. In some species, it's a single membrane, in others a double membrane. [5]

Finally, no common evolutionary origin, or homology, can be proved between the nuclear membrane in planctomycetes and the nuclear membrane in eukaryotes. But, according to micropaleontologists, homology has not been ruled out. At the very least, the existence of the nuclear membrane in planctomycetes should change the definition of "eukaryote" to include complexity beyond the mere presence of a nuclear membrane.

Quantifying the Changing Salinity and Carbonate Budgets

According to geophysicists' Dr. Dorinda Ostermann, Dr. Steve Manganini, Dr. Christopher Brown, and Dr. Steve Piascek, "The salinity budget of the North Atlantic influences the timing and production of globally important North Atlantic Deep Water. These linkages are important because in the Iceland Sea, conditions altering the relative amounts of Polar and Atlantic waters will influence the types of particles produced and their eventual deposition to the deep ocean. These observations are critical to interpreting the paleoclimate records from deep sea cores which contain a history of globally important North Atlantic Deep water production. Long time series work using downcore samples from sediment cores collected from the Arctic Ocean have shown a positive correlation between past Arctic deglacial meltwater events and marine fossil carbonate stable isotopes (Darby et al., 2002). These results imply that changes in Arctic salinity likely played a major role in the variability of North Atlantic overturning circulation which should be recorded in foraminiferal tests. For example, these meltwater events show up as minima in *N. pachyderma* (l) $\delta^{18}O$ and are proportional to the size of the meltwater event. As Arctic sea ice thinned in the past and in the present, we can infer that the waters leaving the Arctic via the Bering, Fram and Davis Straits freshened. But we also need to understand the sensitivity of the circulation system to changing salinity regimes on decadal time scales in order to model the impact these changes may have on the heat budget of the North Atlantic.

"To monitor the long term affect of Arctic sea ice and Greenland glacial thinning on the salinity and carbonate budgets of the Greenland, Iceland, and Norwegian (GIN) Seas, we have continuously deployed a sediment trap NE of Iceland from 1986 to present (68° N 12.4° W, 1848 meters) with colleagues from the Marine Research Institute (MRI), Reykjavik, Iceland. After monitoring the particle flux from this time-series sediment trap for 14 years, a remarkable event was captured in our samples and observed by SeaWiFS satellite imagery. In the summer of 1999 we collected as much carbonate in one two-week interval as in the previous 15 years of collection combined. All but 1% of the carbonate was composed of the coccolithophore *Coccolithus pelagicus*. Concurrent with increased carbonate production, the Iceland Sea freshened markedly (Piascek, unpublished data). We need to understand the possible link between changing hydrographic conditions and the occurrence of *C. pelagicus* at the Iceland Sea mooring location if we are to use it as a paleoproxy indicator of past Polar water intrusion into the GIN Seas. Given the influence of polar regions on global cycles and recent changes in Bering and Barents Sea phytoplankton to predominantly coccolithophorid blooms (Stabeno, 2000), we believe further investigations to monitor and characterize biogeochemical changes in the Iceland Sea are warranted.

"We propose to continue our collection of time-series sediment trap data in the Iceland Sea, 2004–2006, in order to ground-truth SeaWiFS imagery to the plankton group responsible for the high reflectance pixels found in the imagery. We will also use satellite sensor data (color, temperature, dynamic topography, and near surface wind speed) to link plankton blooms (silicate or carbonate) to specific atmospheric and oceanographic conditions. Ostermann has been a welcome participant on the annual MRI research cruise to the trap location. She will recover and redeploy the mooring system in each of the three years of this proposed work. In year one, we will upgrade the mooring system to include a Moored Profiler (MMP), which will continuously monitor the hydrographic conditions of the upper surface waters. Manganini will analyze particulate material including mass flux, organic carbon, nitrogen, inorganic carbon, biogenic silicate, and lithogenic material using the ICP-MS facility at WHOI. The analysis of the liquid portion includes pH, dissolved phosphorus and dissolved nitrogen. Ostermann will analyze the stable isotopic composition of the carbonate particles. We will be working on SeaWiFS data with

Christopher Brown at NOAA, a SeaWiFS project member and will allocate monies each year for imaging and analysis costs. Steve Piascek will work on the GIN hydrographic dataset." [3]

In other words, the game is: how can scientists reconstruct the past? How can they create an accurate "ancient tree of life" that illustrates the evolutionary heritage of life? The traditional answer would be to find fossils at different levels of rock strata, arrange them chronologically, and try to reconstruct the missing intermediates. There have been some successes: Fossil reconstruction has given good insights into the evolutionary ancient tree of life of larger, multicellular organisms that leave durable remains: shells, bones, etc. However, there have been some failures: Fossils of unicellular organisms, organisms without durable remains are sometimes available, but infrequent and often uninformative.

"In a sense," states Dr. Bartley, "all fossils are informative—even if we can't fully interpret the phylogeny of the organism, its presence in a rock tells us something about the organism and its environment. We're learning all the time about organisms that we never thought would be preserved. For example, the discovery of animal embryo fossils in Proterozoic rocks was a tremendous surprise. Now, several workers have been able to find these tiny fossils. Now that we know what to look for, we see many examples of these informative fossils. Maybe the message here is to keep working at the fossil record; it holds more information than we sometimes think! [4]"

Furthermore, claims Dr. Susie Smith (Laboratory of Paleoecology for Northern Arizona University), "The microscopic skeletons of fossil pollen grains preserve in geologic strata because of a mysterious material called sporopollenin, an incredibly tough, chemically unique, polymer. Sporopollenin has been found in widely separated organisms such as algae, fungi, ferns, and flowering plants, in rocks over 3 billion years old, and traces of something that looks like sporopollenin was recovered from meteorites [Brooks, J. and G. Shaw, 1969, Evidence for extraterrestrial life: Identity of sporopollenin with the insoluble organic matter present in the Orgeuil and Murray meteorites and also with some terrestrial microfossils. *Nature* 223:756]. Is this a glimpse of extraterrestrial life?" [6]

The shape of the ancient tree of life has changed, sometimes dramatically, as scientists have learned new ways to play the reconstruction game. Until about 1990, most biologists pictured an ancient tree of life with five major branches—animals, plants, and fungi at the top, protozoa and bacteria at the bottom. Then microbiologists announced that a comparison of the vital molecules that cells use to copy DNA and make proteins pointed to a very different ancient tree of life, with three roughly equal trunks, or domains, emerging from an ancient root.

For example, bacteria (think E. coli) occupy one trunk; the Eucarya (pond scum, people, and everything in between) occupy another as shown in Figure 12.1. A third belongs to the Archaea, exotic microbes that were once lumped with bacteria, but are actually a distinct domain of life more closely related to the Eucarya. (See sidebar, "Three Domains of Life".) As different as people and hydrogen-eating microbes seem, Archaea and Eucarya have the same basic tool set, according to microbiologists. [5]

Three Domains of Life

Bacteria consist of approximately 11 distinct groups. They are at least as varied as more familiar forms of life, but only a tiny fraction of them have been studied.

Bacteria are also prokaryotes (cells without nuclei) in contrast to eukarotes (cells with nuclei)—a group that includes all animals, plants, fungi, and protists. Bacteria, as opposed to humans, are:

- Invisibly small: Because scientists can't see them, they often forget them.
- Incredibly numerous: The most abundant life form on Earth.
- Incredibly fast: Some bacteria can go through a cycle of reproduction in 20 minutes (in contrast to 20 years or more for humans).
- Incredibly sophisticated: Some bacteria can make all their cell materials from any one of hundreds of different chemicals, including mothballs, oil, etc.
- Incredibly diverse: Some bacteria thrive at boiling temperatures, others in the Antarctic ocean; some thrive in concentrated acids, others in strong alkali, etc.

With this in mind, there is a question that begs to be answered: What is the evolutionary relationship of bacteria to other organisms? [5]

According to micropaleontologists, a partial answer to the preceding question .might be: The age of the Earth is about 4.5 billion years. The oldest rocks on Earth are about 3.8 billion years old. Those rocks already contain occasional recognizable fossils of bacteria: Bacteria must have evolved very early in Earth's history. From about two billion years ago, micropaleontologists have found larger stromatolites, fossil remnants of ancient cyanobacteria, the first organisms to produce oxygen. But bacteria are very tiny. They lack morphogical diversity of plants or animals, so it's impossible to reconstruct a history of bacterial evolution based on fossils. On the other hand, eukaryotes only evolved about one billion years ago: The only life on Earth for most of its history was bacteria. So, is it possible to ever figure out the evolutionary tree of bacteria and their true relationship with other life forms?

"To identify the bacteria," claims Dr. Folk, "you have to culture them and see what metabolism they do. [Of the] modern bacteria that are living on your skin or in soil and everywhere around us, only 1% of them are culturable. The other

99% we do not know what they are and we cannot culture them and we cannot classify them. Therefore, in modern day environments, we can only identify 1% of existing bacteria. We cannot possibly identify what a 3.8 billion-years-old bacteria can do. All we do see in rock is their dead bodies. When you see a dead man, you cannot tell whether he was a good person or a bad one. I think that there is no way in present science we can identify what they were doing by means of looking into their metabolism." [2]

Archaea

Archaea consist of three distinct groups. These microbes, once considered bacteria, live in exotic environments like deep-sea hot springs, as well as more prosaic locales. Archaea also means "ancient" because they use ancient energy mechanisms.

As previously mentioned, many archaea are also found in harsh, early earth-like environments and consist of three groups:

1. Halophiles: Halobacterium, which is found only in very concentrated brines, evaporating salt basins, the Dead Sea, etc.
2. Methanogens: Methanococcus jannaschii, which is the first archaeal organism to have its genome sequenced.
3. Extreme thermophiles: Thermocrinus ruber, which grows well at temperatures above boiling!

Eukaryotes

Eukaryotes, or Eucarya—encompassing organisms with cell nuclei, such as plants, animals, and fungi, which are actually closer kin to animals than to plants—consist of four distinct groups:

- Protists: Single-celled organisms
- Fungi: Yeasts, molds, and mushrooms
- Plants: Green, photosynthetic, rooted, producers
- Animals: Usually motile, consumers

NOTE: The same methods of DNA mutational distance can be applied to organisms such as animals for which fossils have already provided good phylogenetic trees. These results depend only on comparisons of DNA sequence for the ribosomal subunit gene; no fossil data is used. They match very well with phylogenetic trees derived from fossil analysis.

Climbing the Branches of the Ancient Tree of Life

Microbiologists are trying to map the luxuriant branches of the ancient tree of life and have found that molecules are not always enough. In the 1970s, molecular biologists were confident that they would soon work out the true ancient tree of life based on differences in genetic sequences. But different genes evolve in different ways, so they often tell conflicting stories about how organisms are related. Microbiologists can't just pick a few organisms and a few genes and say, "Here's my crazy ancient tree of life."

Physical characters, such as anatomical features and cellular structure, have their own problems. Collecting and measuring specimens is time-consuming. Appearances can deceive ancient tree of life builders, as when "convergent evolution" makes look-alikes out of species that actually parted ways millions of years ago, like whales and fish or birds and bats. And, in the vast domains of Bacteria and Archaea, microbes have precious little anatomy to compare.

The ancient tree of life still looks scraggly, with only some 50,000 of the 1.5 million or so known species in place, but researchers think they now have the tools and know-how to fill it out. Different lines of evidence often result in contradictory trees. But when scientists analyze all the evidence—DNA, cellular structure, anatomy, and even behavior such as nest building or seed dispersal—and include both living and fossil relatives, the data tell a more consistent tale. Every character is part of the same story, even if they tell different versions. [5]

Using this total evidence approach, biologists are now building a convincing account of who begot whom, and they are turning up some unlikely relatives. Arrange the birds, crocodiles, and mammals on a ancient tree of life, for example, and birds and mammals look like sister groups. But, if you put in the dinosaurs, you see the birds, crocs, and dinos on one side, with the mammals far away on the other. Similar analyses, based on DNA and newly unearthed fossils, suggest that whales evolved from land animals that looked an awful lot like mangy dogs and are close cousins to the hippopotamus. Whales are so different from other mammals that it wasn't clear where they fit.

According to Vertebrate Paleontologist Dr. Ted Macrini (UT of Austin),

> Morphologists and paleontologists have traditionally reconstructed the amniote (tetrapods exclusive of amphibians) tree with a basically diverging split between the mammal and reptile lineages. The synapsids (mammals and their closest extinct relatives) comprise one of those branches and the reptiles (including turtles, lizards, snakes, birds, and a number of extinct taxa such as non-avian dinosaurs, ichthyosaurs, plesiosaurs, and pterosaurs) make up the other branch.
>
> In the 1980s, some studies using molecular (mainly DNA) data or soft tissue (the parts of organisms that typically do not fossil) data taken exclusively from extant (living) species have challenged this traditional view of the amniote tree. One alternative hypothesis placed birds and mammals as sister taxa in a clade called Haematothermia. Re-analyses of these data and inclusion of fossil taxa and osteological data, overwhelmingly rejected this hypothesis. Inclusion of a single non-mammalian synapsid fossil in the phylogenetic analysis of extant species was enough to reject Haematothermia. In actuality, the synapsid lineage is represented by one of the best fossil records among amniotes, extending back

> over 300 million years. Similarly, the dinosaur lineage is very well represented in the fossil record, leading to the well supported conclusion that birds are in fact dinosaurs.
>
> Whales have been aligned with a number of different groups of carnivoran mammals. Charles Darwin suggested that they are related to bears, but more recently whales have been aligned with an extinct group of mammals called mesonychids. Mesonychids were originally thought to be part of Carnivora but are now considered ungulates. Fossil evidence suggests that whales are the sister taxon to mesonychids and most closely related to artiodactyls (a group of ungulates that includes cattle, pigs, hippos, and deer) among living mammals. Recent molecular studies and some morphological ones suggest that whales are members of Artiodactyla and most closely related to hippos among living mammals. [7]

Just a few minutes tracing the existing ancient tree of life is enough to shatter preconceptions. Fungi are more closely related to animals than to plants, and reptiles have more in common with mammals than with amphibians. Starfish are closer to mammals than they are to shellfish. And speaking of fish (not that oysters and starfish really are fish), many microbiologists believe that all animals with four limbs, including you, your mother, and your ape ancestors, belong to a branch of the fish lineage. According to microbiologists, we are all fish.

Dr. Macrini explains it this way:

> Phylogenetic systematics provides a hierarchical classification of nested groups. For example, all humans belong to a larger group known as Primates, which is part of the larger group of Placentalia which is part of the larger group Mammalia which is part of the still larger group Synapsida. Therefore all humans are primates, placentals, mammals, and synapsids.
>
> Farther down the vertebrate tree, there is a split between Sarcopterygii (the lobe-finned fishes and four-legged vertebrates) and Actinopterygii (the ray-finned fish which includes most of the things traditionally considered to be fish). Sarcopterygii includes

> things like lungfish, coelacanths, and all tetrapods (mammals, birds, amphibians, etc.). Therefore, humans are members of the much larger group Sarcopterygii. We are part of this lineage of vertebrates that includes things traditionally called "fish." [7]

Building a universal ancient tree of life has practical uses, too. Scientists often try to puzzle out what a newly discovered gene does by comparing it with similar genes in other organisms—and they can sharpen those guesses if they know how closely related those organisms are. Fine branches of the ancient tree of life describing how pathogens are related can help epidemiologists track the emergence and spread of diseases, and ecologists trying to predict the impact of an invasive species can often glean clues from its closest relatives in similar ecosystems.

Unified Ancient Tree of Life

Microbiologists are laying plans to coordinate scattered efforts and graft disparate branches onto a unified ancient tree of life. They emphatically cannot go on in an idiosyncratic way. The scientists also need a truly representative sample of life to build a complete ancient tree of life.

Finally, to that end, microbiologists across the country are collecting hundreds of thousands of animal tissue samples, frozen in ten cryogenic vats. DNA from the samples, representing every animal lineage, will be sequenced and the information fed into a supercomputer. Microbiologists also plan to push for a serious effort to catalog Earth's full diversity of life. As much as 90% of living species remain to be discovered; filling in those gaps would provide more clues to helping reconstruct the tree of life, according to microbiologists. [5]

Conclusion

For decades, scientists have used a comprehensive ancient tree of life, showing heat-loving bacteria as the Earth's earliest bacteria. Now, a more accurate reanalysis of the data places those bacteria up among the leaves. The new candidate ancestor, an obscure group of bacteria with an intriguing combination of unique features, is upon the scientific community.

Some microbiology classes routinely culture species from the environment because bacteria are so easy to grow in a lab, or culture. That also applies to commonly studied laboratory strains. But microbiologists' knowledge of bacteria is, in general, very narrow. Only about five percent of all known bacteria have been cultured. In many phyla, no species have been cultured. Scientists know about the so-called uncultured bacteria only from the study of RNA contained in environmental samples. [5]

So where would these phyla emerge in a microbiologists' reanalysis? This is potentially an important problem, since at least 50% of the bacterial phyla contain only uncultured organisms. Microbiologists also plan to study more genes when the complete genomes of some planctomycetes species become available. So, in order to see whether the group is primarily or secondarily adapted to low temperature, another possible approach is to study the genome of a nonhyperthermophilic member of a group that is normally hyperthermophilic. [5]

Finally, the benefits of reconstructing the ancient tree of life could flow back to the natural world, by helping researchers identify unique groups with few neighboring branches that should become the focus of special conservation efforts. Building a universal ancient tree of life is the best way to understand the diversity of life. If scientists work fast enough, it might also be the best way to preserve it. [5]

References

[1] Dr. Peter Koehler, Instructor of Geology, Coconino Community College & Northern Arizona University, Flagstaff, 2004.

[2] Dr. R.L. Folk, Department of Geological Sciences, The University of Texas at Austin, 1 University Station C1100, Austin, Texas 78712-0254, 2004.

[3] Dr. Julie K. Bartley, Assistant Professor, Department of Geosciences, State University of West Georgia, Carrollton, GA 30118.

[4] Dr. Dorinda Ostermann, Dr. Steve Manganini, Dr. Christopher Brown and Dr. Steve Piascek, "Quantifying the Changing Salinity and Carbonate Budgets of the Greenland, Iceland, and Norwegian Seas Using Sediment Traps and Satellites, 1986 to Present," NASA, NRA-04-OES-01 Carbon Cycle Science, Notice of Intent, Woods Hole Oceanographic Institution, Department of Geology & Geophysics, Clark 123, Woods Hole, MA 02543 [WHOI, NASA and NRL], 2004.

[5] "Discoveries Made about Cellular Reaction Processes From Ancient Life," Virginia Polytechnic Institute and State University, Blacksburg, VA 24061, 2003.

[6] Dr. Susie Smith, Laboratory of Paleoecology, Box 6013, Northern Arizona University, Flagstaff, AZ 86011, 2004.

[7] Dr. Ted Macrini, Vertebrate Paleontologist, Department of Geological Sciences, The University of Texas at Austin, 1 University Station C1100, Austin, Texas 78712-0254, 2004.

References

Neuroscience

Free Will

"Democracy is a process by which people have the free will to choose the person who will get the blame."

—Anonymous

Free will! Is there such a thing? This is a burning question for many neuroscientists who have come to feel that, although their ethical life stance both implies and demands the existence of free will, the conclusions of science cast doubt on that very possibility. The dilemma is not new. The "free will–determinism" issue has been key to philosophical discourse in every age, if only because the justice systems of all civilized societies have always been based on the proposition that individuals are responsible for their own behavior. In the end, the position on free will taken by most people in every historical era has come down to their beliefs about the nature of causality: beliefs implied by the worldview prevalent at the time. So, from one of the following three sources comes modern notions about the nature of causality:

- The mechanistic determinism underlying the worldview apparently warranted by the theory and evidence from physics concerning the nature of reality.
- The philosophical dualism justifying most of the world's religious belief systems, which defines

the human as different in kind from other aspects of nature—and thus not subject to nature's regularities as discovered by science.

- The sovereign ego, anti-science type of nondeterminism. [1]

So how is it that humans have the freedom to decide and act? Your brain and sensory organs allow you to respond to your surroundings in a way that is not deterministic. Humans have developed a remarkable ability to make decisions, to plan their actions, and to decide, in part, their fates. But how does this come about? Let's take a look.

Freedom to Decide and Act

The purpose of this chapter is to demonstrate that neuroscientists' confusion and concern is unnecessary with regards to free will. Although neuroscientists have been right in accepting science as the source and final arbiter of any conclusions on the free-will versus determinism question, they have been wrong in their choice of physics as the field to which they turned for guidance. The point is that neither Newtonian nor Einsteinian physics (nor the tool of quantum mechanics) is significant for the issue of human freedom. This is because they do not deal with causal relations at the organic level of existence. There does, in fact, exist a scientific model that allows for a limited concept of human freedom—if not for the sovereign will for which some people yearn. This model is the emerging evolutionary paradigm directing the organic and psychosocial sciences. Those are the studies most directly relevant to human beings.

Science might warrant a limited concept of free will. It needs to concentrate on the universality of cause and effect in nature. Within this all-encompassing process, free will does indeed play a role—in spite of the obvious fact that human behavior, like everything else in the natural world, is to a large extent predictable. All legal and ethical systems are founded on this critical premise, and its denial would seem to contradict common sense. Although each individual act is really the effect of the acting person's character and motives, the person can always choose not to act in any given situation. It is the human process of making inferences on the basis of experience that opens up this possibility.

What Darwin implied about the nature of causality is one of the most important and least clearly comprehended aspects about the role of the human will in the universal causal process. Clearly, the vehicle of natural selection that emerged with life was able to operate in terms of a new kind of cumulative and after-the-fact or contingent causality. Darwin had hypothesized that nothing more mysterious than a gradual accumulation of tiny advantageous changes has produced all of the complex structures of even the highest mammals—including human brains and the so-called spiritual processes of mind, soul, and will. His theory implied that the consequences of the organism's activities within its environment feed back to become, in turn, the causes of subsequent species change. These consequences include both the impact of the organism on its environment and the degree to which it is successful in producing offspring. This makes the organism's actions and resulting environmental impact critical to the entire process of evolutionary change. The approach to the age-old question of free will versus determinism emphasizes new definitions for the sources of these actions (for the first time within an evolutionary framework). It had at last made possible a social-psychological and neuroscientific approach, rather than metaphysical.

If its significance had been widely recognized, the preceding development would have been earthshaking for philosophy and theology. In fact, it was not until near the turn of the century that there occurred the first stirrings of a satisfactory new resolution of the age-old free will dilemma: one abundantly compatible with the findings of modern science. It promised to make the older free-will-versus-determinism argument obsolete and opened up the possibility of a radical reformulation of the origin and role of conscious choice in human affairs.

By using chemicals like hallucinogenic drugs, researchers have discovered that a mystical sensation can be effected mechanically by probing designated areas of the brain. This is a compelling example of how contingent or feedback causality can operate from various emergent levels of interaction (the chemical, neurological, psychological, social, and cultural) to bring about what is interpreted by the experiencer as identical (and equally mysterious and uncaused) feeling states. In many cases, the initiating behavior tends to become compulsive and addictive—that is, it is increasingly less open to the power of reasoned choice. In view of the fact that the person involved invariably feels that an ultimately desirable, mystical consummation has been effected by some extrasensory contact with the individual's sovereign will, all of this should not be surprising, nor should the absolute contact that supposedly can only occur when cognitive activity has been suspended.

What, then, do all of the modern research findings mean for the concept of free will? Is it possible to begin to spell out the shape of an emerging worldview (rooted in current scientifically compelling premises concerning causality) that would be acceptable to scientific humanists? Neuroscientists are convinced that a focus on the role of contingency in human actions points the way to a satisfying resolution of the dilemma. Cognitive neuroscience would seem to provide confirmation within the new paradigm that the human is indeed an open system in a hierarchy of open systems constituting the ongoing stream of evolution. Thus, you do respond to contingencies which, in turn, have resulted from your own thrusts into the totality of your physical and social world. Social scientists believe that among the most powerful of these causal factors are social rituals and the myths that have informed and driven group-created cultures from ancient times. In the words of social behaviorists, you can achieve freedom to alter your previously conditioned behavior (or what you impulsively choose to do) only to the degree that you possess reliable knowledge about the causes and consequences of your actions. This knowledge encourages you to reflect on the consequences of previous actions and to "will" to act in ways that will engender desired future consequences. Thus, the enemies of whatever degree of free will is available to humans, are the cultural beliefs that explain behavior and feelings in terms of unknowable forces and beings.

Neurologist Dr. Miroslav Backonja (Department of Neurology, University of Wisconsin Hospital) explains:

> Free will is a phenomenon, a dynamic process. It arises when an organism faces a problem or an obstacle. The number of solutions for solving the problem is based on the organism's endowment of abilities, inherited (in the form of genetic history and complexity of nervous system) and personal (in the form of life experience) abilities, that will give the impression (to an independent outside observer) of free will.
>
> An example could be illustrated by a question: What is the difference between the free will of a man and the free will of a dog? For either of them to get on the other side of a swollen creek is an obstacle. At any given moment they may have "will" to cross it but how "free" their will is depends on what one considers "free." How and when they will get on the other side will depend on many factors: for a dog it is very limited because in the dog's universe major issues are food (finding one and avoid becoming one), shelter, and once in a while mating, so the incentives for solving this obstacle (we call them degrees of freedom) are limited. For a man there are many tangible motivations (food, shelter, mating) and more

> numerous intangible (social gathering, duty, etc.). So, for a dog, free will is not an issue; the dog will attend to it or go on. For a man it is only a beginning of "planning," which leads to the impression of many choices and therefore many opportunities to exercise different degrees of free will. But what and how man proceeds is still constrained by his inherited and personal abilities. And how free his choices are is detrained by the observer who would judge them within the given points of reference. The number of possibilities and complexities of the interactions is significantly magnified by the number of human interactions that come from social structures and roles. [2]

From the moment of birth, each human being begins to forge a character from what genetic potential makes of experience. There is now considerable evidence that supports this. Neuroscientists also have good reason to conclude that the value system defining that character determines the nature of your choosing. Given the content of your values and knowledge at the moment of choice (and the precipitating circumstances), it is true that you could have done no other than what you did. This does not translate into "sovereignty" of will, but it does mean that your choice is unique. For modern scientific thinkers, free will is meaningful only in terms of this capacity for choice—and a reflective choice at that. No freedom, however limited, can exist if behavior remains merely habitual or impulsive—or if it is driven by demonic notions that your ancestors embedded in the culture. Neuroscientists now know that, as living, culture-creating organisms, humans form an inextricable part of nature's continuous current of cause-and-effect. None of the findings of modern science indicate a possibility of altering or redirecting the process from without. It seems clear that humans can only work from within. The obvious conclusion from all this is that free will, in the sense of the old notion of a sovereign ego operating *upon* a material reality, is merely the stuff of dreams and wishes. But this does not imply determinism as it was formerly understood. Humans, out of necessity, affect nature's direction and velocity, because of humanity's involvement in it's current cultural evolution.

A scientifically supportable concept of free will can be viewed as the potential to redirect the course of nature's continuum of cause-and-effect by achieving conscious control of the consequences of your actions. It means that humans can produce different effects than would otherwise have occurred. Your power both to predict and alter effects has been increased exponentially by the progress of science. The positive message in what modern science implies for free will is that you are not condemned to use that power destructively. To the degree that choices are based on reliable knowledge and reason (combined with the ability to imagine and evaluate possible future consequences), humans can participate wisely in the uniquely human task of guiding the course of cultural evolution. As cultural beings, humans have evolved the potential for choice, and have been both burdened and privileged with the responsibility that choice endows. This is what free will represents for humankind at the beginning of the 21st century.

So, what plays a role in free will? Quantum mechanics must be involved to some extent. If the world were not quantum mechanical, then the future would be determined. The brain (as a complex, dynamical system) is important. Learning and delayed response are two other essential ingredients. But there are more! Let's take a detailed look.

According to professor Dr. Paul J. Bertics (Department of Biomolecular Chemistry, University of Wisconsin):

> If the brain is viewed as a collective of biochemical and physiological processes that are modulated by multiple levels of input, including external factors (e.g., environmental parameters), then as the external cues change with time, these in turn alter the likelihood of specific neurological events from occuring and learning

can be said to have occurred. However, if one also considers that each process is not absolutely defined at each point in the future (i.e., there is a statistical probability that something will occur not an absolute certainty that it will occur) then this scenario allows for biochemical profiles to arise that were unexpected or at least less liklely than other profiles. Hence, unexpected behavior or other neurological activities may appear that may be of value or a detriment to the survival of the individual. [3]

What Actually Plays a Role in Free Will

This part of the chapter discusses what plays a role in free will, the self, and the implications of naturalism for social policy and personal well-being. The first section of this part covers neuroscientists who are skeptical about traditional contra-causal free will; the second section covers some who think that although humans don't have such free will, they should still encourage a belief in it; and, the third covers those who defend traditional accounts of human agency against science and naturalism. What follows in the first section should at least convince those who doubt the viability of naturalism, which suggests that humans don't have free will and that they'd be better off if they made their peace with this fact and adjusted their beliefs and social practices accordingly. In other words, let's very briefly look at neuroscientists who are skeptical about traditional contra-causal free will, and how some explore the personal and social implications of this denial.

The Skeptics

Medical diagnosis was partially responsible for the invention of brain imaging. But its far greater importance is that it may very well confirm, in ways too precise to be disputed, certain theories about the mind, the self, the soul, and free will that are already devoutly believed in by scholars in what is now the hottest field in the academic world: neuroscience!

Neuroscience, the science of the brain and the central nervous system, is on the threshold of a unified theory that will have an impact as powerful as that of Darwinism a hundred years ago. Already, there is a new Darwin, or perhaps an updated Darwin.

According to neuroscientists, every human brain is born not as a blank tablet (a tabula rasa) waiting to be filled in by experience but as an exposed negative waiting to be slipped into developer fluid. You can develop the negative well or you can develop it poorly, but either way you are going to get precious little that is not already imprinted on the film. The print is the individual's genetic history, over thousands of years of evolution, and there is not much anybody can do about it. Furthermore, according to neuroscientists, genetics determine not only things such as temperament, role preferences, emotional responses, and levels of aggression, but also many of your most revered moral choices. These are not choices at all in any free will sense, but tendencies imprinted in the hypothalamus and limbic regions of the brain.

Neuroscience's Strategic High Ground

Here, one can begin to sense the chill that emanates from the hottest field in the academic world. The unspoken and largely unconscious premise of the wrangling over neuroscience's strategic high ground is: You now live in an age in which science is a court from which there is no appeal. And the issue this time around, at the beginning of the 21st century, is not the evolution of the species, which can seem a remote business, but the nature of our own precious inner self.

Neuroscientists are well aware of all this and are cautious—or cautious compared to the new generation. In other words, neuroscientists still hold out the possibility that at some point in evolutionary history, culture began to influence the development of the human brain in ways that cannot be explained by strict Darwinian theory. But the new generation of neuroscientists are not cautious for a second. In private conversations, the bull sessions, as it were, that create the mental atmosphere of any hot new science, neuroscientists express an uncompromising determinism.

Neuroscientists start with the most famous statement in all of modern philosophy, Descartes's Cogito ergo sum (I think, therefore I am), which they regard as the essence of dualism, the old-fashioned notion that the mind is something distinct from its mechanism, the brain and the body. This is also known as the ghost in the machine fallacy, the quaint belief that there is a ghostly self somewhere inside the brain that interprets and directs its operations. Neuroscientists involved in three-dimensional electroencephalography will tell you that there is not even any one place in the brain where consciousness or self-consciousness (Cogito ergo sum) is located. This is merely an illusion created by a medley of neurological systems acting in concert. The young generation takes this yet one step further. Since consciousness and thought are entirely physical products of your brain and nervous system (and since your brain arrived fully imprinted at birth), what makes you think you have free will? Where is it going to come from? What ghost, what mind, what self, what soul, what anything that will not be immediately grabbed by those scornful quotation marks, is going to bubble up your brain stem to give it to you? Neuroscientists have theorized that, given computers of sufficient power and sophistication, it would be possible to predict the course of any human being's life moment by moment, including the fact that the poor devil was about to shake his or her head over the very idea. It is rather doubtful that any Calvinist of the sixteenth century ever believed so completely in predestination as these rational young scientists do at the beginning of the 21st century.

According to Dr. Bertics,

> If one postulates that at any point in time an individual's thoughts and behavior are a consequence of the summation of all of the neurochemical processes and reactions that are occurring at that moment, and that these processes are influenced by both external input (environmental, nutritional, etc.) and by genetic factors, then it is unlikely that it would be possible to predict the course of any human being's life moment by moment, regardless of the computational power available. This philosophy is based on the observation that for each reaction, there is a statistical probability that it will occur at any given time point (some have a high probability and others a low probability). Thus, one can only rank order the probability that a given collection of reactions will occur at each time point—some patterns are very likely and some patterns are highly unlikely, but nonetheless one cannot uniquely or unequivocally determine the single set that will exist at each place on a timeline. If one also factors in that environmental events may alter the likelihood of specific reactions occurring, and that there is a large degree of randomness in the specific combination of environmental cues that will co-exist at each moment, then this further undermines the capacity to definitively state what neurochemical profile will occur at any time. [3]

Is Free Will a Necessary Fiction or Not?

According to some neuroscientists, free will doesn't exist. However, belief in it is necessary to provide essential support for morality, meaning, and the worth of human beings. On the other hand, some neuroscientists argue that you would be better off morally and existentially without believing the falsehood that you have free will. Moreover, having free will could motivate systematic deception about our causal connection to nature, which is neither possible, necessary, nor desirable. Making known the naturalistic truth about ourselves is a far better basis for human flourishing, because free will is not a necessary fiction.

It is thought that some ideas are inherently dangerous. Some pose a threat because they are false, and by misleading you, encourage destructive behavior. Others are dangerous because although true, they might have dire consequences for those who fall under their sway. An example of the former is the belief that human beings are not a significant contributor to global warming, and that therefore there's nothing you can or need do to avert a climate disaster over the next century (see Chapter 19 for a detailed discussion of the dangerous consequences of global warming). Believing such a falsehood will have manifestly destructive consequences for the planet.

Examples of what are considered to be the latter aren't hard to find either. A favorite conservative theme is that telling children the truth about this country's history (like that prostitution was widespread in mining towns during the gold rush) will undercut proper feelings of reverence and respect for the good old U.S. of A. So, rather than risk demoralizing the next generation, it's better to sanitize the past in carefully edited high school textbooks.

In order to protect your moral virtues, it's better to have the false belief that you have free will. Believing that you are ultimately responsible agents (god-like, miniature first causes who choose without being entirely determined to choose) is necessary to supply the requisite motivation to maintain a strong sense of ethical duty and responsibility. Moreover, the very possibility of finding meaning in our lives and truly valuing each other depends on not believing the causal, naturalistic truth about ourselves. Free will, therefore, is a necessary fiction, without which the entire social-psychological fabric of meaning and morality is at risk.

Some neuroscientists want to take exception to this claim and show that humans need not accept the argument from dire consequences. Instead, neuroscientists can safely make known the naturalistic facts about what it is to be human, while keeping and indeed improving the moral commitments and practices. Humans don't need to be misled about their true nature, since a robust morality and quest for meaning are perfectly compatible with naturalism. Once you see that your moral inclinations stem from your shared biological human endowment, honed by culture, you don't need the notion of ultimate responsibility or self-creation exnihilo to back up your sense of ethical duty. In fact, the very belief in ultimate agency arguably creates attitudes and beliefs that reinforce some of the worst human tendencies (for retribution and self-aggrandizement). Humans would be morally better off without it. This is an argument from good consequences, which has the signal virtue of being aligned with the truth about the human situation as most philosophers and scientists see it.

The standard libertarian notion of strong, contra-causal free will, in which humans are in some important respect the first cause of their characters and actions, cannot be sustained in the light of logic or scientific evidence. The libertarian project was worthwhile in attempting to allow a deep moral connection between a given act and the person, and yet not fall into being merely an unfolding of the arbitrarily given, whether determined or random. But it is not possible to find a way in which this can be done.

The Natural Casual Order

Nevertheless, according to neuroscientists, compatibilism is "grimly insufficient," since no one can be "ultimately in control" or "ultimately responsible." The causal perspective given by naturalism shows that it's ultimately all luck, ultimately not one's fault or to one's credit what one becomes. True, but this fact doesn't impugn compatibilism, which can still distinguish between those who are and are not moral agents, and which can justify adequate responses to criminality and moral infractions, all while accepting a naturalistic universal causality.

NOTE

Compatibilism has to do with a person who wholeheartedly accepts the full inclusion of the individual in the causal network, but who nevertheless sees no incompatibility with this and traditional ascriptions of responsibility, with their full complement of punitive implications. This could also be called hard compatibilism: hard, because while it sees causal explanations and ascriptions of responsibility as compatible, it countenances no softening of your reactive dispositions (the desire to retaliate) that you ordinarily think might follow when you discover the full causal story behind behavior. Hard compatibilism is opposed to what seems to be the widely shared, natural inclination to moderate retributive responses (or effusive, fawning praise) directed solely at the individual when it is clearly seen that the causes for behavior lie in external circumstances, which if they had been different, would have made the behavior less likely.

The insufficiency neuroscientists find with compatibilism (and its naturalistic basis) is a matter of judgments made from the libertarian perspective. It isn't the insufficiency of a thoughtfully worked-out compatibilist position on moral responsibility. If someone is punished on compatibilist grounds, for instance to deter, incapacitate, or rehabilitate, this is unjust since the person isn't ultimately responsible for who they are and their actions. It's not really his or her fault. But why, if neuroscientists agree that the libertarian free will is incoherent, since no one could possibly meet the requirements of libertarian agency, can compatibilist notions of justice be judged by such a standard? Why should intuitions about fault or responsibility or control, derived from an impossible contra-causal conception of the person, be thought reasonable criteria for assessing the fairness of compatibilist justice? So, if they haven't been exposed to compatibilist, naturalistic versions of moral responsibility, such intuitions might be someone's knee-jerk response. But humans are not bound to use impossible supernaturalistic standards of agency to assess their moral practices, once the adequacy of naturalism for morality has been explained. [1]

Some neuroscientists claim that compatibilist grounds for moral worth and respect are shallower than libertarian grounds, but since the libertarian view is unreal, why should the compatibilist view be judged shallow in comparison? It's simply what's possible, given the reality of things. It may be shallow, compared to what humans are used to thinking, but once neuroscientists understand that libertarian free will and agency don't exist, then the charge of shallowness ceases to cut any ice against compatibilism, especially once neuroscientists see that compatibilism is perfectly adequate to ground nearly all human intuitions about the moral worth of persons. Libertarians and hard determinists may claim that "real responsibility" and "real agency" must involve contra-causal freedom. But once it is seen that such freedom is an impossibility (which hard determinists agree is the case), then the responsibility and moral agency based on it can't logically be characterized as real. After all, something real has the obligation to exist, or at least be a coherent possibility. While the libertarian agency does not meet these requirements, the compatibilist moral agency does. [1]

Some neuroscientists also charge compatibilism with a complacent compliance with the injustice of not acknowledging the lack of fairness and desert. But again, the notions of fairness and desert here derive from the discredited libertarian perspective. Since some neuroscientists agree that nothing could actually confer such fairness and desert, then it is unfair to find compatibilism lacking on these grounds. True, humans may hanker after such things, but having discovered they don't exist, humans can't fault the truth of the matter for contradicting their hankerings. Instead, humans should give them up. Neuroscientists are straddling two worlds: while admitting one is a fantasy, they render their judgment against compatibilism.

Similarly, the compatibilist practices are in one way unjust, owing to the absence of libertarian free will, which implies that your actions are, on the ultimate level, not up to you. But why should such a sense of injustice be given any weight once you see that it's simply not possible to have authored yourselves? It would only be unjust if the libertarian agency were a live possibility, but it isn't. Just practices reflect what's "up to you" in the compatibilist sense. This is a sense that is perfectly adequate to capture your intuitions about who can be held responsible and who can't. The question is, why do some neuroscientists continue to hold onto this sense of injustice, driven as it is by a definition of freedom that they acknowledge could never be fulfilled? The notion of justice based on libertarian free will simply isn't a fair criterion for assessing compatibilist grounds for their practices once neuroscientists see that it is impossible. [1]

The Presence of Naturalism

There is no reason to claim that the absence of libertarian free will is of no great moral significance. Moreover, to deny the fact that theft without libertarian free will is even a vicious and compatibilistically free criminal who is being punished, is in some important sense a victim of his or her circumstances. The absence of libertarian free will is, one might say, the presence of naturalism, since it is the naturalistic, scientific perspective that establishes the incoherence and empirical implausibility of libertarian, contra-causal, supernatural free will. Because it does just that, naturalism is of great moral significance.

First of all, humans are not victims of circumstance simply because circumstances create them. Humans can only be victims of circumstance in this global sense by supposing that they might conceivably have been masters of circumstance in some global sense (by having libertarian free will). But naturalism forces humans to abandon that perspective as wishful thinking, a traditional cultural myth, a false folk-metaphysics undone by science and logic. The person being punished for his or her crimes under a compatibilist regime is not a victim by virtue of naturalism or universal causality, so there is no intrinsic unfairness to such punishment. If the person punished had been insane, coerced, or in some other sense didn't have the capacity or opportunity to act responsibly, only then would unfairness accrue.

Based on the assumption of the libertarian agency, the moral significance of naturalism (or the absence of libertarian free will) isn't that it makes humans all victims, but that it undercuts retributive and fawning attitudes. Since humans don't create themselves, they do not deserve praise and blame in the ultimate, traditional sense of being the uncaused originators of their actions. Retributive rage and the urge to punish harshly (by imposing the death penalty, permitting rape in prison, or withholding opportunities for rehabilitation) are directly linked to the assumption that persons are self-caused and could have done otherwise in the situations that shaped them. Similarly, the acquiescence to extreme social and economic inequalities is at least partially a function of the belief in the self-made man: those who succeed deserve unlimited rewards since their success is a matter of their self-chosen drive and acumen, not a matter of external conditions. Likewise, those who fail deserve their impoverishment, since they chose not to succeed. If, under naturalism, humans dispense with the self-made self, whether criminal or exemplary, such attitudes lose their primary metaphysical basis, and so might wither in favor of more compassionate, cause-appreciating attitudes. By creating a more enlightened, less punitive culture, this in turn may have a beneficial effect on criminal justice and other social policies.

So, if we reflect upon the fact that many people undergo acute misery, is the fact that those people develop into criminals ultimately beyond their control? Compatibilists, perhaps more than libertarians, tend to be cognizant of the manifold causes that create individuals and societies. And it is precisely the deep appreciation of such causes that motivates less punitive attitudes and practices, which in turn can lessen the unnecessary toll of misery taken by, for instance, the criminal justice system. Thus, the practical moral significance of naturalism partially constitutes such insights.

Similarly, any favorable compatibilist appreciation of persons is necessarily shallow for, in the end, it rests upon factors not under the person's control. Any factor for which one is appreciated, praised, or even loved, is ultimately one's luck. That compatibilists are indifferent to such ultimate arbitrariness. Thus, shallowness and injustice are morally outrageous. The answer here is that compatibilist appreciation for persons isn't shallow, since the deeper sort of appreciation based on a person's capacity for self-choosing isn't possible, given that the capacity doesn't exist. Therefore, it isn't morally outrageous to accept the fact that your appreciation of people's virtues and faults is based on the existence of such characteristics, not on their being self-originated. Even though it may undermine certain attitudes based on traditional notions of agency, knowing that persons are ultimately caused in every respect doesn't erase the moral and aesthetic distinctions one draws between individuals.

Naturalism Risks

Because some neuroscientists think that compatibilism is grimly insufficient to ground our moral practices, humans must protect society against realizing the truth of what has been called "creeping mechanism"—that is, a fully inclusive naturalism that understands human agents as completely caused creatures. But remember that some neuroscientists think this insufficiency isn't a matter of justification, it's that they suppose compatibilism lacks the necessary persuasive power that only the myth of the libertarian agency can provide. Yet, while the justification for these values does not require libertarian free will, in practice they might be at risk were the lack of libertarian free will internalized. In other words, the great unwashed can never be brought safely to such a realization. Meanwhile, neuroscientists and others in the educated, professional classes might be able to understand that compatibilist justifications are all humans need to ground moral practices.

Humans often want a person to blame himself or herself, feel guilty, and even see that he or she deserves to be punished. Such a person is not likely to do all of this if he or she internalizes the ultimate hard determinist perspective. Of course, this is based on what's going on in the actual world, where nothing else could in fact have occurred. In other words, a person could not actually have done anything else except what he or she did do.

A person is unlikely to feel that his or her punishment is appropriate, regardless of whether he or she adopts a completely causal view of his or her behavior. This is an empirical claim needing research, as are all claims concerning the dangers of understanding the impossibility of libertarian free will (or, more positively, accepting naturalism). Naturalism inevitably demoralizes, no matter what sort of education about causality and compatibilism neuroscientists might provide. Such might be the case for some individuals, but the psychological responses and changes in attitudes that might be generated by encountering naturalism really are open questions, not to be prejudged without investigation. Once people see that universal causality does not constitute a blanket excuse—that is, that sufficient reasons exist apart from ultimate, retributive desert to impose sanctions on offenders (namely social protection, deterrence, rehabilitation, and restitution)—then they quickly understand that not everything is permitted under naturalism. They might indeed feel that their punishment is deserved in the compatibilist sense of being necessary for guiding good behavior and for achieving other agreed-upon social goals. In any case, neuroscientists need to actually find out what sorts of responses people have to naturalism, not imagine the worst and then restrict knowledge about it on that assumption. [1]

The sentiment supporting common intuitions of justice should be almost instinctive, according to neuroscientists. But of course it is instinctive; in fact it's so hard-wired that it's probable that ethical instincts drive the libertarian assumption of contra-causal agency, not the other way around. This suggests that humans don't need to protect such instincts from naturalism. Intuitions about justice and morality derive from deeply embedded emotional responses to injury, threat, cooperation, and mutual aid that most likely conferred survival advantages upon the individuals (and groups, perhaps) who manifested them. This is why it is highly unlikely that naturalism will undermine such intuitions, although it may moderate retributive impulses commonly justified by free will. So, in order to prevent wholesale demoralization, despite some neuroscientists' claim to the contrary, it seems unlikely that humans need to continue inculcating the myth of a contra-causal, supernatural agency.

Some neuroscientists also claim that understanding that you don't have free will is an obvious danger to your moral motivation, but this is true only if compatibilist reasons for ethics aren't appreciated, which they can be quite easily. It also suggests that our values only function from the standpoint of wanting and imposing strong, ultimate credit and blame as defined by supernaturalistic and dualistic views of agency, when in fact they derive from your natural inclinations to cherish and protect. So, in order to function adequately as moral agents, humans don't need contra-causal freedom, nor need they believe in it. A strong parallel exists between neuroscientists' claim and the claim that human beings need to believe in God to be good. But just as atheists are perfectly capable of goodness, and indeed aren't handicapped by religious absolutism in achieving tolerance of other views, those who don't believe in free will are perfectly capable of being moral agents. Thorough-going naturalists may even have an advantage in achieving certain moral virtues, since they are perhaps less susceptible to the radical individualism and egoism generated by belief in libertarian agency.

NOTE

In cultures where the notion of free will is unheard of or is considerably muted (in some Eastern societies), moral codes nevertheless thrive apace.

According to neuroscientists, illusory beliefs are in place concerning free will and moral responsibility, and the role they play is largely positive. Humanity is fortunately deceived on the free will issue, and this seems to be a condition of civilized morality and personal value. The benefit neuroscientists see in maintaining the fiction of strong, originative free will is to sustain morality and meaning as they currently exist for perhaps the majority of citizens in Western society. But is the "civilized" status quo what neuroscientists really want, given the psychological and social pathologies generated by assigning ultimate credit and blame, and the excessive rewards and punishments justified by libertarian free will? Also, should neuroscientists continue to ignore the causes in social and personal dysfunction, violence, and crime, by continuing to believe that such evils ultimately derive from uncaused human choices? Furthermore, is the fiction of free will really necessary to rally support for perfectly viable compatibilist moral and criminal justice practices that can be justified on their own terms, are consistent with scientific understanding of the world, and avoid the punitive excesses driven by the belief in libertarian agency? And do humans need the assumption that they are self-caused choosers in order to perceive themselves and others as truly valuable? The answer to all these questions, according to some neuroscientists, is no! Instead, appreciating and accepting a naturalistic view of ourselves and your behavior, while adopting an explicitly compatibilist morality, far from demoralizing us, will be a vast improvement over the status quo, both for morals and meaning. [1]

Is Big Brother Watching?

Humans would be better off ignorant of naturalism. It would be better for morality and meaning if scientific findings showing human beings as caused creatures, were not widely disseminated or understood. In fact, if neuroscientists took illusionism seriously, it might well motivate an Orwellian project of doublethink, in which an increasingly manifest truth (our causal interconnectedness with nature) is perpetually denied.

With this in mind, to make a convincing case that you need the free will illusion, neuroscientists have to prove that people would indeed lose their moral compass, that under naturalism they would succumb to "pragmatic consequentialist temptations" (punishing the innocent), or be driven to "unprincipled nihilism." But, of course, it would take considerable empirical research to show that such horrors would necessarily follow the dissemination of naturalism, instead of the consequences that neuroscientists foresee. This would be a lessening of ego-driven, punitive, and fawning attitudes, and an increase in knowledge about the actual causes of crime and social and personal dysfunction that neuroscientists can use to reduce human suffering. The case for motivated obscurity regarding free will (that is, shielding people from naturalism), is not *prima facie.* This is because people can understand compatibilist grounds for the criminal justice practices, and they don't necessarily fall into a moral panic when they discover they've never had libertarian free will but still possess all their real causal powers.

In short, for an adequate moral and personal reality, it's not at all clear that humans need the illusion of free will. Free will is not a necessary fiction. Neuroscientists raise issues that suggest research is needed on beliefs about agency, responsibility, and the psychological consequences of being disabused about libertarian free will. But must humans remain deceived about the truth of their full causal connection to nature? Nothing suggests such deceit is necessary, possible, or desirable. The only and best way forward is in the light of our best collective understanding about who and what we ultimately are, which is delivered by continuing exploration of how we fit into the natural scheme of things.[1]

Believers

The problem of free will engages humans deeply because it seems central to their conception of who they are, their place in the world, and their moral intuitions. To take a position on whether humans have free will, and what sort of freedom this is, is to take positions on a host of other fundamental and necessarily interlocking issues. This consists of what humans ultimately consist of as selves, the relation of mind to body, the role of consciousness in behavior, the proper methods of scientific and phenomenological inquiry, the need for foundations in ethics, and the possibility of the supernatural, among many other questions. To define free will, or volition, and argue that this definition captures the truth of the matter, is to invoke an entire worldview, which must stand against its competitors.

Dr. Backonja explains:

> Free will and consciousness are two biological phenomena that were first recognized and described in philosophy and only recently, very recently, received the needed attention of science. Since free will and consciousness were in the domain of philosophy, we have to respect the richness of the philosophical tradition, but with advances of cognitive neurosciences we can readily find the origins of those two phenomena in biology. Both of them present a so-called inverse problem, common to scientific process, during which by knowing the manifestation of the phenomenon one tries to understand the underlying mechanisms, so by measuring the changes in the light spectrum one measures the type and distance of the light source. The primary constraint that makes these two subjects very difficult is the language; each of these terms is ascribed to a wide range of biological, psychosocial, and philosophical entities. Therefore to talk about any of them, one has to qualify which one it is the one is talking about.
>
> The issue that closely follows is the issue of context. If taken out of context and discussed in most abstract terms, then they can be anything. But for the scientific process they are nothing: a phrase without meaning, an empty shell. [2]

And competition there is, since worldviews have consequences in policy and politics. When U.S. cigarette companies, defending themselves in class action suits, claim that smokers could have chosen to quit at any time, they are appealing to a particular notion of will and its relation to behavior. As the Truth and Reconciliation Commission completed its work in South Africa, some have questioned whether justice is served by foregoing punishment in return for the true story of one's misdeeds. Shouldn't retribution against the freely willing despot trump the need for social reconstruction? Or, conversely, might not the retributive impulse be softened by understanding the causal histories of political criminals? Such questions inevitably get answered using background assumptions at the heart of the free will debate. [1]

As previously mentioned, the argument over free will is in a very real sense the successor, and to some extent the companion, of the centuries-old argument about God. Since the Enlightenment, Western theistic views, placing power and our ultimate fate in the hands of an almighty deity, have to some extent given way to a secularism in which human destiny is understood to lie largely within our own hands. Humans have usurped God's power, or at least a good share of it, but many continue to believe that the capacity for humans to shape themselves still supercedes the rest of nature and its physical laws. Free will (on the libertarian incompatibilist account), in which humans are ultimately responsible for themselves and their acts, makes the self more or less a first cause. In other words, it makes the self an unmoved mover: Humans could have willed otherwise in the radical sense that the will is not the explicable or predictable result of any set of conditions that held at the moment of choice. The question of whether humans actually have such free will thus recapitulates in the domain of human metaphysics the question of the existence of God. That many believe humans stand above nature in some essential respect suggests that the Enlightenment was more successful in its glorification of the individual than in its challenge to the supernatural. [1]

Libertarians tend to suppose that not only do humans have such freedom, but that without it, human moral intuitions and institutions are at risk. The ultimately responsible, self-originating self is thought necessary to ground ethical judgments and social justice, and a fully explanatory and inclusive science of human behavior threatens the status of such a self. Other such libertarians are pessimists regarding the compatibility of (current) science and personal responsibility. On the other hand, compatibilists and determinists, such as some neuroscientists, are optimists who cheerfully accept that the will is entirely a function of antecedent conditions, and find that no capacity for ultimate self-determination is either conceivable or necessary to found human moral intuitions.

Some neuroscientists believe that free will is an essential aspect of humanity that may eventually be confirmed by science, perhaps in radical departures from current theory. On the other side, neuroscientists are taking more or less the compatibilist position: Volition is no more (or less) than the brain in action. And, this sort of volition is sufficient for human moral purposes.

In any event, the notion of free will has ramifications for the most fundamental conceptions of self, agency, and responsibility, and to these humans are deeply attached. Humans do want credit; they do enjoy imposing just deserts; they can't, realistically, give up entirely those reactive attitudes such as gratefulness and resentment that seem to point to the instigatory self. As much as free will exposes humans to the threat of unlimited retaliation for wrong-doing, it nevertheless compensates them by making them the lords of their little domains, the micro-gods of their minds. By making humans finally responsible for themselves, contra-causal freedom pays them the ultimate compliment, even if sometimes it exacts the ultimate price.

There is the truth of matter about free will. It therefore must be approached with all due humility—and then there is the practical issue of whether the truth (in this particular case) is something humans are ready to face. Some neuroscientists have argued that freedom is directly perceived in the irrefutable experience of choice, but others take the truth about free will to be a matter of where a scientific understanding of themselves leads, and experience is simply data added to the mix. To decide between these two approaches is to decide between intuitionism and empiricism, or between personal modes of knowing, and collective and experimental modes. There is no final arbitration of this issue, except to point out that one can pitch intuitionist arguments successfully to those, like much of the public, that are prepared to buy them, but to convince scientists and (most) philosophers, you'd better have a theory linked to some institutional, peer-reviewed wisdom. So libertarians will have to come up with a plausible naturalistic model for an unmoved mover (not likely) if they want to get funded for research. And libertarians first have to change attitudes about what counts as evidence (not likely) if compatibilists want to change the public's conception of free will.

Libertarians too feel the empiricist pull, since they search science to confirm their interpretation of the experience of volition. They, just as much as compatibilists, want the satisfying unification of intuition and observation, the personal and the impersonal perspectives. However, as much as libertarians want validation by science, it will never be forthcoming since the very notion of the ultimately responsible self is inherently opposed to scientific objectives of explanation and prediction. In short, there's a conceptual conflict at the deepest level that blocks this sort of cognitive unification for libertarians.

Thus, the commonsense concept of personal liberty should evolve to become compatible with determinism. Whatever the truth about free will may be, too much is at stake in terms of human social and legal institutions, their self-esteem, and their personal power and creativity to relinquish this most central of assumptions: that humans alone choose themselves and their futures. Better to finesse the science indefinitely by clinging to the straw that just because there's no good evidence for free will, that doesn't constitute proof that it doesn't exist. So, why not persist in human libertarian convictions, since science can't ever pronounce the definitive death of ultimate freedom?[1]

Mere absence of evidence is perhaps the least persuasive reason for those convinced that the *causa sui* self simply must exist. But there are many other reasons: First, the fear of mechanism must be diffused by showing that a causally embedded self can be a moral agent, responsive to the value-reinforcing effects of social sanctions and rewards (indeed, only such a self is fully responsive; a *causa sui* self is, by definition, not). Responsibility doesn't have to be ultimate to justify praise and blame, which means a causal understanding of voluntary acts isn't tantamount to excusing them, as is often supposed. Nor does the fair application of sanctions need a basis beyond the fact that behavior is indeed modified by its consequences. Because they are not the ones needing correction, punishing others for your sins is unjust. Also, because the punishment fits no crime of theirs that needs extinguishing in the Skinnerian sense, their suffering is undeserved (nonfunctional and needless).

Compatibilists must show that determinism is no threat to human efficacy and creativity beyond the moral issues. Their personal power derives not from having some mysterious causal priority over circumstances, but from exploiting the causal context of which humans are so inextricably a part. Being ultimate self-choosers would merely tie humans in knots as they tried to discover some basis for taking that first step towards self-definition. Being proximate self-choosers, on the other hand, is all humans need to fulfill the desires of nature and nurture bequeath them in such variety.

There also may well be benefits flowing from a thoroughly naturalistic conception of the self and its choices. The retributive impulse, cut off from its metaphysical justification in free will, might soften, leading to a less punitive culture. More attention might be paid to improving the social conditions shaping individuals, and no longer will policy makers so blithely blame the victim (remember the multitudes in the U.S. who chose to be homeless during the 1980s?). On the personal level, dethroning the supervisory "I" might help us become less self-conscious, more playful, and less likely to wallow in excessive self-blame, pride, envy, or resentment. [1]

Finally, might we become less ambitious, once we see that we don't ultimately choose ourselves or our projects, and that our successes (and failures) result from thousands of combining circumstances? Perhaps, but this might be all to the good, given that the unfettered accumulation of wealth seems likely to compromise the long-term sustainability of resources, or at least concentrate them in a very few hands. And, after all, you need not worry that putting the self in its natural, causal context will extinguish desire, any more than you need worry that it will undermine your rights and liberties. Ourselves, physically embodied, are virtually constituted by desire, and real freedom lies in having the opportunity to pursue our motives as we discover them arising in us. Ultimate responsibility for oneself may well lead to the more responsible use of such freedom, by seeing that the self neither has, nor needs.

Conclusion

So far, all scientific evidence suggests that the atoms comprising human beings are not exempt from the laws of physics. Since physical systems are deterministic, with the exception of quantum-mechanical considerations (since human behavior seems to be determined by desires and reasoning with the exception that humans feel that humans are free to choose their actions), quantum mechanics has been seen by some (by process of elimination) to be a window to the willful conscious mind. This chapter has very briefly examined the various arguments that have been made for and against quantum mechanical explanations of free will, whether they are necessary, and what implications these arguments have for responsibility.

Finally, quantum indeterminacy is little more than a red herring, a way to avoid what humans see as an unacceptable conclusion that their free will is an illusion. In a sense, a person is only free when viewed from the standpoint of his or her own consciousness. From the standpoint of the rest of humankind, the only evidence that she or he has free will is his or her own reporting that he or she has it.

References

[1] Timothy O'Connor, "Free Will," *Stanford Encyclopedia of Philosophy*, The Metaphysics Research Lab, Center for the Study of Language and Information, Ventura Hall, Stanford University, Stanford, CA 94305-4115, 2002.

[2] Dr. Miroslav Backonja, Neurology, Madison Wisconsin, Department of Neurology, H6/570 University of Wisconsin Hospital, 600 Highland Avenue, Madison, WI 53792.

[3] Dr. Paul J. Bertics, Professor, Department of Biomolecular Chemistry, 571 Medical Sciences Bldg, UW Medical School, University of Wisconsin,1300 University Avenue, Madison, WI 53706, USA.

Consciousness

"Dignity consists not in possessing honors, but in the consciousness that we deserve them."
—Aristotle (384–322 BC)

The most baffling problem in the science of the mind is consciousness. There is nothing that neuroscientists know more intimately than conscious experience, but there is nothing that is harder to explain. All sorts of mental phenomena have yielded to scientific investigation in recent years, but consciousness has stubbornly resisted. Many have tried to explain it, but the explanations always seem to fall short of the target. Some have been led to suppose that no good explanation can be given and that the problem is intractable.

Neuroscientists have to confront consciousness directly in order to make progress on the problem. This chapter isolates the truly hard part of the problem, separating it from more tractable parts and giving an account of why it is so difficult to explain. For example, some recent work has used reductive methods to address consciousness. However, some neuroscientists argue that such methods inevitably fail to come to grips with the hardest part of problem. Once this failure is recognized, the door to further progress is opened. Later in chapter, a discussion will ensue about a new kind of nonreductive explanation: how a naturalistic account of consciousness can be given. A discussion will also ensue about a nonreductive theory based on principles of structural coherence, organizational invariance, and a double-aspect view of information.

The Problems

Consciousness doesn't consist of just one problem. Consciousness is an ambiguous term, referring to many different phenomena. Each of these phenomena needs to be explained, but some are easier to explain than others. At the start, it is useful to divide the associated problems of consciousness into hard and easy problems. The easy problems of consciousness are those that seem directly susceptible to the standard methods of cognitive science, whereby a phenomenon is explained in terms of computational or neutral mechanisms. The hard problems are those that seem to resist those methods. The easy problems of consciousness include those of explaining the following phenomena:

- The ability to discriminate, categorize, and react to environmental stimuli
- The ability of a system to access its own internal states
- The deliberate control of behavior
- The difference between wakefulness and sleep
- The focus of attention
- The integration of information by a cognitive system
- The reportability of mental states

Neurologist Miroslav Backonja (Department of Neurology, University of Wisconsin Hospital) explains:

> Like with everything in science and math, the starting point is everything; it determines were we go and how far we can go. The same refers to consciousness.

> Consciousness is a fundamental characteristic of every living organism, and it is the ability of the organism to distinguish itself from the surrounding environment and react to it. As such, organisms have to have the ability to monitor their own status, starting with homeostasis all the way to the abilities to recruit their reactions based on past "experiences." Complexity of the system that subserves this function increases with the complexity of the phylogenetic development of the organism, culminating in the human conscious experience, and all of that is the function of evolutionary development.
>
> Human conscious experience is characterized and distinguished from other organisms by its complexity of neural bases to which it evolved to and the ability to discuss itself. The hard problem/question could be then phrased: If human conscious experience is the product of primarily electrochemical brain activity (as are the rest of the human cognitive functions), what is the form of consciousness of the lower organisms? Chemical, and which chemical? Electrical?" [1]

All of these phenomena are associated with the notion of consciousness. For example, neuroscientists sometimes say that a mental state is conscious when it is verbally reportable, or when it is internally accessible. Sometimes a system is said to be conscious of some information when it has the ability to react on the basis of that information, or, more strongly, when it attends to that information, or when it can integrate that information and exploit it in the sophisticated control of behavior. Neuroscientists sometimes say that an action is conscious precisely when it is deliberate. They also say that an organism is awake by saying that it is conscious.

Whether these phenomena can be explained scientifically is not the issue here. All of them are straightforwardly vulnerable to explanation in terms of computational or neural mechanisms. To explain access and reportability, for example, neuroscientists need only to specify the mechanism by which information about internal states is retrieved and made available for verbal report. To explain the integration of information, neuroscientists need only exhibit mechanisms by which information is brought together and exploited by later processes. For an account of sleep and wakefulness, all appropriate neurophysiological accounts of the processes responsible for organisms' contrasting behavior in those states will suffice. An appropriate cognitive or neurophysiological model can clearly do the explanatory work.

Consciousness would not be much of a problem if these phenomena were all there was to it. Although neuroscientists do not yet have anything close to a complete explanation of these phenomena, they have a clear idea of how they might go about explaining them. This is why neuroscientists call these problems the easy problems. Of course, easy is a relative term. Getting the details right will probably take a century or two of difficult empirical work. Still, the methods of cognitive science and neuroscience will succeed. There is every reason to believe that.

The problem of experience is really the hard problem of consciousness. When neuroscientists think and perceive, there is a whir of information processing, but there is also a subjective aspect. In other words, there is something it is like to be a conscious organism. This subjective aspect is experience. When neuroscientists see, for example, they experience visual sensations: the felt quality of redness, the experience of dark and light, the quality of depth in a visual field. Other experiences go along with perception in different modalities: the sound of a clarinet, the smell of mothballs. Then there are bodily sensations, from pains to orgasms; mental images that are conjured up internally; the felt quality of emotion; and the experience of a stream of conscious thought. What unites all of these states is that there is something it is like to be in them. All of them are states of experience. [2]

Some organisms are subjects of experience, but the question of how it is that the systems are subjects of experience is perplexing. Why is it that when your cognitive systems engage in visual and auditory information processing, you have visual or auditory experience: the quality of deep blue, the sensation of middle C? How can neuroscientists explain why you are able to entertain a mental image, or to experience an emotion? It is widely agreed that experience arises from a physical basis, but they have no good explanation of why and how it so arises. Why should physical processing give rise to rich inner life at all? It seems objectively unreasonable that it should, and yet it does. [2]

If any problem that qualifies as the problem of consciousness, it is this one. In this central sense of consciousness, an organism is conscious if there is something it is like to be that organism, and a mental state is conscious if there is something it is like to be in that state. Sometimes terms such as phenomenal consciousness and qualia are also used here, but neuroscientists find it more natural to speak of conscious experience or simply experience. Another useful way to avoid confusion is to reserve the term consciousness for the phenomenon of experience, using the less loaded term awareness for the more straightforward phenomenon described earlier. Communication would be much easier if such a convention were widely adopted. Those who talk about consciousness are frequently talking past each other, as things stand.

Both philosophers and scientists who are writing on consciousness often exploit the ambiguity of the term. It is common to see literature on consciousness begin with an invocation of the mystery of consciousness, noting the strange intangibility and ineffability of subjectivity, and worrying that so far neuroscientists have no theory of the phenomenon. Here, the topic is clearly the hard problem—the problem of experience. Later in the chapter, the tone becomes more optimistic. Upon examination, this tone turns out to be a theory of one of the more straightforward phenomena—of reportability, of introspective access, or whatever. Toward the end of the chapter, a discussion ensues about how consciousness has turned out to be tractable after all, but you the reader might be left feeling like a victim of a bait-and-switch. The hard problem remains untouched.

Cognitive Abilities and Functions

Why is the hard problem hard, and the easy problems easy? The easy problems are easy precisely because they concern the explanation of cognitive abilities and functions. To explain a cognitive function, neuroscientists need only specify a mechanism that can perform the function. The methods of cognitive science are well suited for this sort of explanation, and so are well suited to the easy problems of consciousness. By contrast, the hard problem is hard precisely because it is not a problem about the performance of functions. So, even when the performance of all the relevant functions is explained, the problem still persists. [2]

NOTE

Here function is not used in the narrow teleological sense of something that a system is designed to do, but in the broader sense of any causal role in the production of behavior that a system might perform.

For instance, to explain reportability is just to explain how a system could perform the function of producing reports on internal states. To explain internal access, neuroscientists need to explain how a system could be appropriately affected by its internal states and use information about those states in directing later processes. To explain integration and control, neuroscientists need to explain how a system's central processes can bring information contents together and use them in the facilitation of various behaviors. These are all problems about the explanation of functions. [2]

So, how do neuroscientists explain the performance of a function? By specifying a mechanism that performs the function. Here, neurophysiological and cognitive modeling are perfect for the task. If neuroscientists want a detailed low-level explanation, they can specify the neural mechanism that is responsible for the function. If neuroscientists want a more abstract explanation, they can specify a mechanism in computational terms. Either way, a full and satisfying explanation will result. For example, the bulk of their work in explaining reportability is over once neuroscientists have specified the neural or computational mechanism that performs the function of verbal report.

The point is trivial in a way. It is a conceptual fact about these phenomena that their explanation only involves the explanation of various functions, as the phenomena are functionally definable. All it means for reportability to be instantiated in a system is that the system has the capacity for verbal reports of the internal information. All it means for a system to be awake is for it to be appropriately receptive to information from the environment and for it to be able to use this information in directing behavior in an appropriate way.

NOTE

To see that this sort of thing is a conceptual fact, someone who says you have explained the performance of the verbal report function, but you have not explained reportability, is making a trivial conceptual mistake about reportability. All it could possibly take to explain reportability is an explanation of how the relevant function is performed. The same goes for the other phenomena questions.

Reductive explanation works in just this way throughout the higher-level sciences. To explain the gene, for instance, neuroscientists needed to specify the mechanism that stores and transmits hereditary information from one generation to the next. It turns out that DNA performs this function. Once neuroscientists explain how the function is performed, they have explained the gene. To explain life, neuroscientists ultimately need to explain how a system can reproduce, adapt to its environment, metabolize, and so on. All of these are questions about the performance of functions, and so are well suited to reductive explanation. The same holds for most problems in cognitive science. To explain learning, neuroscientists need to explain the way in which a system's behavioral capacities are modified in light of environmental information, and the way in which information can be brought to bear in adapting a system's actions to its environment. If neuroscientists show how a neural or computational mechanism does the job, they have explained learning. Neuroscientists can say the same for other cognitive phenomena, such as perception, memory, and language. Sometimes the relevant functions need to be characterized quite subtly, but it is clear that insofar as cognitive science explains these phenomena at all, it does so by explaining the performance of functions.

This sort of explanation fails when it comes to conscious experience. What makes the hard problem hard and almost unique is that it goes beyond problems about the performance of functions.

NOTE

Even when neuroscientists have explained the performance of all the cognitive and behavioral functions in the vicinity of experience (perceptual discrimination, categorization, internal access, verbal report), there may still remain a further unanswered question: Why is the performance of these functions accompanied by experience? This question is left open by a simple explanation of the functions.

In the explanation of genes, or of life, or of learning, there are no analogous or further questions. If someone says, "I can see that you have explained how DNA stores and transmits hereditary information, one generation to the next, but you have not explained how it is a gene," then they are making a conceptual mistake. All it means to be a gene is to be an entity that performs the relevant storage and transmission function. But if someone says "I can see that you have explained how information is discriminated, integrated, and reported, but you have not explained how it is experienced," they are not making a conceptual mistake. This is a nontrivial further question. [2]

With regards to the problem of consciousness, that is the key question. Why doesn't all this information-processing go on in the dark, free of any inner feel? Why is it that when electromagnetic waveforms impinge on a retina and are discriminated and categorized by a visual system, this discrimination and categorization is experienced as a sensation of vivid red? Neuroscientists know that conscious experience does arise when these functions are performed, but the very fact that it arises is the central mystery. There is an explanatory gap between the functions and experience, and neuroscientists need an explanatory bridge to cross it. The materials for the bridge must be found elsewhere, because a mere account of the functions stays on one side of the gap. [2]

This is not to say that experience has no function. Perhaps it will turn out to play an important cognitive role. But, for any role it might play, there will be more to the explanation of experience than a simple explanation of the function. Perhaps it will even turn out that in the course of explaining a function, neuroscientists will be led to the key insight that allows an explanation of experience. If this happens, though, the discovery will be an extra explanatory reward. There is no cognitive function such that neuroscientists can say in advance that explanation of that function will automatically explain experience.

Neuroscientists need a new approach to explain experience. The usual explanatory methods of cognitive science and neuroscience do not suffice. These methods have been developed precisely to explain the performance of cognitive functions, and they do a good job of it. But, as these methods stand, they are only equipped to explain the performance of functions. The standard approach has nothing to say when it comes to the hard problem.

The Conscious Experience

In order to account for conscious experience, neuroscientists have seen that there are systematic reasons why the usual methods of cognitive science and neuroscience fail. These are simply the wrong sort of methods: Nothing that they give to neuroscientists can yield an explanation. To account for conscious experience, neuroscientists need an extra ingredient in the explanation. This makes for a challenge to those who are serious about the hard problem of consciousness. What is your extra ingredient, and why should that account for conscious experience? [2]

There is no shortage of extra ingredients to be had. Some propose an injection of chaos and nonlinear dynamics. Some think that the key lies in nonalgorithmic processing. Some appeal to future discoveries in neurophysiology. Some suppose that the key to the mystery will lie at the level of quantum mechanics. (See sidebar, "Human Consciousness Shapes Not Only the Present but the Past As Well".) It is easy to see why all these suggestions are put forward. None of the old methods work, so the solutions must lie with something new. Unfortunately, these suggestions all suffer from the same old problems.

Human Consciousness Shapes not Only the Present but the Past as Well

Why does the universe exist? The quest for an answer to that question inevitably entails wrestling with the implications of one of the strangest aspects of modern physics: According to the rules of quantum mechanics, neuroscientists' observations influence the universe at the most fundamental levels. The boundary between an objective world out there and your own subjective consciousness that seemed so clearly defined in physics before the eerie discoveries of the 20th century blurs in quantum mechanics. When physicists look at the basic constituents of reality—atoms and their innards, or the particles of light called photons, what they see depends on how they have set up their experiment. A physicist's observations determine whether an atom behaves like a fluid wave or a hard particle, or which path it follows in traveling from one point to another. From the quantum perspective the universe is an extremely interactive place. [3]

The universe is built like an enormous feedback loop, a loop in which you contribute to the ongoing creation of not just the present and the future, but the past as well. It illustrates a key principle of quantum mechanics: Light has a dual nature. Sometimes light behaves like a compact particle, a photon. Sometimes it seems to behave like a wave spread out in space, just like the ripples in a pond. In the experiment, light (a stream of photons) shines through two parallel slits and hits a strip of photographic film behind the slits. The experiment can be run two ways: with photon detectors right beside each slit that allow physicists to observe the photons as they pass, or with detectors removed, which allows the photons to travel unobserved. When physicists use the photon detectors, the result is unsurprising: Every photon is observed to pass through one slit or the other. The photons, in other words, act like particles (see Chapter 9 for a detailed discussion of the light experiment).

However, something weird occurs when the photon detectors are removed. One would expect to see two distinct clusters of dots on the film, corresponding to where individual photons hit after randomly passing through one slit or the other. Instead, a pattern of alternating light and dark stripes appears. Such a pattern could be produced only if the photons are behaving like waves, with each individual photon spreading out and surging against both slits at once, like a breaker hitting a jetty. The film shows where crests from those waves overlap, through alternating bright stripes in the pattern. Dark stripes indicate that a crest and a trough have canceled each other out.

So, what the physicists try to measure is what the outcome of the experiment depends on: If they set up detectors beside the slits, the photons act like ordinary particles, always traversing one route or the other, not both at the same time. In that case the striped pattern doesn't appear on the film. But, if the physicists remove the detectors, each photon seems to travel both routes simultaneously like a tiny wave, producing the striped pattern.

Where the classic experiment demonstrates that physicists' observations determine the behavior of a photon in the present, a neuroscientist's observation in the present can affect how a photon behaved in the past. For example, imagine a quasar—a very luminous and very remote young galaxy. Now imagine that there are two other large galaxies between Earth and the quasar. The gravity from massive objects like galaxies can bend light, just as conventional glass lenses do. In experiments conducted by scientists, the two huge galaxies substitute for the pair of slits. The quasar is also the light source. Just as in the two-slit experiment, light—photons—from the quasar can follow two different paths, past one galaxy or the other.

Suppose that on Earth, some astronomers decide to observe the quasars. In this case, a telescope plays the role of the photon detector in the two-slit experiment. If the astronomers point a telescope in the direction of one of the two intervening galaxies, they will see photons from the quasar that were deflected by that galaxy. They also would get the same result by looking at the other galaxy. But the astronomers could also mimic the second part of the two-slit experiment. By carefully arranging mirrors, they could make photons arriving from the routes around both galaxies strike a piece of photographic film simultaneously. Identical to the pattern found when photons pass through the two slits, alternating light and dark bands appear on the film.

Here's the odd part. The quasar could be very distant from Earth, with light so faint that its photons hit the piece of film only one at a time. But the results of the experiment wouldn't change. The striped pattern would still show up, meaning that a lone photon not observed by the telescope traveled both paths toward Earth, even if those paths were separated by many light-years. And that's not all.

The photon could have already journeyed for billions of years, long before life appeared on Earth, by the time the astronomers decide which measurement to make (whether to pin down the photon to one definite route or to have it follow both paths simultaneously). The measurements made *now* determine the photon's past. In one case the astronomers create a past in which a photon took both possible routes from the quasar to Earth. Even though the photon began its jaunt long before any detectors existed, alternatively, physicists can retroactively force the photon onto one straight trail toward their detector.

In 1984, physicists at the University of Maryland set up a tabletop version of the delayed-choice scenario. Even though those measurements were made after the photons had already left the light source and begun their circuit through the course of mirrors (using a light source and an arrangement of mirrors to provide a number of possible photon routes), the physicists were able to show that the paths the photons took were not fixed until they made their measurements.

Humans are part of a universe that is a work in progress. They are tiny patches of the universe looking at itself and building itself. It's not only the future that is still undetermined, but the past as well, and physicists' present observations select one out of many possible quantum histories for the universe by peering back into time, even all the way back to the Big Bang.

Does this mean humans are necessary to the existence of the universe? While conscious observers certainly partake in the creation of the participatory universe, they are not the only, or even primary, way by which quantum potentials become real. Ordinary matter and radiation play the dominant roles. Most physicists like to use the example of a high-energy particle released by a radioactive element like radium in Earth's crust. The particle, as with the photons in the two-slit experiment, exists in many possible states at once, traveling in every possible direction, not quite real and solid until it interacts with something, say a piece of mica in Earth's crust. When that happens, one of those many different probable outcomes becomes real. In this case the mica, not a conscious being, is the object that transforms what might happen into what does happen. The trail of disrupted atoms left in the mica by the high-energy particle becomes part of the real world. [3]

The entire universe is filled with such events, at every moment. This is where the infinite variety inherent in quantum mechanics manifests as a physical cosmos, and where the possible outcomes of countless interactions become real. And physicists see only a tiny portion of that cosmos. Most of the universe consists of huge clouds of uncertainty that have not yet interacted either with a conscious observer or even with some lump of inanimate matter. In other words, where the past is not yet fixed, the universe is a vast arena that contains realms.

This is a mind-stretching idea. It's not even really a theory, but more of an intuition about what a final theory of everything might be like. It's a tenuous lead, a clue that the mystery of creation may lie not in the distant past, but in the living present. This point of view is what gives neuroscientists hope that the question, "How come existence?" can be answered.

How Come Existence?

The problem of existence goes back to the earliest days of quantum mechanics and was formulated most famously by the Austrian physicist Erwin Schrodinger, who imagined a Rube Goldberg-type of quantum experiment with a cat. Put a cat in a closed box, along with a vial of poison gas, a piece of uranium, and a Geiger counter hooked up to a hammer suspended above the gas vial. During the course of the experiment, the radioactive uranium may or may not emit a particle. If the particle is released, the Geiger counter will detect it and send a signal to a mechanism controlling the hammer, which will strike the vial and release the gas, killing the cat. If the particle is not released, the cat will live. Schrodinger asked, "Before opening the box, what could be known about the cat?" [3]

The answer would be simple, if there were no such thing as quantum mechanics: Depending of course, on whether a particle hit the Geiger counter, the cat is either alive or dead. But in the quantum world, things are not so straightforward. The particle and the cat now form a quantum system consisting of all possible outcomes of the experiment. One outcome includes a dead cat; another, a live one. Neither becomes real until someone opens the box and looks inside. With that observation, an entire consistent sequence of events (the particle jettisoned from the uranium, the release of the poison gas, the cat's death) at once becomes real, giving the appearance of something that has taken weeks to transpire. Some physicists believe this quantum paradox gets to the heart of nature in the universe: The principles of quantum mechanics dictate severe limits on the certainty of your knowledge.

Before you start looking at it, you may ask whether the universe really existed. That's the same Schrodinger cat question. And a physicist's answer would be that the universe looks as if it existed before you started looking at it. When you open the cat's box after a week, you're going to find either a live cat or a smelly piece of meat. You can say that the cat looks as if it were dead or as if it were alive during the whole week. Likewise, when physicists look at the universe, the best they can say is that it looks as if it were there 10 billion years ago.

The conscious observers are an essential component of the universe and cannot be replaced by inanimate objects. The universe and the observer exist as a pair. You can say that the universe is there only when there is an observer who can say, "Yes, I see the universe there." These small words (it *looks* like it was here) for practical purposes may not matter much, but for you as a human being, you do not know any sense in which you could claim that the universe is here in the absence of observers. You are together, the universe and you. The moment you say that the universe exists without any observers, physicists cannot make any sense out of that. They cannot imagine a consistent theory of everything that ignores consciousness. A recording device cannot play the role of an observer, because who will read what is written on this recording device? In order for physicists to see that something happens, and say to one another that something happens, you need to have a universe, you need to have a recording device, and you need to have physicists. It's not enough for the information to be stored somewhere, completely inaccessible to anybody. It's necessary for somebody to look at it. You need an observer who looks at the universe. Our universe is dead in the absence of observers. [3]

So, how come existence? Will this question ever be answered? Is human intelligence capable of answering that question? Neuroscientists don't expect dogs or ants to be able to figure out everything about the universe. And, in the sweep of evolution, neuroscientists doubt that humans are the last word in intelligence. There might be higher levels later. So why should neuroscientists think humans are at the point where they can understand everything? At the same time, neuroscientists think it's great to ask the question and see how far you can go before you bump into a wall. [3]

You know, if you say that humans are smart enough to figure everything out, that is a very arrogant thought. If you say that humans are not smart enough, that is a very humiliating thought.

So, how come existence? If the question was meaningless, it did not stop neuroscientists from asking it. [3]

Will neuroscientists ever understand why the universe came into being? Or, at least how? Absolutely![3]

Because of the role it might play in the process of conscious mathematical insight, nonalgorithmic processing, for example, is put forward by neuroscientists. The arguments about mathematics are controversial, but even if they succeed and an account of nonalgorithmic processing in the human brain is given, it will still only be an account of the functions involved in mathematical reasoning and the like. For a nonalgorithmic process as much as an algorithmic process, the question is left unanswered: Why should this process give rise to experience? There is no special role for nonalgorithmic processing in answering this question. [2]

So, for nonlinear and chaotic dynamics, the same goes. These might provide a novel account of the dynamics of cognitive functioning, quite different from that given by standard methods in cognitive science. But from dynamics, one only gets more dynamics. The question about experience here is as mysterious as ever. The point is even clearer for new discoveries in neurophysiology. These new discoveries may help neuroscientists make significant progress in understanding brain function, but for any neural process neuroscientists isolate, the same question will always arise. It is difficult to imagine what a proponent of new neurophysiology expects to happen, over and above the explanation of further cognitive functions. It is not as if neuroscientists will suddenly discover a phenomenal glow inside a neuron [2]!

Quantum mechanics is perhaps the most popular extra ingredient of all. The attractiveness of quantum theories of consciousness may stem from a Law of Minimization of Mystery: Consciousness is mysterious and quantum mechanics is mysterious, so maybe the two mysteries have a common source. Nevertheless, quantum theories of consciousness suffer from the same difficulties as neural or computational theories. Quantum phenomena have some remarkable functional properties, such as nondeterminism and nonlocality. It is natural to speculate that these properties may play some role in the explanation of cognitive functions, such as random choice and the integration of information, and this hypothesis cannot be ruled out *a priori*. But, when it comes to the explanation of experience, quantum processes are in the same boat as any other. Entirely unanswered is the question of why these processes should give rise to experience. [2]

NOTE

One special attraction of quantum theories is the fact that on some interpretations of quantum mechanics, consciousness plays an active role in collapsing the quantum wave function. Such interpretations are controversial, but in any case they offer no hope of explaining consciousness in terms of quantum processes. Rather, these theories assume the existence of consciousness, and use it in the explanation of quantum processes. At best, these theories tell neuroscientists something about a physical role that consciousness may play. They tell neuroscientists nothing about how it arises.

The same criticism applies to any purely physical account of consciousness at the end of the day. For any physical process, neuroscientists specify there will be an unanswered question: Why should this process give rise to experience? Given any such process, it is conceptually coherent that it could be instantiated in the absence of experience. It follows that no mere account of the physical process will tell neuroscientists why experience arises. The emergence of experience goes beyond what can be derived from physical theory. [2]

A purely physical explanation is well suited to the explanation of physical structures. Explaining macroscopic structures in terms of detailed microstructural constituents is important, and it provides a satisfying explanation of the performance of functions, thus accounting for these functions in terms of the physical mechanisms that perform them. This is because a physical account can entail the facts about functions: Once the internal details of the physical account are given, the structural and functional properties fall out as an automatic consequence. But the structure and dynamics of physical processes yield only more structure and dynamics, so structures and functions are all neuroscientists can expect these processes to explain. The facts about experience cannot be an automatic consequence of any physical account, as it is conceptually coherent that any given process could exist without experience. Experience is not entailed by the physical, but may arise from the physical.

You can't explain conscious experience on the cheap. This is the moral of all this. It is a fact that reductive methods (methods that explain a high-level phenomenon wholly in terms of more basic physical processes) work well in so many domains. In a sense, one can explain most biological and cognitive phenomena on the cheap, in that these phenomena are seen as automatic consequences of more fundamental processes. It would be wonderful if reductive methods could explain experience, too: Neuroscientists have hoped for a long time that they might. Unfortunately, there are systematic reasons why these methods must fail. Reductive methods are successful in most domains because what needs explaining in those domains are structures and functions, and these are the kind of thing that a physical account can entail, so these methods are impotent when it comes to a problem over and above the explanation of structures and functions.

This might seem reminiscent of the vitalist claim that no physical account could explain life, but the cases are disanalogous. What drove vitalist skepticism was doubt about whether physical mechanisms could perform the many remarkable functions associated with life, such as complex adaptive behavior and reproduction. The conceptual claim that explanation of functions is what is needed was implicitly accepted, but lacking detailed knowledge of biochemical mechanisms, vitalists doubted whether any physical process could do the job and put forward the hypothesis of the vital spirit as an alternative explanation. Vitalist doubts melted away once it turned out that physical processes could perform the relevant functions.

On the other hand, with experience, physical explanation of the functions is not in question. The key is instead the conceptual point that the explanation of functions does not suffice for the explanation of experience. This basic conceptual point is not something that further neuroscientific investigation will affect. In a similar way, experience is disanalogous to the *elan vital.* The vital spirit was put forward as an explanatory posit in order to explain the relevant functions, and could therefore be discarded when those functions were explained without it. Experience, then, is not a candidate for this sort of elimination. It is also not an explanatory posit, but an explanandum in its own right.

NOTE

All sorts of puzzling phenomena have eventually turned out to be explainable in physical terms, but each of these were problems about the observable behavior of physical objects, coming down to problems in the explanation of structures and functions. Because of this, these phenomena have always been the kind of thing that a physical account *might* explain, even if at some points there have been good reasons to suspect that no such explanation would be forthcoming. The tempting induction from these cases fails in the case of consciousness, which is not a problem about physical structures and functions. The problem of consciousness is puzzling in an entirely different way. An analysis of the problem shows neuroscientists that conscious experience is just not the kind of thing that a wholly reductive account could succeed in explaining.

The Unconscious Consciousness Experience

Holding that they will never have a theory of conscious experience, at this point, some neuroscientists are tempted to give up. The problem is too hard for neuroscientists' limited minds: they are cognitively closed with respect to the phenomenon. Others have argued that conscious experience is outside the domain of scientific theory altogether. [2]

Some neuroscientists think this pessimism is premature. This is not the place to give up. It is the place where things get interesting. When simple methods of explanation are ruled out, neuroscientists need to investigate the alternatives. Nonreductive explanation is the natural choice when reductive explanation fails.

This is not universal, given that a remarkable number of phenomena have turned out to be explicable wholly in terms of entities simpler than themselves. In physics, it occasionally happens that an entity has to be taken as fundamental. Fundamental entities are not explained in terms of anything simpler. Instead, one takes them as basic, and gives a theory of how they relate to everything else in the world. For example, in the nineteenth century, it turned out that electromagnetic processes could not be explained in terms of the wholly mechanical processes that previous physical theories appealed to, so scientists introduced electromagnetic charge and electromagnetic forces as new fundamental components of a physical theory. To explain electromagnetism, the ontology of physics had to be expanded. In order to give a satisfactory account of the phenomenon, new basic properties and basic laws were needed.

Mass and space–time are other features that physical theory takes as being fundamental. No attempt is made to explain these features in terms of anything simpler, but this does not rule out the possibility of a theory of mass or of space–time. There is an intricate theory of how these features interrelate, and of the basic laws they enter into. At a much higher level, these basic principles are used to explain many familiar phenomena concerning space and time.

Some neuroscientists suggest that a theory of consciousness should take experience as fundamental. Neuroscientists know that a theory of consciousness requires the addition of something fundamental to human ontology, as everything in physical theory is compatible with the absence of consciousness. Neuroscientists might add some entirely new nonphysical feature from which experience can be derived, but it is hard to see what such a feature would be like. More likely, neuroscientists will take experience itself as a fundamental feature of the world, alongside mass, charge, and space–time. Neuroscientists can go about the business of constructing a theory of experience if they take experience as being fundamental.

There are fundamental laws where there is a fundamental property. A nonreductive theory of experience will add new principles to the furniture of the basic laws of nature. These basic principles will ultimately carry the explanatory burden in a theory of consciousness. Just as neuroscientists explain familiar phenomena involving experience in terms of more basic principles involving experience and other entities, they might explain familiar high-level phenomena involving mass in terms of more basic principles involving mass and other entities.

In particular, a nonreductive theory of experience will specify basic principles telling neuroscientists how experience depends on physical features of the world. These psychophysical principles will not interfere with physical laws, as it seems that physical laws already form a closed system. Rather, they will be a supplement to a physical theory. A physical theory gives a theory of physical processes, and a psychophysical theory tells neuroscientists how those processes give rise to experience. Neuroscientists know that experience depends on physical processes, but they also know that this dependence cannot be derived from physical laws alone. So, the extra ingredient that neuroscientists need to build an explanatory bridge is regarded as the new basic principle postulated by a nonreductive theory.

Of course, there is a sense in which this approach does not tell neuroscientists why there is experience in the first place by taking experience as being fundamental. But, this is the same for any fundamental theory. Nothing in physics tells scientists why there is matter in the first place, but they do not count this against theories of matter. Certain features of the world need to be taken as fundamental by any scientific theory. A theory of matter can still explain all sorts of facts about matter, by showing how they are consequences of the basic laws. The same goes for a theory of experience.

As this position postulates basic properties over and above the properties invoked by physics, it qualifies as a variety of dualism. But it is an innocent version of dualism, entirely compatible with the scientific view of the world. Nothing in this approach contradicts anything in physical theory. Neuroscientists need to add further bridging principles to explain how experience arises from physical processes. There is nothing particularly spiritual or mystical about this theory—its overall shape is like that of a physical theory, with a few fundamental entities connected by fundamental laws. It expands the ontology slightly, to be sure. Indeed, the overall structure of this position is entirely naturalistic, allowing that ultimately the universe comes down to a network of basic entities obeying simple laws, and allowing that there may ultimately be a theory of consciousness cast in terms of such laws. A good choice might be naturalistic dualism, if the position is to have a name.

In some ways, a theory of consciousness will have more in common with a theory in physics than a theory in biology, if the preceding view is right. Biological theories involve no principles that are fundamental in this way, so biological theory has a certain complexity and messiness to it. But theories in physics, insofar as they deal with fundamental principles, aspire to simplicity and elegance. The fundamental laws of nature are part of the basic furniture of the world, and physical theories are telling neuroscientists that this basic furniture is remarkably simple. If a theory of consciousness also involves fundamental principles, then neuroscientists should expect the same. The principles of simplicity, elegance, and even beauty that drive physicists' search for a fundamental theory will also apply to a theory of consciousness.

NOTE

Some philosophers argue that even though there is a conceptual gap between physical processes and experience, there need be no metaphysical gap, so that experience might in a certain sense still be physical. Usually this line of argument is supported by an appeal to the notion of *a posteriori* necessity. Some neuroscientists think that this position rests on a misunderstanding of *a posteriori* necessity, however, or else requires an entirely new sort of necessity that they have no reason to believe in. In any case, this position still concedes an explanatory gap between physical processes and experience. For example, the principles connecting the physical and the experiential will not be derivable from the laws of physics, so such principles must be taken as explanatorily fundamental. So, even on this sort of view, the explanatory structure of a theory of consciousness will be much as it has been described in this chapter.

Consciousness Theory

It is not too soon to begin work on a theory. Neuroscientists are already in a position to understand certain key facts about the relationship between physical processes and experience, and about the regularities that connect them. Neuroscientists can lay those facts on the table once reductive explanation is set aside. This is so that they can play their proper role as constraints on the basic laws that constitute an ultimate theory and as the initial pieces in a nonreductive theory of consciousness.

The paucity of objective data is an obvious problem that plagues the development of a theory of consciousness. Conscious experience is not directly observable in an experimental context, so neuroscientists cannot generate data about the relationship between physical processes and experience at will. Nevertheless, neuroscientists all have access to a rich source of data in their own case. Many important regularities between experience and processing can be inferred from considerations about one's own experience. There are also good indirect sources of data from observable cases, as when one relies on the verbal report of a subject as an indication of experience. Neuroscientists have more than enough data to get a theory off the ground, even though these methods have their limitations.

Also useful in getting value out of the data neuroscientists have is philosophical analysis. This sort of analysis can yield a number of principles relating consciousness and cognition, thereby strongly constraining the shape of an ultimate theory. The method of thought-experimentation can also yield significant rewards, as you will see. Finally, the fact that neuroscientists are searching for a fundamental theory means that they can appeal to such nonempirical constraints as simplicity, homogeneity, and the like in developing a theory. Neuroscientists must make the inference to the simplest possible theory that explains the data while remaining a plausible candidate to be part of the fundamental furniture of the world. They must also seek to systematize the information they have and to extend it as far as possible by careful analysis.

Because of the impossibility of conclusive intersubjective experimental tests, such theories will always retain an element of speculation that is not present in other scientific theories. Still, neuroscientists can certainly construct theories that are compatible with the data that they have and evaluate them in comparison to each other. Even in the absence of intersubjective observation, there are numerous criteria available for the evaluation of such theories: simplicity, internal coherence, coherence with theories in other domains, the ability to reproduce the properties of experience that are familiar from neuroscientists' own case, and even an overall fit with the dictates of common sense. Perhaps there will be significant indeterminacies remaining even when all these constraints are applied, but neuroscientists can at least develop plausible candidates. Neuroscientists will be able to evaluate candidate theories only when they have been developed.

A nonreductive theory of consciousness will consist in a number of psychophysical principles. These are principles connecting the properties of physical processes to the properties of experience. Neuroscientists think of these principles as encapsulating the way in which experience arises from the physical. Ultimately, these principles should tell neuroscientists what sort of physical systems will have associated experiences, and for the systems that do, they should tell neuroscientists what sort of physical properties are relevant to the emergence of experience, and just what sort of experience neuroscientists should expect any given physical system to yield. But there is no reason why neuroscientists should not get started, even though this is a tall order.

Now, let's look at the following three psychophysical principles candidates presented by neuroscientists that might go into a theory of consciousness:

1. Structural coherence
2. Organizational invariance
3. Double-aspect theory of information

The first two of these are nonbasic principles—systematic connections between processing and experience at a relatively high level. These principles can play a significant role in developing and constraining a theory of consciousness, but they are not cast at a sufficiently fundamental level to qualify as truly basic laws. The final principle is a candidate for a basic principle that might form the cornerstone of a fundamental theory of consciousness. This final principle is particularly speculative, but it is the kind of speculation that is required if neuroscientists are ever to have a satisfying theory of consciousness.

Structural Coherence

Structural coherence is a principle of coherence between the structure of consciousness and the structure of awareness. Recall that awareness was used earlier to refer to the various functional phenomena that are associated with consciousness. It's being used here to refer to a somewhat more specific process in the cognitive underpinnings of experience. In particular, the contents of awareness are to be understood as those information contents that are accessible to central systems and brought to bear in a widespread way in the control of behavior. Briefly put, neuroscientists can think of awareness as direct availability for global control. The contents of awareness are the contents that are directly accessible and potentially reportable (at least in a language-using system) to a first approximation.

Awareness is nevertheless intimately linked to conscious experience, and it is a purely functional notion. In familiar cases, wherever neuroscientists find consciousness, they find awareness. Wherever there is conscious experience, there is some corresponding information in the cognitive system that is available in the control of behavior and available for verbal report. Conversely, it seems that whenever information is available for report and for global control, there is a corresponding conscious experience. Thus, there is a direct correspondence between awareness and consciousness.

The correspondence can be taken further. It is a central fact about experience that it has a complex structure. The visual field has a complex geometry, for instance. There are also relations of similarity and difference between experiences, and relations in such things as relative intensity. Every subject's experience can be at least partly characterized and decomposed in terms of these structural properties: similarity and difference relations, perceived location, relative intensity, geometric structure, and so on. There is also a corresponding feature, in the information-processing structure of awareness, to each of these structural features.

For example, take color sensations: For every distinction between color experiences, there is a corresponding distinction in processing. The different phenomenal colors that humans experience form a complex three-dimensional space, varying in hue, saturation, and intensity. The properties of this space can be recovered from information-processing considerations: Examination of the visual systems shows that waveforms of light are discriminated and analyzed along three different axes, and it is this three-dimensional information that is relevant to later processing. The three-dimensional structure of phenomenal color space therefore corresponds directly to the three-dimensional structure of visual awareness. This is precisely what neuroscientists would expect. After all, every color distinction corresponds to some reportable information, and therefore to a distinction that is represented in the structure of processing.

The geometric structure of the visual field is directly reflected in a structure that can be recovered from visual processing in a more straightforward way. Every geometric relation corresponds to something that can be reported and is therefore cognitively represented. Neuroscientists could not directly observe that agent's visual experiences, but they could nevertheless infer those experiences' structural properties. Of course, this could only occur if neuroscientists were given only the story about information-processing in an agent's visual and cognitive system.

Any information that is consciously experienced will also be cognitively represented in general. The fine-grained structure of the visual field will correspond to some fine-grained structure in visual processing. The same goes for experiences in other modalities, and even for nonsensory experiences. Internal mental images have geometric properties that are represented in processing. Even emotions' structural properties, such as relative intensity, correspond directly to a structural property of processing. So, where there is greater intensity, neuroscientists find a greater effect on later processes. Thus, those properties will be directly represented in the structure of awareness, precisely because the structural properties of experience are accessible and reportable.

What constitutes the principle of structural coherence is the isomorphism between the structures of consciousness and awareness. This principle reflects the central fact that even though cognitive processes do not conceptually entail facts about conscious experience, consciousness and cognition do not float free of one another, but cohere in an intimate way.

This principle has its limits. It allows neuroscientists to recover structural properties of experience from information-processing properties, but not all properties of experience are structural properties. There are properties of experience, such as the intrinsic nature of a sensation of red, that cannot be fully captured in a structural description. The very intelligibility of inverted spectrum scenarios, where experiences of red and green are inverted, but all structural properties remain the same, show that structural properties constrain experience without exhausting it. Nevertheless, the very fact that neuroscientists feel compelled to leave structural properties unaltered when they imagine experiences inverted between functionally identical systems, shows how central the principle of structural coherence is to human conception of their mental lives. After all, it is not a logically necessary principle. Humans can imagine all the information processing occurring without any experience at all, but it is nevertheless a strong and familiar constraint on the psychophysical connection.

In terms of physical processes, the principle of structural coherence allows for a very useful kind of indirect explanation of experience. For example, neuroscientists can use facts about neural processing of visual information to indirectly explain the structure of color space. The facts about neural processing can entail and explain the structure of awareness. If neuroscientists take the coherence principle for granted, the structure of experience will also be explained. Empirical investigation might even lead neuroscientists to better understand the structure of awareness within a bat, shedding indirect light on the vexing question of what it is like to be a bat. This principle provides a natural interpretation of much existing work on the explanation of consciousness, although it is often appealed to inexplicitly, so it is taken for granted by almost everybody that is familiar with it. And in the cognitive explanation of consciousness, it is a central plank.

A natural interpretation of work in neuroscience directed at isolating the substrate (or the neural correlate) of consciousness is allowed by the coherence between consciousness and awareness. Various specific hypotheses have been put forward. For example, some neuroscientists suggest that 40-Hz oscillations may be the neural correlate of consciousness, whereas other neuroscientists suggest that temporally extended neural activity is central. If neuroscientists accept the principle of coherence, the most direct physical correlate of consciousness is awareness: the process whereby information is made directly available for global control. The different specific hypotheses can be interpreted as empirical suggestions about how awareness might be achieved. For example, some neuroscientists suggest that 40-Hz oscillations are the gateway by which information is integrated into working memory and thereby made available to later processes. Similarly, it is natural to suppose that temporally extended activity is relevant precisely because only that sort of activity achieves global availability. The same applies to other suggested correlates such as the global workspace, the high-quality representations, and the selector inputs to action systems. All of these can be seen as hypotheses about the mechanisms of awareness: the mechanisms that perform the function of making information directly available for global control.

So, it follows that a mechanism of awareness will itself be a correlate of conscious experience, given the coherence between consciousness and awareness. The question of just which mechanisms in the brain govern global availability is an empirical one. Perhaps there are many such mechanisms, but neuroscientists have reason to believe that the processes that explain awareness will at the same time be part of the basis of consciousness, if they accept the coherence principle. [2]

Organizational Invariance

The principle of organizational invariance states that any two systems with the same fine-grained functional organization will have qualitatively identical experiences. If the causal patterns of neural organization were duplicated in silicon, for example, with a silicon chip for every neuron and the same patterns of interaction, then the same experiences would arise. According to this principle, what matters for the emergence of experience is not the specific physical makeup of a system, but an abstract pattern of causal interaction between its components. This principle is controversial, of course. Some neuroscientists have thought that consciousness is tied to a specific biology, so that a silicon isomorph of a human need not be conscious. However, by the analysis of thought experiments, some neuroscientists believe that the principle can be given significant support.

Very briefly: Suppose (for the purposes of a *reductio ad absurdum)* that there could be two functionally isomorphic systems with different experiences and that the principle is false? Perhaps only one of the systems is conscious, or perhaps both are conscious, but they have different experiences. For the purposes of illustration, let's say that one system is made of neurons and the other of silicon, and that one experiences red where the other experiences blue. The two systems have the same organization, so neuroscientists can imagine gradually transforming one into the other, perhaps replacing neurons one at a time by silicon chips with the same local function. Neuroscientists thus gain a spectrum of intermediate cases, each with the same organization, but with slightly different physical makeup and slightly different experiences. Along this spectrum, there must be two systems, A and B, between which neuroscientists replace less than one tenth of the system, but whose experiences differ. Except for the fact that a small neural circuit in A has been replaced by a silicon circuit in B, these two systems are physically identical.

The key step in the thought experiment is to take the relevant neural circuit in A, and install alongside it a causally isomorphic silicon circuit, with a switch between the two. What happens when neuroscientists flip the switch? By hypothesis, the system's conscious experiences will change. This is from red to blue, say, for the purpose of illustration. This follows from the fact that the system after the change is essentially a version of B, whereas before the change it is just A. [2]

There is no way for the system to notice the changes, given the assumptions! Its causal organization stays constant so that all of its functional states and behavioral dispositions stay fixed. As far as the system is concerned, nothing unusual has happened. There is no room for the thought, "Hmm! Something strange has just happened!" In general, the structure of any such thought must be reflected in processing, but the structure of processing remains constant here. If there were to be such a thought, it must float entirely free of the system and would be utterly impotent to affect later processing.

NOTE

If it affected later processing, the systems would be functionally distinct, contrary to hypothesis.

Neuroscientists might even flip the switch a number of times, so that experiences of red and blue dance back and forth before the system's inner eye. The system can never notice these dancing qualia, according to the hypothesis.

Neuroscientists take this to be *a reductio* of the original assumption. It is a central fact about experience. From the neuroscientists' own case (that whenever experiences change significantly and they are paying attention), they notice the change: If this were not the case, they would be led to the skeptical possibility that their experiences are dancing before their eyes all the time. This hypothesis has the same status as the possibility that the world was created five minutes ago:

perhaps it is logically coherent, but it is not plausible. Given the extremely plausible assumption that changes in experience correspond to changes in processing, neuroscientists are led to the conclusion that the original hypothesis is impossible, and that any two functionally isomorphic systems must have the same sort of experiences. While logically possible, the philosophical hypotheses of absent qualia and inverted qualia, are empirically and nomologically impossible.[2]

NOTE

Some neuroscientists may worry that a silicon isomorph of a neural system might be impossible for technical reasons. That question remains open. The invariance principle says only that if an isomorph were possible, then it would have the same sort of conscious experience.

This gives the basic flavor, even though there is more to be said here. Once again, this thought experiment draws on familiar facts about the coherence between consciousness and cognitive processing to yield a strong conclusion about the relation between physical structure and experience. If the argument goes through, neuroscientists know that the only physical properties directly relevant to the emergence of experience are organizational properties. Based on a theory of consciousness, this acts as a further strong constraint.

Theory of Double-Aspect of Information

The two preceding principles have been nonbasic principles. They involve high-level notions such as awareness and organization, and therefore lie at the wrong level to constitute the fundamental laws in a theory of consciousness. Nevertheless, they act as strong constraints. What is further needed are basic principles that might ultimately explain and fit these constraints.

The basic principle that neuroscientists suggest here centrally involves the notion of information. Where there is information, there are information states embedded in an information space. An information space has its basic structure of difference relations between its elements, characterizing the ways in which different elements in a space are similar or different, possibly in complex ways. An information space is an abstract object, but neuroscientists can see information as physically embodied when there is a space of distinct physical states, the differences between which can be transmitted down some causal pathway. The states that are transmitted can be seen as themselves constituting an information space. Thus, physical information is a difference that makes a difference.

The double-aspect principle stems from the observation that there is a direct isomorphism between certain physically embodied information spaces and certain phenomenal (or experiential) information space. From the same sort of observations that went into the principle of structural coherence, neuroscientists can note that the differences between phenomenal states have a structure that corresponds directly to the differences embedded in physical processes. In particular, this is in regards to those differences that make a difference down certain causal pathways implicated in global availability and control. That is, neuroscientists can find the same abstract information space embedded in conscious experience and physical processing.

Information (or at least some information) has two basic aspects, a physical aspect and a phenomenal aspect. This has the status of a basic principle that might underlie and explain the emergence of experience from the physical. Experience arises by virtue of its status as one aspect of information when the other aspect is found embodied in physical processing.

This principle is lent support by a number of considerations, which can only be discussed briefly here. First of all, consideration of the sort of physical changes that correspond to changes in conscious experience suggests that such changes are always relevant by virtue of their role in constituting informational changes—differences within an abstract space of states that are divided up precisely according to their causal differences along certain causal pathways. Second, if the principle of organizational invariance is to hold, then neuroscientists need to find some fundamental organizational property for experience to be linked to, and information is an organizational property par excellence. Third, this principle offers some hope of explaining the principle of structural coherence in terms of the structure present within information spaces. Fourth, analysis of the cognitive explanation of neuroscientists' judgments and claims about conscious experience—judgments that are functionally explainable but nevertheless deeply tied to experience itself—suggests that explanation centrally involves the information states embedded in cognitive processing. It follows that a theory based on information allows a deep coherence between the explanation of neuroscientists' judgments and claims about it, and the explanation of experience.

Information is fundamental to the physics of the universe. According to this "it from bit" doctrine, the laws of physics can be cast in terms of information, postulating different states that give rise to different effects without actually saying what those states are. It is only their position in an information space that counts. If so, then information is a natural candidate to also play a role in a fundamental theory of consciousness. Neuroscientists are led to a conception of the world in which information has two basic aspects (corresponding to the physical and the phenomenal features of the world) and in which information is truly fundamental.

Of course, by leaving a number of key questions unanswered, the double-aspect principle is extremely speculative and is also underdetermined. An obvious question is whether all information has a phenomenal aspect. One possibility is that neuroscientists need a further constraint on the fundamental theory, indicating just what sort information has a phenomenal aspect. The other possibility is that there is no such constraint. If not, then experience is much more widespread than neuroscientists might have believed, as information is everywhere. This is counterintuitive at first, but some neuroscientists think the position gains a certain plausibility and elegance. Where there is simple information processing, there is simple experience, and where there is complex information processing, there is complex experience. A mouse has a simpler information-processing structure than a human and has correspondingly simpler experience. So, does a thermostat (a maximally simple information-processing structure) have a maximally simple experience? Indeed, if experience is truly a fundamental property, it would be surprising for it to arise only every now and then. Most fundamental properties are more evenly spread. Some neuroscientists believe that the position is not as implausible as it is often thought to be. In any case, this is very much an open question.

Finally, the door is opened to some grander metaphysical speculation concerning the nature of the world, once a fundamental link between information and experience is on the table. For example, it is often noted that physics characterizes its basic entities only extrinsically, in terms of their relations to other entities, which are themselves characterized extrinsically, and so on. The intrinsic nature of physical entities is left aside. Some argue that no such intrinsic properties exist, but then one is left with a world that is pure causal flux (a pure flow of information) with no properties for the causation to relate. If one allows that intrinsic properties exist, a natural speculation is that the intrinsic properties of the physical (the properties that causation ultimately relates) are themselves phenomenal properties. Neuroscientists might say that phenomenal properties are the internal aspect of information. This could answer a concern about the causal relevance of experience—a natural worry, given a picture in which the physical domain is causally closed, and on which experience is supplementary to the physical. The informational view allows neuroscientists to understand how experience might have a subtle kind of causal relevance in virtue of its status as the intrinsic nature of the physical. But, in addressing some philosophical issues, metaphysical speculation is quite suggestive. This is probably best ignored for the purposes of developing a scientific theory.

Conclusion

The theory neuroscientists have presented in this chapter is speculative, but it is a candidate theory. The principles of structural coherence and organizational invariance will be planks in any satisfactory theory of consciousness. The status of the double-aspect theory of information is less certain. Indeed, right now it is more of an idea than a theory. To have any hope of eventual explanatory success, it will have to be specified more fully and fleshed out into a more powerful form. Still, reflection on just what is plausible and implausible about it, on where it works and where it fails, can only lead to a better theory.

Finally, most existing theories of consciousness either deny the phenomenon, explain something else, or elevate the problem to an eternal mystery. To make further progress, neuroscientists will need further investigation, more refined theories, and more careful analysis. The hard problem is a hard problem, but there is no reason to believe that it will remain permanently unsolved. [2]

References

[1] Dr. Miroslav Backonja, Neurology, Madison Wisconsin, Department of Neurology, H6/570 University of Wisconsin Hospital, 600 Highland Avenue, Madison, WI 53792.

[2] Francis Crick and Christof Koch, "The Problem of Consciousness," *Scientific American,* 415 Madison Ave., New York, NY 10017, Vol. 12, No. 1, 2002.

[3] "Intuition: Consciousness—How fundamental is it?" *Physics-Intuition-Applications,* [http://www.p-i-a.com/], 2004.

References

Geology

The Dynamics of the Inner Earth

"Now I see the secret of the making of the best persons. It is to grow in the open air and to eat and sleep with the earth."

—Walt Whitman (1819–1892)

Recently, a team of researchers found that the elastic properties of iron are quite different at extremely high temperatures than at low temperatures. This is the result of work that promises to advance understanding and answer questions about the origin and dynamics of Earth's iron-rich inner core, mantle, and the generation of the planet's magnetic field. The surprising finding calls into question previous interpretations of seismic data that were based on iron's low-temperature properties.

"What we know of the deep Earth is primarily from studying the recordings (seismograms) of sound waves generated by earthquakes that travel through the Earth," explains geophysicist Dr. Ronald Cohen (Geophysical Laboratory, Carnegie Institution of Washington), "and from the free oscillations of the Earth, (the Earth rings like a bell). The data can be inverted to give the velocity and density structure of the Earth. The correct interpretation of this data is key to understanding the structure of earth's interior." [1]

According to Professor Dr. Lars Stixrude (Dept. of Geological Sciences, University of Michigan), "What is remarkable about the discovery of structure in the inner core is that it has given us our first glimpse into the dynamics and evolution of the deepest part of our planet. By understanding the origin of the structure that seismologists observe, we hope to elucidate the rate and extent of cooling of the earth's interior, which we believe led to the formation of the inner core, and the role that the inner core may play in the generation of the geomagnetic field." [2]

"Seismology cannot tell us what is the meaning of the changes in sound velocity and density that are observed," states Dr. Cohen. "Mineral physics provides data and theory to interpret these data, by giving the sound velocities and densities of rocks and minerals as functions of pressure and temperature.

"Data at pressures of up to about 25 Gpa (equivalent to depths of 700 km in the Earth (1 Gpa is 10,000 atmospheres or 150,000 lbs/sq.in.) can be obtained in multianvil apparati, and higher pressures over the pressures of the Earth's core (364 Gpa is the pressure in the center of the Earth) can be obtained in the diamond anvil cell. New methods are being developed to measure elastic constants in the diamond anvil cell, but at this time it is impossible to measure elastic constants, and thus, the sound speeds under the pressures (135–364 Gpa) and temperatures (2,000–6,000K) of the Earth's core (note that the Earth's core is hotter than the surface of the sun!)." [1]

According to Dr. Cohen, the "first-principles theory provides a route to estimate elastic constants of materials from fundamental physics, using no essential experimental data. The only input in a first principles calculation are the atomic (nuclear) positions and charges for any material. Density functional theory relates the ground state properties (such as energy) to the charge density, and the correct density is the one that minimizes the energy for any given crystal structure. Elastic constants (at zero temperature) can be obtained by computing the energy as a function of lattice strain (deformation). This is a computational challenge (though now straightforward for simple crystals structures) since great precision is needed in the energies as they change only little as a function of small strains. This was done for iron under core densities, but first at zero temperature [3] [4]; the results seemed consistent with seismic observations only if the inner core of the Earth (the inner core is crystalline, and is surrounded by a molten outer core) was a giant single crystal, or perfectly aligned crystals of high-pressure, hexagonal-clode packed (hcp) iron. It seemed unlikely that this result would change by including temperature in the computations, since at zero pressure elastic constants for transition metals such as iron tend not to change qualitatively with temperature." [1]

"It is much more difficult to compute the effects of temperature on elastic constants from first principles," explains Dr. Cohen. "One must include the vibrations of atoms, and how this changes with strain. There are different approximations, but no tractable exact way to do this, even with density functional theory. Using a model called the Particle in a Cell (PIC) model, Steinle-Neumann et al. computed the effects of temperature on the structure and elasticity of hcp iron. [5] In the hcp structure, there is only one structural degree of freedom at a given crystal volume (density), the c/a ratio, which is how much the crystal structure is stretched out. Steinle-Neumann and colleagues found, according to their theory, that the c/a ratio increased with temperature, and this in turn impacted the elastic constants. The net result was that the seismic data for the Earth's inner core could be understood with a reasonable amount (30%) of crystal alignment. The question remains open, however, because experiments so far have observed more moderate increases in c/a with temperature." [6]

Scientists have long known that Earth's core, which is responsible for generating the planet's magnetic field, is primarily composed of iron. Its center portion (the inner core) is a solid sphere, which over the course of Earth's history has grown to its present size of 1,200 km (about 745 miles).

"The earth's magnetic field is generated by turbulent flow of the liquid iron in the Earth's outer core," explains Dr. Cohen. "The turbulent flow acts like a dynamo. The flow carries charge, so there is a net electric current. The current sets up a magnetic field, which in turn generates current in the fluid. This process takes thermal energy (from the temperature difference across the core) and converts it into electromagnetic energy. It has been shown that only turbulent flow can give rise to a self-perpetuating dynamo. Geophysicists have tried to study this effect by numerical simulations and by laboratory experiments with rotating metals. So far, it has not been possible to reach the parameter regimes that are believed present for the geodynamo, though much has been learned about the properties and requirements for a perpetuating and spontaneously reversing dynamo such as is present in the Earth." [1]

However, researchers have been puzzled by observations, based on seismological measurements, that elastic waves generated by earthquakes travel through the inner core faster along directions parallel to Earth's polar axis than in the equatorial plane. The cause of this difference has not been well understood, partly because the elastic properties of iron at the high pressure and temperature of Earth's center are not known (it is impossible to take direct measurements at the core, and arid core conditions are difficult to duplicate in the laboratory).

"It is hard for us to imagine how different conditions are at the center of the earth from any conceivable direct human experience," explains Dr. Lars Stixrude. "The pressure at the center is known to be more than 3,000 times greater than that in the deepest ocean trench, while the temperature is thought to be approximately the same as the surface of the sun. Even under these conditions, however, we believe that the basic rules of physics, as embodied in the theory of quantum mechanics, are essentially unchanged from those that govern our surface existence." [2]

By performing simulations on supercomputers, researchers can, however, predict the properties of iron at core conditions. Rather than relying on direct measurements or experimental data, these simulations are based on fundamental physics. Researchers used supercomputer simulations to study changes in the crystal structure of iron at high pressure and very high temperatures of 4,000–7,000K (6,740–12,140°F).

"The structure of iron at the conditions of the earth's interior is quite different from room conditions," states Dr. Stixrude. "The atoms are arranged in hexagonal planes, and these planes are stacked one on top of the other. As temperature increases, it appears that the planes move farther apart from each other, which causes seismic waves to travel more slowly perpendicular to the planes, and more rapidly parallel to the planes. Supposing that the inner core were made of many such crystals and these crystals showed a tendency to align so that the planes were parallel to the earth's rotation axis, the speed of elastic waves in different directions would match seismological observations very well." [2]

Changes in the basic hexagonal prism shape of the iron crystals directly influence elastic properties. To the researchers' surprise, the elastic properties of iron were quite different at high temperatures than at low temperatures, an observation that has led to new interpretations of seismic images. According to geophysicist Dr. Gerd Steinle-Neumann (Bayerisches Geoinstitut–University Bayreuth), "These computational findings on the elastic properties, however, have been obtained with various approximations, and will need additional confirmation from experimental studies or even more accurate modeling approaches." [7]

The results support the hypothesis that the directional behavior in seismic wave propagation reflects the alignment of crystals in the inner core. Stresses acting on the inner core influence the alignment process, and various models have been proposed to explain how that occurs. This chapter very briefly surveys a number of those models, such as the development of a simple model of inner-core structure, in which the hexagonal bases of the crystals tend to align with Earth's polar axis.

Researchers were led to infer that the temperature in the center of the Earth was the result of a strong temperature dependence of the average seismic wave velocity in iron, and an almost perfect agreement of such properties with those of the inner core (at a temperature of 5,700K, or 9,800°F). The researchers hope that other scientists will use this new understanding of the high-temperature elasticity of iron to refine models of the dynamics in Earth's inner core. These refined models should help to finally explain why seismic waves travel faster in particular directions. Geophysicists are characterizing the material properties, which ultimately will help them understand the dynamic processes that underlie the differences in seismic wave velocity.

Theoretical studies involving simulations of the earth's core and mantle dynamics are increasingly complementing experimental and observational work in the Earth sciences, material physics, and chemistry. They have successfully provided insight into the microscopic causes of many physical phenomena, predicted material properties, and expanded the range of conditions under which materials have been studied.

As geophysicists try to uncover the Earth's deepest secrets, they must study the crust first, then the mantle, and finally the core. Let's take a look.

Revealing Earth's Deepest Secrets

About 300 years ago, the English scientist Isaac Newton calculated, from his studies of planets and the force of gravity, that the average density of the Earth is twice that of surface rocks and therefore the Earth's interior must be composed of much denser material. Geophysicists' knowledge of what's inside the Earth has improved immensely since Newton's time, but his estimate of the density remains essentially unchanged. Geophysicists' current information comes from studies of the paths and characteristics of earthquake waves travelling through the Earth, as well as from laboratory experiments on surface minerals and rocks at high pressure and temperature. Other important data on the Earth's interior come from geological observations of surface rocks and studies of the Earth's motions in the solar system, its gravity and magnetic fields, and the flow of heat from inside the Earth.

The planet Earth is made up of three main shells: the very thin, brittle crust; the mantle; and the core, as shown in Figure 15.1.[8] The mantle and core are each divided into two parts. Although the core and mantle are about equal in thickness, the core actually forms only 15% of the Earth's volume, whereas the mantle occupies 84%. The crust makes up the remaining 1%. Geophysicists' knowledge of the layering and chemical composition of the Earth is steadily being improved by earth

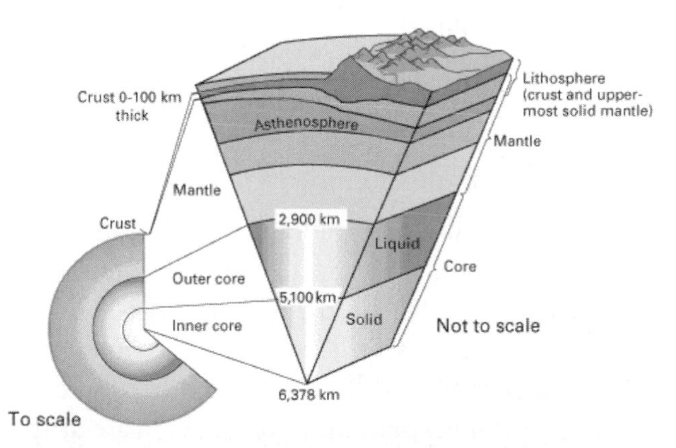

Figure 15.1 Cutaway views showing the internal structure of the Earth. This view is drawn to scale. It demonstrates that the Earth's crust literally is only skin deep. It also shows the Earth's three main layers (crust, mantle, and core) in more detail.

scientists doing laboratory experiments on rocks at high pressure and analyzing earthquake records on computers.

Crust

Because the crust is accessible to geophysicists, its geology has been extensively studied, and therefore much more information is known about its structure and composition than about the structure and composition of the mantle and core. Within the crust, intricate patterns are created when rocks are redistributed and deposited in layers through the geologic processes of eruption and intrusion of lava, erosion and consolidation of rock particles, and solidification and recrystallization of porous rock.

By the large-scale process of plate tectonics, about twelve plates, which contain combinations of continents and ocean basins, have moved around on the Earth's surface through much of geologic time. The edges of the plates are marked by concentrations of earthquakes and volcanoes. Collisions of plates can produce mountains like the Himalayas, the tallest range in the world. The plates include the crust and part of the upper mantle, and they move over a hot, yielding upper mantle zone at very slow rates of a few centimeters per year, slower than the rate at which fingernails grow. The crust is much thinner under the oceans than under continents (see Figure 15.1).

The boundary between the crust and mantle is called the Mohorovicic discontinuity (or Moho). It is named in honor of the man who discovered it, the Croatian scientist Andrija Mohorovicic. No one has ever seen this boundary, but it can be detected by a sharp increase downward in the speed of earthquake waves there. The explanation for the increase at the Moho is presumed to be a change in rock types. Drill holes to penetrate the Moho have been proposed, and a Soviet hole on the Kola Peninsula has been drilled to a depth of 12 kilometers

(7.5 miles), but drilling expense increases enormously with depth, and Moho penetration is not likely very soon.

The Mantle

The size of the Earth—about 12,750 kilometers (km) in diameter—was known by the ancient Greeks, but it was not until the turn of the 20th century that scientists determined that your planet is made up of three main layers: crust, mantle, and core, as shown in Figure 15.1. This layered structure can be compared to that of a boiled egg. The crust, the outermost layer, is rigid and very thin compared with the other two. Beneath the oceans, the crust varies little in thickness, generally extending only to about 5 km (3.1 miles). The thickness of the crust beneath continents is much more variable, but averages about 30 km (18.8 miles); under large mountain ranges, such as the Alps or the Sierra Nevada, however, the base of the crust can be as deep as 100 km (62.5 miles). Like the shell of an egg, the Earth's crust is brittle and can break.

Below the crust is the mantle, a dense, hot layer of semi-solid rock approximately 2,900 km (1,812.5 miles) thick. The mantle, which contains more iron, magnesium, and calcium than the crust, is hotter and denser because temperature and pressure inside the Earth increase with depth. As a comparison, the mantle might be thought of as the white of a boiled egg. With this in mind, this part of the chapter will take a very brief look at the Earth's mantle dynamics.

Geophysicists' knowledge of the upper mantle, including the tectonic plates, is derived from analyses of earthquake waves, heat flow, magnetic and gravity studies, and laboratory experiments on rocks and minerals. Between 100 and 200 kilometers (62.5 and 125 miles) below the Earth's surface, the temperature of the rock is near the melting point; molten rock erupted by some volcanoes originates in this region of the mantle. (See sidebar, "U.S. Geological Survey Research Casts Major Doubts on Geologic Theory of How Volcanic Regions Are Created: Mantle Plumes May Be Nonexistent after All.")

U.S. Geological Survey Research Casts Major Doubts on Geologic Theory of How Volcanic Regions Are Created: Mantle Plumes May Be Nonexistent after All

Standard fare in geology textbooks and school classrooms across the world is that the hot springs, geysers and volcanoes of Yellowstone National Park, Hawaii, Iceland, and many other volcanic regions were created by plumes of hot rock that rise from near the Earth's core. New results from recently published U.S. Geological Survey (USGS) research hint, astonishingly, that such plumes may not exist at all.

Recent results from seismic tomography (a method that uses earthquake waves to CAT scan the Earth's secretive goings-on), suggest that the magma system beneath Yellowstone is only skin deep (shallower than 120 miles), far less than the 1,750 miles scientists would expect if the magma arose from near the Earth's molten core as has been thought for decades. The Yellowstone results have implications that range much farther than just the local geology of America's first National Park. A progressively older trail of volcanic rock stretches westward across Wyoming, Idaho, and Nevada more than 200 miles from Yellowstone, following the eastern Snake River Plain. Geophysicists have interpreted this trail as the geologic remains of past volcanism left behind as the North American plate slowly drifted over the fixed Yellowstone plume.

Similar trails elsewhere have also been attributed to mantle plumes. In particular, the spectacular chain of volcanic islands that stretches from the Big Island of Hawaii, northwest along the Hawaiian island chain, is commonly considered compelling evidence for a plume and is the textbook example of how mantle plumes work to create volcanic islands and other geologic features above the Earth's surface.

If Yellowstone can leave a volcano trail without a plume, then other hotspots might also. The implication is that Hawaii may not be underlain by a plume after all. Seismic tomography has also been conducted at Iceland, traditionally considered to be a landform created by a mantle plume.

The results from Yellowstone and Iceland are drawing wide attention in the geological community because many other kinds of geologic data have only been able to be interpreted in terms of the plume model with great difficulty. A growing number of Earth scientists are thus taking a second look at their own data, and are starting to cast around for alternatives to one of their favorite ideas. [8]

This zone of extremely yielding rock has a slightly lower velocity of earthquake waves and is presumed to be the layer on which the tectonic plates ride. Below this low-velocity zone is a transition zone in the upper mantle; it contains two discontinuities caused by changes from less dense to more dense minerals. The chemical composition and crystal forms of these minerals have been identified by laboratory experiments at high pressure and temperature. The lower mantle, below the transition zone, is made up of relatively simple iron and magnesium silicate minerals, which change gradually with depth to very dense forms. Going from mantle to core, there is a marked decrease (about 30%) in earthquake wave velocity and a marked increase (about 30%) in density.

Core

The core was the first internal structural element to be identified. It was discovered in 1906 by R.D. Oldham, from his study of earthquake records, and it helped to explain Newton's calculation of the Earth's density. The outer core is presumed to be liquid, because it does not transmit shear (S) waves and because the velocity of compressional (P) waves that pass through it is sharply reduced. The inner core is considered to be solid because of the behavior of P and S waves passing through it.

Finally, data from earthquake waves, rotations and inertia of the whole Earth, magnetic-field dynamo theory, and laboratory experiments on melting and alloying of iron, all contribute to the identification of the composition of the inner and outer core. The core is presumed to be composed principally of iron, with about 10% alloy of oxygen or sulfur or nickel, or perhaps some combination of these three elements.

Conclusion

As previously discussed, seismology is an attempt by geophysicists to listen to the ring and vibration of the Earth to discover its content. Seismology has become the principle method used in studying Earth's interior. Seismology on Earth deals with the study of vibrations that are produced by earthquakes, the impact of meteorites, or artificial means such as an explosion. On these occasions, a seismograph is used to measure and record the actual movements and vibrations within the Earth and on the ground.

Scientists categorize seismic movements into four types of diagnostic waves that travel at speeds ranging from 3 to 15 kilometers (1.9 to 9.4 miles) per second. Two of the waves travel around the surface of the Earth in rolling swells. The other two, Primary (P) or compression waves and Secondary (S) or shear waves, penetrate the interior of the Earth. Primary waves compress and dilate the matter they travel through (either rock or liquid) similar to sound waves. They also have the ability to move twice as fast as S waves. Secondary waves propagate through rock, but are not able to travel through liquid. Both P and S waves refract or reflect at points where layers of differing physical properties meet. They also reduce speed when moving through hotter material. These changes in direction and velocity are the means of locating discontinuities.

Finally, seismic discontinuities aid in distinguishing divisions of the Earth into inner core, outer core, lower mantle, transition region, upper mantle, and crust (oceanic and continental). Lateral discontinuities also have been distinguished and mapped through seismic tomography, but are beyond the scope of this chapter.

References

[1] Dr. Ronald Cohen, Geophysical Laboratory, Carnegie Institution of Washington, 5251 Broad Branch Rd. N.W., Washington, D.C. 20015.

[2] Dr. Lars Stixrude, Professor, Dept. of Geological Sciences, University of Michigan, 425 E. University Ave., Ann Arbor, MI 48109-1063.

[3] Lars Stixrude and R.E. Cohen, *Science,* 267, 1972 (1995).

[4] Gerd Steinle-Neumann, Lars Stixrude, and R. Cohen, *Physical Review, B 60,* 791 (1999).

[5] Gerd Steinle-Neumann, Lars Stixrude, R. E. Cohen, et al., *Nature, 413,* 57 (2001).

[6] J.-F. Lin, D. L. Heinz, A. J. Cambell, et al., *Geophysical Research Letters, 29,* 10, 1029/2002GL015089 (2002).

[7] Dr. Gerd Steinle-Neumann, Bayerisches Geoinstitut, University Bayreuth, 95440 Bayreuth, Germany.

[8] "Inside the Earth," U.S. Geological Survey, MS119 National Center, Reston, VA 20192, 2003.

Earthquake Predicting

"You can no more win a war than you can win an earthquake."
—Jeannette Rankin (1880–1973)

Many of the world's major cities (Los Angeles, Mexico City, Tokyo, and even New York) are located in earthquake zones. People living there have learned to live with minor tremors. But sooner or later, all of these cities will be hit by the Big One—a devastating quake that may claim millions of lives, as dramatized by the recent NBC two-part mini series *10.5*. Seismologists (the scientists whose job it is to predict earthquakes) now face a race against time. In their quest to find a way to give at least some warning, the seismologists are looking to space for help.

NOTE

There is no chance of running away from an earthquake. The fastest waves, called *P waves*, roar along at three miles per second.

Shake, Rattle, and Roll Watch

Almost every day, there is an earthquake somewhere in the world. Most of them occur in areas known as *earthquake zones.* Earthquakes happen because the Earth's solid outer crust, or *lithosphere,* is broken into giant fragments called *tectonic plates* that float on the molten rock below. Earthquake zones lie along the cracks between these plates, and earthquakes are triggered by the plates moving past each other. As the plates grind together, the rock on either side of the crack bends, stretches, and then snaps. This sudden jolt sends shock, or *seismic,* waves shuddering through the ground to cause earthquakes.

NOTE

The P (primary) waves arrive first. The slower S (secondary) waves strike seconds later, shaking down already weakened buildings.

The problem with earthquake prediction is knowing just when the rock is going to snap. (See sidebar, "Why is Earthquake Prediction so Difficult.") In fact, if you live in an earthquake zone and you have not had an earthquake for some time, expect one soon. The longer the time since the last quake, the bigger the next quake will be. This is because there has been more time for the strain in the rocks to build up. Big quakes might occur every century or so, smaller ones every few decades. [1]

"More generally," explains Dr. Paul G. Silver (Senior Staff Scientist for Carnegie Institution of Washington), "I would say that the three most significant problems in our field involve the questions that arise from the plate tectonic revolution:

1. What is the pattern of convective flow in the mantle that accompanies plate tectonics, and what are the forces that move the plates?
2. How did plate tectonics begin? What was the tectonic regime in the earliest era of Earth's history?
3. Can we predict earthquakes? Plate tectonics shows us the regions where earthquakes are likely

to occur (along plate boundaries). Can we predict the place and time?" [2]

Why is Earthquake Prediction So Difficult?

"Earthquake predictability has been the subject of several recent, controversial articles within the seismological community. The intensity of the controversy is understandable, since it stems both from our present inability to predict earthquakes and from the potentially great value that prediction could have for society. Our difficulty in predicting earthquakes is partly due to the inherent characteristics of earthquakes and seismic waves and partly to an incomplete understanding of the earthquake process. In order to gain some perspective on this issue, it is useful to place the earthquake problem in the context of other natural hazards. Nearly all other natural hazards, from hurricanes to wildfires to volcanic eruptions, are predictable to some extent. The predictions are based on precursors, defined here as the non-threatening, initial phase of a natural hazard. This precursory phase consists of two parts, the preparation of the disturbance itself and the propagation of that disturbance to population centers. For example, a hurricane represents an atmospheric disturbance that develops at sea in the tropics and subsequently moves slowly toward population centers, at which time it becomes a threat. The precursory phase, including the preparation and propagation times, lasts hours to days. Other weather disturbances, such as tornadoes, occur on much shorter time scales, although the conditions under which tornadoes are highly probable can usually be recognized. In the case of earthquakes, we have yet to observe a reliable preparation phase, and the propagation time is very short, on the order of seconds. Tsunamis possess much longer propagation times, so that forecasting is possible, in the absence of an observed preparation phase. Probably the most closely related hazard to earthquakes is volcanic eruptions. Prediction must often be based solely on identifying a preparation phase because in many cases the propagation time is very short. For volcanic eruptions, however, the preparation phase has a known physical basis, namely the pre-eruption upward transport of magma. This magma transport has several observable manifestations, including crustal deformation, microseismicity, changes in the gas chemistry and increases in the temperature of hydrothermal fluids. Volcano prediction is a reality. Perhaps the most successful prediction was the June, 1991 eruption of Mount Pinatubo (in the Philippines), which led to the evacuation of 80,000 people and saved billions of dollars in U.S. aircraft that were moved from Clark Air Force Base. This volcano had not erupted in 400 years but was predictable from a variety of precursory signals. It is likely that improved monitoring of crustal deformation, seismicity, and other magma-transport indicators will ultimately lead to the routine prediction of volcanic eruptions.

"Our poor success record in earthquake prediction has understandably produced a shift in emphasis to other aspects of natural disaster reduction. There are clearly things that can be done to reduce the vulnerability to an earthquake hazard, in the absence of predictive capability, and we have made much progress in these areas. We have taken significant steps forward both in mitigation, the long-term actions that reduce the vulnerability to hazards, and preparedness, the short-term actions taken around the time of an event. For example, there have been important efforts to identify those areas that are most prone to sustaining significant earthquake damage (through the generation of hazard maps), so that informed decisions can be made about land use and building codes. Warning systems that detect and immediately broadcast the occurrence of an earthquake help with preparedness and in guiding the emergency response to a disaster, and in special cases can give a few seconds of advanced warning. For the other significant natural hazards, short-term forecasting is an integral component of preparedness. With hurricanes and volcanic eruptions, for example, buildings can be secured, equipment can be removed, emergency services can be put on alert, and populations can be evacuated, if necessary. It is often said in the seismological community that earthquake forecasting would not be valuable, even if it were possible. All we need are stronger buildings to withstand earthquakes. I believe that this sentiment is misguided. Clearly, advanced warning has been extremely valuable for other natural hazards, and such information in the case of earthquakes would be equally valuable. We must ultimately admit that much of this sentiment stems from our frustration over the surprising difficulty of the earthquake prediction problem.

"Why are earthquakes different from these other hazards? Why are earthquakes the last of the natural hazards to be predictable? For one thing, the short propagation time means that prediction must be based on the existence of a preparation phase. It is clear that we have yet to detect, on a reliable basis, such a preparation phase. Is this because there is no such phase in the case of earthquakes, or because we have not yet observed it? This question is at the heart of the present debate on the predictability of earthquakes.

"The notion that slow tectonic deformation might precede significant earthquakes, and be detectable by seismic instrumentation, has been around for decades. This still remains, in my opinion, the most likely form of an earthquake preparation phase. Progress, however, has been slow in evaluating this hypothesis and more generally in understanding the deformational context of earthquake occurrence. It is the knowledge of this deformational environment that I believe will fill a major gap in our understanding of earthquakes. In the broadest sense, plate tectonic theory has provided us with the underlying cause of most earthquakes, as due to the relative motion of plates along their boundaries. Yet, we have only begun to explore this relationship, and the most important questions remain unanswered. How does steady plate motion ultimately lead to the occurrence of individual seismic events? Are there transients in plate boundary deformation, as suggested by recent strain/geodetic observations, and if so, what are their spatial and temporal characteristics? Do transients propagate? How do the individual faults within a fault system interact? How do earthquakes interact? And finally, is there an observable preparation phase to earthquakes that may form the basis for prediction? It is becoming increasingly apparent that earthquakes are only the most visible part of a complex system of interactions that we have only begun to explore. In order to more fully understand this system that defines the plate-motion/earthquake relationship, I believe it is necessary to characterize and understand its most easily observable manifestation: plate-boundary deformation.

"How, then, should we proceed? We can gain insight from other fields that study complex natural phenomena. Without exception, these fields are primarily data driven. Major advances in understanding have followed major increases in monitoring capability. For example, recent advances in meteorology generally, and weather forecasting particularly, were in large part due to the deployment of a multi-billion-dollar satellite system that allows for continuous, global monitoring of atmospheric disturbances. Our greatest limitation in the study of plate-boundary deformation is the lack of adequate monitoring capability. Of course, we have thousands of seismometers that perform the important task of monitoring seismic activity. But these are primarily intended for studying the earthquake process itself, rather than its deformational environment.

"The earthquake science community needs an adequate facility for the semi-permanent monitoring of the plate-boundary deformation field. A plate boundary deformation network (PBDN) ought to be established that is capable of monitoring deformation along the roughly 1000 km by 200 km segment of the Pacific-North American Plate boundary zone that is dominated by the San Andreas fault system. Such a network should be capable of detecting surface strain spanning the spatial/temporal range defined by plate motion at one end and earthquake rupture at the other: seconds to decades and meters to 100s of kilometers. At present, there is no one seismic/geodetic technique that covers this broad range with adequate sensitivity and dynamic range, and at least two would be required. For example, GPS (or SAR) can cover the long-period (one-month to decades), long-wavelength (>10 km) part of this spectrum, while point-strain measurements, such as those obtained by borehole tensor strainmeters, could be used at shorter period (one-hour to one-month) and wavelength, where they enjoy orders of magnitude of greater sensitivity. Seismometers can adequately cover periods shorter than one hour. The required instrumentation has already been developed for such a network. GPS (and SAR) technology is now standard. Also, several types of borehole tensor strainmeters are either operational or are in final stages of development. The PBDN should also be able to monitor strain at seismogenic depths, as well as at the surface. This is more relevant to the problem of earthquake occurrence, but far more difficult, because strain cannot be measured directly. Instead, strain indicators, such as microseismicity and temporal variations in elastic properties, must be monitored and 'calibrated' in some way. Many of these indicators can be observed using three-component seismometers.

"The PBDN would ideally consist of 1,000–2,000 sites covering the plate boundary zone at roughly 10 km spacing, and each could include, for example, a GPS receiver, a borehole strainmeter, a borehole broadband three-component seismometer and additionally a strong-motion accelerometer for covering the high-frequency, high-amplitude end of deformation. It is encouraging that the GPS/seismometer components of such a network are presently being deployed in Southern California as part of the SCIGN network. In order to adequately fill the sensitivity gap in the one-hour-to-one-month period band, however, it would be necessary to augment this configuration with strainmeters. Indeed, most of the published studies of strain transients have used data from these instruments.

"The major obstacle to deploying such a monitoring network is not technical or logistical, but insufficient resources. It is estimated that the cost of each site would be about $100,000, including a 200-meter-deep hole for the borehole instruments, or about $100 million for the entire network. If viewed as a 20-year experiment, with $10 million/year for maintenance and another $10 million/year for research, this would average out to $25 million/year over the life of the experiment. While this may sound like a large sum to many, it should be put in perspective. As a large facility/research program in the earthquake sciences, it would have a budget comparable to other large programs, such as NEHRP or IRIS. The people of California could entirely support such a program with a contribution of less than a dollar from each resident per year!

"There has been justifiable concern, which I share, that basing a major earthquake science program solely on earthquake predictability would be very risky, given our lack of success to date in achieving this goal. For this reason, earthquake prediction should be embedded within a rich scientific problem that will generate significant results regardless of whether prediction is ultimately achieved. Plate boundary deformation constitutes such a problem. In its own right, it is an important and relatively unexplored area of plate tectonics that is at the foundation of nearly all active tectonics. Geophysics, and particularly seismology, would play a leading role in such a broad endeavor." [3]

NOTE

When an earthquake hit Mexico City in 1985, it was the so-called "earthquake-proof" buildings that suffered the worst damage by far!

Scientists now get help from space in making high-precision surveys that monitor the risk areas for signs of slight horizontal and vertical movement. The margin for error is still wide, but even a slight increase in accuracy could one day save millions of lives.

NOTE

One sign of tectonic plates on the move may be changes in ground water levels, because the stress in the rock squeezes the water upward. A group of Chinese seismologists claim success with predictions based on water levels in wells and boreholes. Another sign may be increased levels of radon gas in the ground, forced upward by built-up pressure. Changes in the electrical resistance of the ground, or in its magnetism, may also hold clues.

Tectonic Plates on the Move

Scientists now have a fairly good understanding of how the tectonic plates move and how such movements relate to earthquake activity. Most movement occurs along narrow zones between plates, where the results of plate-tectonic forces are most evident. There are four types of plate boundaries as shown in Figure 16.1:

- Divergent boundaries, where new crust is generated as the plates pull away from each other
- Convergent boundaries, where crust is destroyed as one plate dives under another
- Transform boundaries, where crust is neither produced nor destroyed as the plates slide horizontally past each other
- Plate boundary zones, broad belts in which boundaries are not well defined and the effects of plate interaction are unclear. [1]

NOTE

Most deaths in earthquakes are caused by collapsing buildings and roadways, so in earthquake-prone San Francisco and in many Japanese cities, all new buildings must be built to particular safety guidelines. Ideas on the best construction for earthquake resistance differ, but regulations insist on deep, firm foundations of steel and concrete. Wide-based pyramid structures are also encouraged. But even such buildings do not come with a guarantee.

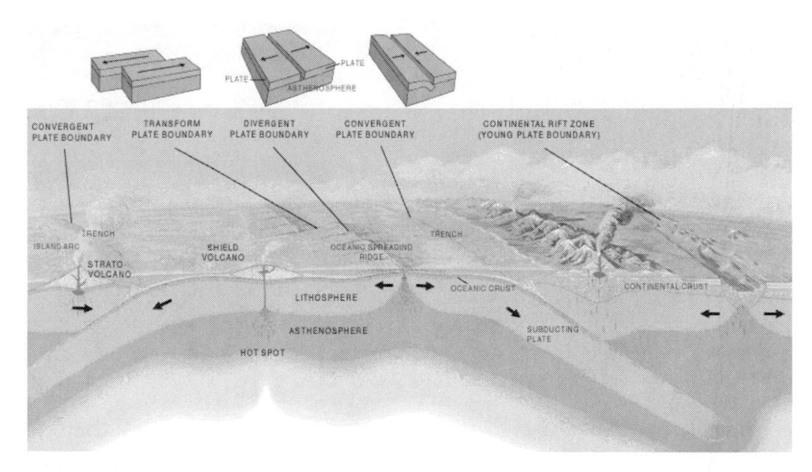

Figure 16.1 The main types of plate boundaries. East African Rift Zone is a good example of a continental rift zone. [1]

Divergent Boundaries

Divergent boundaries occur along spreading centers where plates are moving apart and new crust is created by magma pushing up from the mantle. Picture two giant conveyor belts, facing each other, but slowly moving in opposite directions as they transport newly formed oceanic crust away from the ridge crest.

NOTE

Along California's San Andreas fault, two giant tectonic plates are slipping past each other. Since the great San Francisco earthquake of 1906, movement along this fault has been monitored. Some sections show almost continuous microearthquakes, which release tension. Other sections, seismic gaps, show no signs of activity. This is where the "big one" will occur, as the pressure builds up to a point where it can only be released by a massive shift. But when?

Perhaps the best known of the divergent boundaries is the Mid-Atlantic Ridge, as shown in Figure 16.2.[1] This sub-

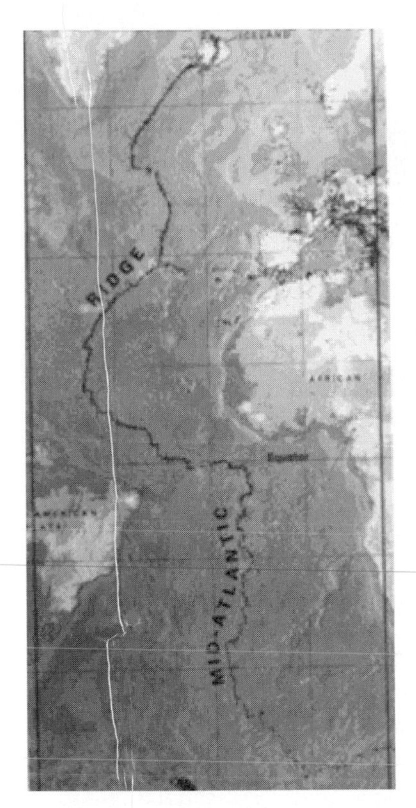

Figure 16.2 The Mid-Atlantic Ridge, which splits nearly the entire Atlantic Ocean north to south, is probably the best-known and most-studied example of a divergent plate boundary.

merged mountain range, which extends from the Arctic Ocean to beyond the southern tip of Africa, is but one segment of the global mid-ocean ridge system that encircles the Earth. The rate of spreading along the Mid-Atlantic Ridge averages about 2.5 centimeters per year (cm/yr), or 25 km in a million years. This rate may seem slow by human standards, but because this process has been going on for millions of years, it has resulted in plate movement of thousands of kilometers. Sea-floor spreading over the past 100 to 200 million years has caused the Atlantic Ocean to grow from a tiny inlet of water between the continents of Europe, Africa, and the Americas into the vast ocean that exists today. [1]

NOTE

On average, plates slip past each other an inch or so a year. In a slip that triggers a major quake, the plates may move a few yards or more. A major earthquake can easily lift an entire mountain yards in the air. Shallow earthquakes originate 0–40 miles below ground. These are the ones that do the most damage. Intermediate earthquakes originate 40–180 miles below ground. Deep ones originate below 180 miles down. The deepest recorded quake started 450 miles down.

The volcanic country of Iceland, which straddles the Mid-Atlantic Ridge, offers scientists a natural laboratory for studying on land the processes also occurring along the submerged parts of a spreading ridge. Iceland is splitting along the spreading center between the North American and Eurasian Plates, as North America moves westward relative to Eurasia (see Figure 16.3). [1]

NOTE

The worst earthquake disaster to hit America in recent years was the deadly quake that struck without warning in the Los Angeles suburb of Northridge on January 19, 1994. In just 30 seconds of total mayhem, 11,000 homes were destroyed, dozens of freeways reduced to rubble, and scores of people killed or injured. Yet before the fatal disaster, no one knew of the existence of the branch of the fault line whose suddenly shifting tectonic plates triggered the devastation.

The consequences of plate movement are easy to see around Krafla Volcano (see Figure 16.4) [1], in the northeastern part of Iceland. Here, existing ground cracks have widened and

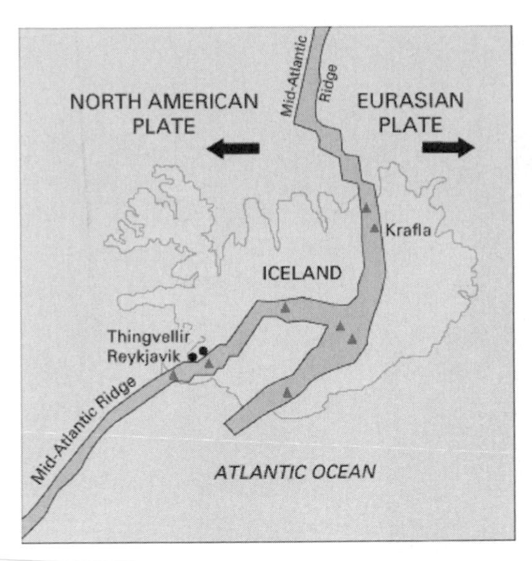

Figure 16.3 Map showing the Mid-Atlantic Ridge splitting Iceland and separating the North American and Eurasian Plates. The map also shows Reykjavik, the capital of Iceland, the Thingvellir area, and the locations of some of Iceland's active volcanoes (triangles), including Krafla.

Figure 16.4 Lava fountains (10 meters high) spouting from eruptive fissures during the October 1980 eruption of Krafla Volcano.

new ones appear every few months. From 1975 to 1984, numerous episodes of rifting (surface cracking) took place along the Krafla fissure zone (see Figure 16.5).[1] Some of

Figure 16.5 Aerial view of the area around Thingvellir, Iceland, showing a fissure zone (in shadow) that is an on-land exposure of the Mid-Atlantic Ridge. Left of the fissure, the North American Plate is pulling westward away from the Eurasian Plate (right of fissure). This photograph encompasses the historical tourist area of Thingvellir, the site of Iceland's first parliament, called the Althing, founded around the year A.D. 930. Large building (upper center) is a hotel for visitors.

these rifting events were accompanied by volcanic activity; the ground would gradually rise one to two meters before abruptly dropping, signaling an impending eruption. Between 1975 and 1984, the displacements caused by rifting totaled about seven meters.

In East Africa (see Figure 16.6) [1], spreading processes have already torn Saudi Arabia away from the rest of the African continent, forming the Red Sea. The actively splitting African Plate and the Arabian Plate meet in what geophysicists call a triple junction, where the Red Sea meets the Gulf of Aden. A new

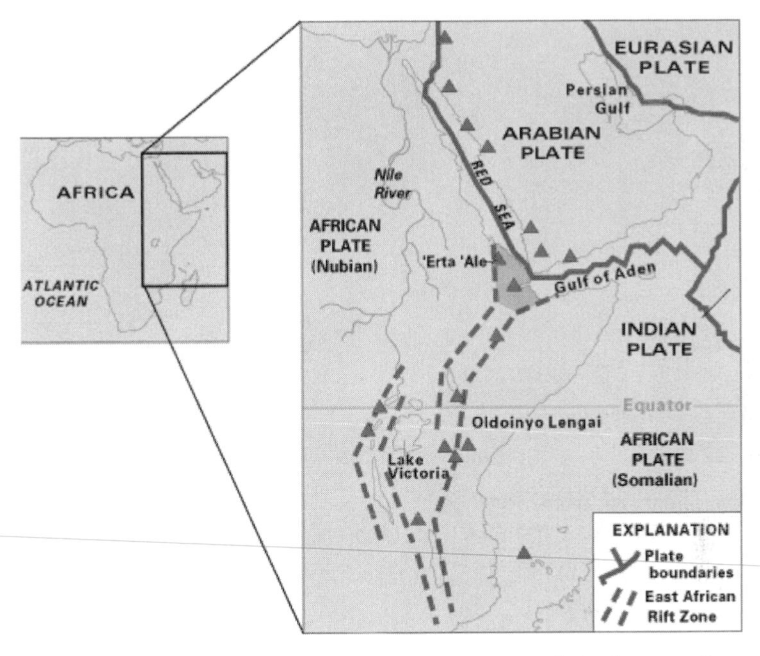

Figure 16.6 Map of East Africa showing some of the historically active volcanoes (triangles) and the Afar Triangle (shaded, center)—a so-called triple junction (or triple point), where three plates are pulling away from one another: the Arabian Plate, and the two parts of the African Plate (the Nubian and the Somalian) splitting along the East African Rift Zone.

spreading center may be developing under Africa along the East African Rift Zone. When the continental crust stretches beyond its limits, tension cracks begin to appear on the Earth's surface. Magma rises and squeezes through the widening cracks, sometimes to erupt and form volcanoes. The rising magma, whether or not it erupts, puts more pressure on the crust to produce additional fractures and, ultimately, the rift zone.

East Africa may be the site of the Earth's next major ocean. Plate interactions in the region provide scientists an opportunity to study first-hand how the Atlantic may have begun to form about 200 million years ago (see Figures 16.7 and 16.8). [1] Geophysicists believe that, if spreading continues, the

Figure 16.7 Helicopter view (in February 1994) of the active lava lake within the summit crater of Erta Ale (Ethiopia), one of the active volcanoes in the East African Rift Zone. Two helmeted volcanologists (observing the activity from the crater rim) provide scale.

Figure 16.8 Oldoinyo Lengai, another active volcano in the East African Rift Zone, erupts explosively in 1966.

three plates that meet at the edge of the present-day African continent will separate completely, allowing the Indian Ocean to flood the area and making the easternmost corner of Africa (the Horn of Africa) a large island.

Convergent Boundaries

The size of the Earth has not changed significantly during the past 600 million years, and very likely not since shortly after its formation 4.6 billion years ago. The Earth's unchanging size implies that the crust must be destroyed at about the same rate as it is being created. Such destruction (recycling) of crust takes place along convergent boundaries where plates are moving toward each other, and sometimes one plate sinks (is subducted) under another. The location where sinking of a plate occurs is called a subduction zone.

The type of convergence (called by some a very slow "collision") that takes place between plates depends on the kind of lithosphere involved. Convergence can occur between an oceanic and a largely continental plate, or between two largely oceanic plates, or between two largely continental plates.

Oceanic–Continental Convergence

If by magic you could pull a plug and drain the Pacific Ocean, you would see a most amazing sight—a number of long narrow, curving trenches thousands of kilometers long and 8 to 10 km deep cutting into the ocean floor. Trenches are the deepest parts of the ocean floor and are created by subduction (see Figure 16.9). [1]

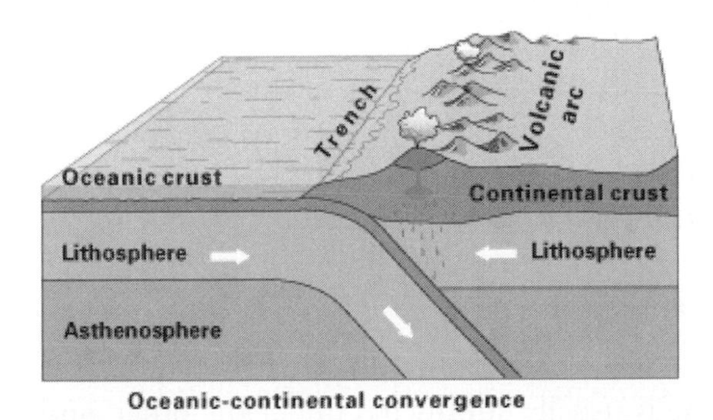

Figure 16.9 Oceanic–Continental Convergence.

Off the coast of South America along the Peru–Chile trench, the oceanic Nazca Plate is pushing into and being subducted under the continental part of the South American Plate. In turn, the overriding South American Plate is being lifted up, creating the towering Andes mountains, the backbone of the continent. Strong, destructive earthquakes and the rapid uplift of mountain ranges are common in this region. Even though the Nazca Plate as a whole is sinking smoothly and continuously into the trench, the deepest part of the subducting plate breaks into smaller pieces that become locked in place for long periods of time before suddenly moving to generate large earthquakes. Such earthquakes are often accompanied by uplift of the land by as much as a few meters.

On June 9, 1994, a magnitude-8.3 earthquake struck about 320 km (200 miles) northeast of La Paz, Bolivia, at a depth of 636 km (398 miles). This earthquake, within the subduction zone between the Nazca Plate and the South American Plate, was one of deepest and largest subduction earthquakes recorded in South America. Fortunately, even though this powerful earthquake was felt as far away as Minnesota and Toronto, Canada, it caused no major damage because of its great depth. [1]

Oceanic–continental convergence also sustains many of the Earth's active volcanoes, such as those in the Andes and the Cascade Range in the Pacific Northwest. The eruptive activity is clearly associated with subduction, but scientists vigorously debate the possible sources of magma: Is magma generated by the partial melting of the subducted oceanic slab, or the overlying continental lithosphere, or both? [1]

Oceanic–Oceanic Convergence

As with oceanic–continental convergence, when two oceanic plates converge, one is usually subducted under the other, and in the process a trench is formed. The Marianas Trench (paralleling the Mariana Islands), for example, marks where the fast-moving Pacific Plate converges against the slower moving Philippine Plate. The Challenger Deep, at the southern end of the Marianas Trench, plunges deeper into the Earth's interior (nearly 11,000 m) than Mount Everest, the world's tallest mountain, rises above sea level (about 8,854 m). [1]

Subduction processes in oceanic–oceanic plate convergence also result in the formation of volcanoes. Over millions of years, the erupted lava and volcanic debris pile up on the ocean floor until a submarine volcano rises above sea level to form an island volcano. Such volcanoes are typically strung out in chains called island arcs. As the name implies, volcanic island arcs, which closely parallel the trenches, are generally curved. The trenches are the key to understanding how island arcs such as the Marianas and the Aleutian Islands have formed and why they experience numerous strong earthquakes. Magmas that form island arcs are produced by the partial melting of the descending plate and/or the overlying oceanic lithosphere. The descending plate also provides a source of stress as the two plates interact, leading to frequent moderate to strong earthquakes.

Continental–Continental Convergence

The Himalayan mountain range dramatically demonstrates one of the most visible and spectacular consequences of plate tectonics. (See sidebar, "The Himalayas.") When two continents meet head-on, neither is subducted because the continental rocks are relatively light and, like two colliding icebergs, resist downward motion. Instead, the crust tends to buckle and be pushed upward or sideways. The collision of India into Asia 50 million years ago caused the Eurasian Plate to crumple up and override the Indian Plate convergence. After the collision, the slow continuous convergence of the two plates over millions of years pushed up the Himalayas and the Tibetan

Plateau to their present heights. Most of this growth occurred during the past 10 million years. The Himalayas, towering as high as 8,854 m above sea level, form the highest continental mountains in the world. Moreover, the neighboring Tibetan Plateau, at an average elevation of about 4,600 m, is higher than all the peaks in the Alps except for Mont Blanc and Monte Rosa, and is well above the summits of most mountains in the United States. [1]

The Himalayas

"Among the most dramatic and visible creations of plate-tectonic forces are the lofty Himalayas, which stretch 2,900 km along the border between India and Tibet. This immense mountain range began to form between 40 and 50 million years ago, when two large landmasses, India and Eurasia, driven by plate movement, collided. Because both these continental landmasses have about the same rock density, one plate could not be subducted under the other. The pressure of the impinging plates could only be relieved by thrusting skyward, contorting the collision zone, and forming the jagged Himalayan peaks.

"About 225 million years ago, India was a large island still situated off the Australian coast, and a vast ocean (called Tethys Sea) separated India from the Asian continent. When Pangaea broke apart about 200 million years ago, India began to forge northward. By studying the history (and ultimately the closing) of the Tethys, scientists have reconstructed India's northward journey. About 80 million years ago, India was located roughly 6,400 km south of the Asian continent, moving northward at a rate of about 9 m a century. When India rammed into Asia about 40 to 50 million years ago, its northward advance slowed by about half. The collision and associated decrease in the rate of plate movement are interpreted to mark the beginning of the rapid uplift of the Himalayas.

"The Himalayas and the Tibetan Plateau to the north have risen very rapidly. In just 50 million years, peaks such as Mt. Everest have risen to heights of more than 9 km. The impinging of the two landmasses has yet to end. The Himalayas continue to rise more than 1 cm a year—a growth rate of 10 km in a million years! If that is so, why aren't the Himalayas even higher? Scientists believe that the Eurasian Plate may now be stretching out rather than thrusting up, and such stretching would result in some subsidence due to gravity.

"Fifty kilometers north of Lhasa (the capital of Tibet), scientists found layers of pink sandstone containing grains of magnetic minerals (magnetite) that have recorded the pattern of the Earth's flip-flopping magnetic field. These sandstones also contain plant and animal fossils that were deposited when the Tethys Sea periodically flooded the region. The study of these fossils has revealed not only their geologic age, but also the type of environment and climate in which they formed. For example, such studies indicate that the fossils lived under a relatively mild, wet environment about 105 million years ago, when Tibet was closer to the equator. Today, Tibet's climate is much more arid, reflecting the region's uplift and northward shift of nearly 2,000 km. Fossils found in the sandstone layers offer dramatic evidence of the climate change in the Tibetan region due to plate movement over the past 100 million years.

"At present, the movement of India continues to put enormous pressure on the Asian continent, and Tibet in turn presses on the landmass to the north that is hemming it in. The net effect of plate-tectonics forces acting on this geologically complicated region is to squeeze parts of Asia eastward toward the Pacific Ocean. One serious consequence of these processes is a deadly "domino" effect: Tremendous stresses build up within the Earth's crust, which are relieved periodically by earthquakes along the numerous faults that scar the landscape. Some of the world's most destructive earthquakes in history are related to continuing tectonic processes that began some 50 million years ago when the Indian and Eurasian continents first met." [1]

Transform Boundaries

The zone between two plates sliding horizontally past one another is called a transform-fault boundary, or simply a transform boundary. The concept of transform faults originated with Canadian geophysicist J. Tuzo Wilson, who proposed that these large faults or fracture zones connect two spreading centers (divergent plate boundaries) or, less commonly, trenches (convergent plate boundaries). Most transform faults are found on the ocean floor. They commonly offset the active spreading ridges, producing zig-zag plate margins, and are generally defined by shallow earthquakes. However, a few occur on land—for example, the San Andreas fault zone in California. This transform fault connects the East Pacific Rise, a divergent boundary to the south, with the South Gorda (Juan de Fuca) Explorer Ridge, another divergent boundary to the north.

The San Andreas fault zone, which is about 1,300 km long and in places tens of kilometers wide, slices through two thirds of the length of California. Along it, the Pacific Plate has been grinding horizontally past the North American Plate for 10 million years, at an average rate of about 5 cm/yr. Land on the west side of the fault zone (on the Pacific Plate) is moving in a northwesterly direction relative to the land on the east side of the fault zone (on the North American Plate).[1]

Oceanic fracture zones are ocean floor valleys that horizontally offset spreading ridges; some of these zones are hundreds to thousands of kilometers long and as much as 8 km deep. Examples of these large scars include the Clarion, Molokai, and Pioneer fracture zones in the Northeast Pacific off the coast of California and Mexico. These zones are presently inactive, but the offsets of the patterns of magnetic striping provide evidence of their previous transform-fault activity.[1]

Plate-Boundary Zones

Not all plate boundaries are as simple as the main types discussed so far. In some regions, the boundaries are not well defined because the plate movement deformation occurring there extends over a broad belt (called a plate-boundary zone). One of these zones marks the Mediterranean–Alpine region between the Eurasian and African Plates, within which several smaller fragments of plates (microplates) have been recognized. Because plate-boundary zones involve at least two large plates and one or more microplates caught up between them, they tend to have complicated geological structures and earthquake patterns.

Rates of Motion

Seismologists can measure how fast tectonic plates are moving today, but how do scientists know what the rates of plate movement have been over geologic time? The oceans hold one of the key pieces to the puzzle. Because the ocean floor magnetic striping records the flip-flops in the Earth's magnetic field, scientists, knowing the approximate duration of the reversal, can calculate the average rate of plate movement during a given time span. These average rates of plate separations can range widely. The Arctic Ridge has the slowest rate (less than 2.5 cm/yr), and the East Pacific Rise near Easter Island in the South Pacific, about 3,400 km west of Chile, has the fastest rate (more than 15 cm/yr). [1]

Evidence of past rates of plate movement also can be obtained from geologic mapping studies. If a rock formation of known age (with distinctive composition, structure, or fossils) mapped on one side of a plate boundary can be matched with the same formation on the other side of the boundary, then measuring the distance that the formation has been offset can give an estimate of the average rate of plate motion. This simple but effective technique has been used to determine the rates of plate motion at divergent boundaries, such as the Mid-Atlantic Ridge, and transform boundaries, such as the San Andreas Fault.

Current plate movement can be tracked directly by means of ground-based or space-based geodetic measurements; geodesy is the science of the size and shape of the Earth. Ground-based measurements are taken with conventional but very precise ground-surveying techniques, using laser-electronic instruments. However, because plate motions are global in scale, they are best measured by satellite-based methods. The late 1970s witnessed the rapid growth of space geodesy, a term applied to space-based techniques for taking precise, repeated measurements of carefully chosen points on the Earth's surface separated by hundreds to thousands of kilometers. The three most commonly used space-geodetic techniques (very long baseline interferometry [VLBI], satellite laser ranging [SLR], and the Global Positioning System [GPS]) are based on technologies developed for military and aerospace research, notably radio astronomy and satellite tracking.

Among the three techniques, to date the GPS has been the most useful for studying the Earth's crustal movements. Twenty-one satellites are currently in orbit 20,000 km above the Earth as part of the NavStar system of the U.S. Department of Defense. These satellites continuously transmit radio signals back to Earth. To determine its precise position on Earth (longitude, latitude, elevation), each GPS ground site must simultaneously receive signals from at least four satellites, recording the exact time and location of each satellite when its signal was received. By repeatedly measuring distances between specific points, geophysicists can determine if there has been active movement along faults or between plates. The separations between GPS sites are already being measured regularly around the Pacific basin. By monitoring the interaction between the Pacific Plate and the surrounding, largely continental plates, scientists hope to learn more about the events building up to earthquakes and volcanic eruptions in the circum-Pacific Ring of Fire. Space-geodetic data have already confirmed that the rates and direction of plate movement, averaged over several years, compare well with rates and direction of plate movement averaged over millions of years.

The Future: Using Satellites to Detect Ground Movement

As previously explained, most seismologists pin their hopes for the future of earthquake prediction on accurate measurement of ground movements using space technology. There are already 45 observatories in more than 30 countries using lasers and satellites to make amazingly accurate measurements—so precise that they can pick up movements of a fraction of an inch over thousands of miles. These systems work by bouncing laser beams off reflectors on satellites orbiting the Earth and measuring how long they take to return.

A linked network of four laser-satellite stations called Keystone have come into operation around Tokyo Bay in Japan. A fifth linked station is airborne. This linked network is able to track satellite passes and make measurements in earthquake zones over 700 times a month. The system provides seismologists with more accurate information about ground movements almost instantly. Armed with this new technology, the scientists are able to map the slight, but telltale, shifts that indicate an earthquake is on its way. It may not be the best answer, but it is better than nothing. [1]

NOTE

Using laser beams bounced off satellites, it is possible to detect the tiny movements in the Earth's crust that precede a quake. But even if you could predict that an earthquake was about to hit Los Angeles, you would still have the problem of evacuating the city.

Lubricating the Cracks in the Earth's Crust

Finally, earthquakes are set off by friction between the rock on either side of a fault line. Some people argue that seismologists might be able to reduce the severity of an earthquake by injecting lubricating fluids into the fault so that the rock plates slide past each other more smoothly. This could possibly avert the build-up of tension that results in major quakes. In the past, attempts have been made to lubricate fault lines with water and mud. More recently, it has been suggested that the tension could be reduced by forcing the sides of the cracks apart using high-pressure steam.

Conclusion

The goal of earthquake predicting is to give warning of potentially damaging earthquakes early enough to allow appropriate response to the disaster, enabling people to minimize loss of life and damage. There is great interest in predicting earthquakes because of their devastating potential. According to geophysicists, there is currently no reliable way to predict the days or months when an event will occur in any specific location. However, there is a great amount of information to inform geophysicists where earthquakes are likely to occur. Ultimately, scientists would like to be able to specify a high probability for a specific earthquake within a particular year. There are two ways to predict earthquakes: by studying the history of large earthquakes in a specific area and analyzing the rate at which strain accumulates in the rock.

Scientists study the past frequency of large earthquakes in order to determine the future likelihood of similar large shocks. According to the geophysicists, this is called long term prediction. In places where earthquakes are known to occur repeatedly, like along plate boundaries, it is possible to detect a regular pattern of recurring large quakes. To do this, seismologists have to look at prehistoric earthquakes. After studying these earthquakes, scientists can look at the intervals between these major quakes. With this information, it may be possible to predict where and when a large quake may happen next. Scientists have found that prehistoric earthquakes happened along the Pacific Rim and so they get quite a bit of attention. For example, if a region has experienced four magnitude 7 or larger quakes during 200 years of recorded history, then scientists assign a 50% probability to the occurrence of another magnitude 7 or larger quake during the next 50 years. [1]

Another way to estimate the likelihood of future quakes is to study how fast strain accumulates. Prediction of this kind is short-term. Most research on short-term earthquake prediction involves monitoring changes in the properties of elastically strained rocks. Tilting or bulging of the ground and slow rises and falls in elevation are signs that strains are building up. Cracks and fractures are significantly important in the strained rock. When the strained rocks come to the critical level, like when a rubber band is pulled too tight, the rocks will break and slip into a new position. Scientists then measure how much strain accumulates along a fault segment each year, how much time has passed since the last earthquake along the segment, and how much strain was released in the last earthquake. This information is then used to calculate the time required for the accumulating strain to build to levels that result in earthquakes. This method is rare because it is hard to find such detailed information about faults and strained rock.

For example, the Pacific Northwest faults are complex, and it is not yet possible to forecast when any particular fault segment in Washington or Oregon will break. However, seismologists can look at recurring patterns and roughly predict when an earthquake is going to occur. The Pacific Northwest has only been monitored for a couple of decades, not long enough to allow seismologists to see what patterns, if any, exist here. Seismologists are still trying to understand what types of earthquakes are possible there, and what kind of shaking you will experience from future earthquakes (depending on the earthquake location and size, and the site geology and topography).

Finally, scientists' understanding of earthquakes is of vital importance to the world. As the population increases, expanding urban development and construction works encroach upon areas susceptible to earthquakes. With a greater understanding of causes and effects of earthquakes, seismologists may be able to reduce damage and loss of life from this destructive phenomenon. [1]

References

[1] U.S. Geological Survey, MS119 National Center, Reston, VA 20192, 2003.

[2] Dr. Paul G. Silver, Senior Staff Scientist, Carnegie Institution of Washington, Department of Terrestrial Magnetism, 5241 Broad Branch Rd. NW, Washington, DC 20015.

[3] Dr. Paul G. Silver, "Why is earthquake prediction so difficult?," *Seismol. Res. Lett.*, 69, Carnegie Institution of Washington, Department of Terrestrial Magnetism, 5241 Broad Branch Rd. NW, Washington, DC 20015, 1998, 111–113.

Chemistry

How Microscopic Atomic Forces Produce Various Macroscopic Behaviors

"The release of atomic energy has not created a new problem. It has merely made more urgent the necessity of solving an existing one."

—Albert Einstein (1879–1955)

The violation of universal natural law is the main contributor to all problems. Problems of national health, crime, the economy, education, and the environment—all problems plaguing individual and national life have their origin with the widespread violation of universal natural law by the whole population. However, it is first necessary to understand what universal natural law is and how it governs the universe before you can really understand this sweeping statement.

Universal Natural Laws

Universal natural law governs the entire universe. Universal natural laws are the orderly principles that govern physical events and processes. Science defines the universal law of nature. It is a precise description of how nature behaves under specific circumstances.

Modern science, over the last several centuries has identified myriad universal laws of nature governing behavior at all levels of the physical universe: on subatomic, atomic, molecular, biological, geological, astrophysical, and cosmological scales. Indeed, according to science, universal natural law pervades everything. So, from the motion of a particle, to the evolution of life, to the functioning of entire societies, nothing happens that is not the direct result of universal natural law.

Humans rely on universal natural law. From the most modern technologies harnessing fundamental universal laws of nature at the electronic and nuclear levels, to the most mundane act, such as eating, which utilizes countless laws governing metabolism and digestion, everything that humans accomplish is achieved by stimulating and enlisting the support of universal natural laws.

In other words, universal natural law governs every blink of the eye, every breath, and every heartbeat. Universal natural law is as intimate to humans as life itself.

Universal Natural Law Scope

From the microscopic atomic forces to various macroscopic behaviors (which will be discussed briefly later in the chapter)—and from the subatomic world of elementary particles to the large-scale structure of the universe, modern science has discovered that the universe is structured in layers of existence. Science has also learned that these various levels are distinguished not only by vastly different time and distance scales, but that every level of the physical universe has its own set of laws. For example, early attempts to understand the atom relied on planetary models, with the massive sun corresponding to the heavy nucleus and the planets in their orbits analogous to electrons. At that time, it seemed natural and intuitive that the same basic universal laws of nature would apply—even at vastly different scales. This assumption proved false. The planetary model thus failed utterly to account for even the most basic properties of the atom—its structure and stability.

Associate Professor, Dr. Andrew Cooksy (Department of Chemistry, San Diego State University) explains:

> While it's true that for practical purposes we rely on different equations at the macroscopic level than at the microscopic level, I believe most of us expect and hope that there is actually a single set of fundamental laws that describe the universe at all of these levels. We use quantum mechanics to predict the properties of matter and energy at extremely small scales, because we've proven that classical mechanics doesn't work there. But the reverse is not true: We can demonstrate that as the scale of the system increases, those same quantum mechanical formulas do predict the observed, classical behavior. The practical consideration appears to be simply that the quantum mechanical approach is too demanding (and too detailed in the resulting information) for use on large systems. One sees the same behavior in the laws of relativity: The relativistic equations we apply to very energetic systems smoothly segue into the familiar classical laws as the systems approach more familiar energy scales. [1]

Instead, in order to describe physics at the atomic scale, a whole new conceptual framework, known as quantum mechanics, was required. An entirely new language had to be introduced, involving new mathematics based on infinite dimensional complex vector spaces. All of this led scientists to conclude, reluctantly, that the microscopic world is entirely different from the macroscopic world of everyday events. The universe began to be understood as comprised of worlds within worlds, where each world has its own set of laws, its own qualitative flavor, its own quantitative characteristics—and each world self-consistent within itself.

According to Professor of Chemistry, Dr. James F. Harrison (Department of Chemistry, Michigan State University),

Quantum chemistry is the application of the laws of quantum mechanics to the solution of chemical problems. These problems may be conceptual, such as understanding the nature of the chemical bond, or more pragmatic, such as predicting the geometry and properties of a molecule from first principles. In 1929, the famous physicist P.A.M. Dirac said regarding quantum mechanics: "The underlying physical laws necessary for the mathematical theory of a large part of physics and the whole of chemistry are thus completely known, and the difficulty is only that the exact application of these laws leads to equations much too complicated to be soluble." Dirac was alluding to the fact that the equations of quantum mechanics, essentially the Schrodinger equation, could be written down for any atom or molecule or solid and contained only the fundamental constants of nature as input. These constants include Planck's constant, the masses and electrical charge of the nuclei and the electron. The idea was that one could predict, *ab initio*, from these constants and computation alone, the result of a laboratory experiment. This was an astonishing view as it implied that, in principle, chemistry could be predicted from the laws of quantum mechanics. The intervening 70 plus years have witnessed incredible progress in our ability to apply the laws of quantum mechanics with ever increasing accuracy to the problems of chemistry. Before the advent of computers, most research in quantum chemistry focused on applying the concepts of quantum mechanics to understand how the experimentally derived laws of chemistry related to the mathematical structure of quantum mechanics. This was necessary, as the complex equations of quantum mechanics resisted all but very simple approximate solutions. This was a remarkably creative period in the development of chemistry, and the success obtained led most researchers to the view that chemistry was indeed contained in quantum mechanics. In the early 1960s when commercial computers were becoming available, several visionaries began to explore the possibility that computers could enable one to solve, quantitatively, the equations of quantum mechanics as they apply to atoms, molecules, and solids. The success of these pioneers, coupled with the incredible increase in computer power over the last four decades, has resulted in quantum chemistry being an essential part of chemistry. One often speaks of theory in science as having a speculative connotation. It is important to realize that, to the extent that quantum mechanics is correct, there is no speculation. Its predictions over the past 75 years have been verified by countless experiments, and whenever a dispute has arisen, quantum mechanics has emerged victorious. It is now safe to say that, in many areas, quantum mechanics provides an alternative to traditional laboratory experiments.

I am interested in using quantum mechanics to understand the electronic structure of atoms, molecules, and solids. Presently my focus is on the electronic structure of small molecules containing an element from the first transition series of the periodic table and one or more main-group elements. These small, highly reactive molecules are characterized by having several low-energy, excited states with high-spin multiplicities, and they are believed to be intermediates in many chemical reactions involving transition metals. The nature of the chemical bond in these systems is poorly understood, and theoretical studies provide the most direct and least ambiguous means of study.

Presently one is able to compute the properties of small molecules (5–10 atoms) with an accuracy that is comparable to or occasionally exceeds experiment. However, the computational techniques used scale poorly with system size. For example, a very popular technique for obtaining accurate solutions to the Schrodinger equation is called the coupled-cluster method, and this scales like the 6th power of the system size. So if it takes one hour to calculate the properties of a molecule and one doubles the size of the molecule (system), then the same accuracy calculation will take 2**6, or 64 hours. Since many systems of chemical, physical, or biological interest are very large, this scaling time could amount to years if not centuries and effectively precludes the application of these theoretical techniques to many systems. For example, a small protein may contain 1,000 atoms, and if it was a factor of 100 larger than a system that took 1 second to compute, a comparable calculation could take 100**6 seconds, or 100 million seconds, or 1,160 days, or 3 years. So there is a critical need to

> develop theories and computer algorithms that scale more reasonably. For example, if one had a method that scales linearly, then our 1,000-atom protein would take 1,000 seconds, or 17 minutes, rather than 3 years. Breakthroughs in this area would allow the power of quantum mechanics to be applied to biological problems and would result in unprecedented insights into the diseases that afflict humanity. [2]

This early 20th-century discovery of the quantum world has been followed in recent years by the rapid unfolding of scientific knowledge of even deeper layers of nature's functioning. These subatomic (nuclear and subnuclear) levels reveal deeply hidden symmetries of nature that unite the fundamental forces and particles—forces and particles that appear distinctly different at more superficial levels.

In 1979, the Nobel prize was awarded to Professors Weinberg, Salam, and Glashow for their theory uniting the weak and electromagnetic forces (two of the four fundamental forces) at a distance scale 100 times smaller than the atomic nucleus. The remarkable success of this unified electroweak theory gave profound support to this scientific process of inner exploration of deeper levels of nature and led to the rapid unfolding of still more unified theories of the fundamental forces.

These deeper layers of nature's functioning are characterized by unparalleled levels of energy and dynamism. For the same reason that nuclear transformations liberate a million-fold more energy than chemical transformations, the quantum uncertainty principle similarly guarantees that still more fundamental levels of natural law are vastly more energetic. The characteristic energy is so enormous that the energy density in a single cubic centimeter of empty space exceeds, by orders of magnitude, the entire mass energy of the visible universe at the deepest level of nature's dynamics.

Nonetheless, deeper levels of universal natural law are increasingly comprehensive and holistic in addition to being more energetic. The more superficial levels of nature can be seen as fragmented expressions of the more unified laws governing deeper levels. Macroscopic behaviors are merely superficial manifestations of more fundamental underlying processes. For example, the behavior of classical objects can be derived from the basis of their atoms (just as thermodynamics can be derived from statistical mechanics), whereas the converse is not the case. In other words, deeper levels are more fundamental. The outer is based upon the inner.

Dr. Andrew Cooksy states,

> I have a fondness, certainly not shared by all of my colleagues, for the planetary model of atoms. It is indeed inaccurate, but I wouldn't say that it failed utterly. The Bohr model of the atom is fundamentally a planetary model, and was nonetheless remarkably successful (in my opinion) in predicting the observed spectra of one-electron atoms. That is its only scientific success, but it was a critical one in advancing the understanding of quantum mechanics in chemical systems. Its failures are many, including an incorrect equation for the angular momentum and the intrinsic inability of a planetary model to reproduce the actual, three-dimensional shapes of the electron "clouds." On the other hand, just try to get a typical chemist to avoid falling back into this very classical image of how the atom works. As an aside, this is a typical case of how chemists in particular often use quantum mechanics. We often keep the more intuitive, classical picture in our heads but correct it by imposing some of the constraints that come from quantum mechanics. For example, the Bohr model was adequate for predicting the spectra of one-electron atoms because it took the planetary model and then enforced the nonclassical restriction that the circumference of the orbit be limited to certain values (later shown to be multiples of the de Broglie wavelength of the electron). [1]

Chemists can see, then, that the whole, seemingly inert, manifest universe is a superficial, partial, and fragmented reflection of the concentrated, comprehensive intelligence and dazzling symmetries that exist at fundamental scales—the tip of the iceberg of universal natural law. The culmination of this inward march of modern science has been the recent discovery of completely unified field theories. These theories, based on the superstring theory, identify a single, universal field of nature's intelligence at the basis of all forms and phenomena in the universe—the fountainhead of all known laws of nature. At this superunified scale (the Planck scale, or 10^{33} cm), these theories describe a total transformation of universal natural law, involving higher dimensions of space–time, infinite towers of massive particles, and a spectacular, nonlinear dynamics that is so intricate and subtle that it transcends current mathematical description.

Supreme Administration

It is helpful to recognize that the governance of nature through universal natural laws is not different in principle from the governance of society through national laws. The reason for this recognition is to bring this abstract discussion of laws of nature into the familiar realm of human experience. A national government administers society through the formulation and enforcement of specific laws governing various forms of human behavior and human interaction—financial, social, environmental, etc. Similarly, chemists have seen that nature's government administers all universal natural events and processes through specific universal natural laws. Interestingly, both national law and universal natural law have a hierarchical structure, with deeper, more foundational laws underlying more specific laws. National law has its ultimate foundation in the Constitution, on the basis of which the constitutionality or legitimacy of the more specific and temporal laws and regulations are derived. As you will see, in the same way, the hierarchical structure of nature's government has a foundational level (a Constitution of the Universe). Furthermore, this relates to the laws governing the dynamics of the unified field itself, which underlie and give rise to the diversified universal laws of nature governing every level of the physical universe.

What are the properties of nature's government? And how does it administer the immensely vast and unfathomably complex universe? [3]

First of all, nature's government is rich with profound order. This order stems from the symmetrical structure of a snowflake, to the profound symmetries displayed in the elementary particles; from the orderly structure of the atom, to the precise elliptical orbits of the planets; and from the regular structure of crystals, to the highly ordered states of open thermodynamic systems.

Second, nature's government is maximally efficient. The most pervasive law governing the whole of macroscopic behavior is the Principle of Least Action. This principle states that whatever nature accomplishes, nature accomplishes with maximum efficiency and economy: That all physical events unfold along the most economical path (the path where the action) a mathematically defined measure of energy and complexity—is minimized. This principle explains, for example, why a ray of light refracts in its propagation from air to water. The resulting bent path is actually the path where the propagation time (the time required to go from source to goal) is minimized. Thus, based on this simple principle of maximum economy of nature, all classical behavior can be derived. [3]

"Another pervasive law governing macroscopic behavior is the second law of thermodynamics," explains Dr. Cooksy:

> Whereas the principle of least action dictates the path taken by a process from one point to another the second law can be said to explain why the process is going between those two points (and

> especially why it goes in the direction it does.) The second law is a marvelous example of microscopic properties driving macroscopic activity, stating simply that entropy always increases during any change (except for idealized cases where it can stay the same). Bolzmann's law defines the entropy to be essentially a count of the number of microscopic quantum states accessible to your macroscopic system. One finds that all processes evolve in such a way that the total number of available quantum states in the world is always increasing. [1]

The principle of conservation of energy is another example of nature's absolute efficiency. In all physical processes, nothing is wasted—energy is neither created nor destroyed. And somewhat surprisingly, at the deepest level, nature is even energy-creating. In modern inflationary models of cosmology, the universe undergoes a period of vast, exponential growth during which, due to a delicate balance of gravitational and material energy, the universe expands by many orders of magnitude while maintaining a constant energy density. The result is that the entire universe, with its unimaginable mass and energy, emerges, essentially, from nothing.

Finally, universal natural law is inherently life-supporting. It has given birth to and sustains the existence of over eight million species on earth. Recent evidence suggests that this evolutionary quality is not, in any sense, peculiar to earth. Wherever marginally hospitable environments exist, living organisms are likely to be found. Remarkable and varied life forms emerge, even deep within the earth, or in fissures deep within the ocean with temperatures above boiling.

Indeed, evolution is the very nature of life. Wherever chemists see life, they see growth, evolution: growth of size, complexity, capability, power, knowledge. Thus, the nature of life is to grow, even with the most cursory examination of life.

There is a prevailing impression that nature is full of violence and cruelty. Certainly, violence is found in nature. When you consider the entire lifespan of an organism, that organism inevitably dies. This death is an isolated incident that occurs but once in a lifetime, spent growing in size, experience, capability, power, and reproducing itself. Death is a part of life, in other words, but a small part, an isolated event in the course of a lifetime, and it ultimately plays a natural role in the larger evolutionary process of sustaining and nourishing millions of organisms in a complex ecosystem.

Thus, nature as inherently life supporting is a more accurate global perspective. It is, after all, the universal laws of nature that have given birth to eight million species and sustain the complex ecosystems in which these myriad species mutually nourish and support one another. The natural world is teeming with life. This includes healthy topsoil, which contains thousands of microorganisms in every square inch of it.

From this, chemists conclude that nature's government is life-supporting, orderly, maximally efficient, and evolutionary. Moreover, science reveals that these qualities are fundamental—they have their origin in the deepest levels of universal natural law.

For example, chemists have previously seen that the orderliness displayed at macroscopic levels is just a partial reflection of deeply hidden symmetries of nature present at more fundamental scales. Even the symmetry of a snowflake is a macroscopic reflection of the angular symmetry of the water molecule. Nature's absolute efficiency, embodied in the principle of least action governing all classical phenomena, actually derives from the deeper quantum mechanical principle of democracy of paths. [3]

NOTE

The path of least action is just the macroscopic outcome of the simultaneous superposition of multiple coexisting paths at the microscopic level.

Similarly, the evolutionary quality, other fundamental properties of nature, and conservation of energy have their origins at deeper levels. Ultimately, all of nature's fundamental qualities are rooted in the Constitution of the Universe. These are the laws governing the behavior of the unified field itself. [3]

One can only marvel at the perfection of nature's government, which administers the immensely complex universe without problems. Given a choice, any head of state would like his or her government to enjoy the same success, orderliness, efficiency, and life-supporting quality as nature's government. The scientific knowledge and practical methodologies to achieve this are available now, as you will see.

Universal Natural Law Application

How can society draw upon the immense organizational intelligence displayed throughout the physical universe? There is a long tradition of society making use of universal natural laws, both through technologies and through other means. Technology is the practical application of the laws of nature discovered by science. The electric light, which has profoundly revolutionized our lifestyle, is a simple application of the laws of electrodynamics. Medicines are applications of the laws of chemistry and human physiology. Other advanced applications of nuclear, chemical, and electronic technologies are transforming the face of human civilization in a similar way. [3]

The use of universal natural law is far more pervasive and intimate to life, while technology is the advanced application of specific laws discovered by science. From the time of birth, children learn to take advantage of the universal natural laws governing their environment. Even without the intellectual understanding of those laws, the simple act of walking is an immensely complex process, which stimulates and utilizes myriad laws governing electrochemical, neuromuscular, and mechanical processes.

Indispensable to every aspect of life is the efficient application of universal natural law. Some proficiency in the applications of universal natural law is gained through the natural process of maturation and life experience, as exemplified in the process of walking. In a developed society, such proficiency is also gained through education and through the scientific discovery of new universal laws of nature and their application through technology. With such scientific development, one can take advantage of universal natural laws that may be otherwise hidden or inaccessible. For instance, through the application of electrical and chemical energy, one can learn to generate heat and warmth in the winter.

Everything humans accomplish, from the simple act of digestion to putting a man on the moon, they accomplish through the skillful application of universal natural law. Conversely, as you will see next, everything you fail to accomplish is due to the failure to apply universal natural law effectively.

The Source of All Problems

As you have just seen, it is possible to enlist the support of universal natural law to fulfill a particular desire or to accomplish a specific purpose. This successful application of laws of nature may be either through the sophisticated application of universal natural laws through technology or instinctive, learned behavior.

It is also possible that universal laws of nature may be used ineffectively—or even destructively, without sufficient knowledge of universal natural law. Ineffective use of universal natural law means not taking maximal advantage of universal laws of nature to accomplish one's ends most effectively. For instance, a simple task, like the displacement of a boulder, becomes arduous and complex in the absence of the knowledge of the lever.

Nonetheless, whether accidental or intentional, the misapplication of universal natural law can have severe negative repercussions. A young child, unaware of the implications of the laws of gravity, can crawl off a bed and cause injury to himself. Doctors, unaware of the laws governing radiation and its carcinogenic effect on the body, prescribed radiation to remove facial hair in the 1950s, resulting in widespread incidence of skin cancer. Scientists, unaware of the deeper laws governing the delicate balance of ecosystems, advocated worldwide usage of DDT, resulting in serious health repercussions for humans, extinction of numerous species, and soil toxicity.

What chemists term a violation of universal natural law is the misapplication of universal natural law, with all its negative repercussions. Thus, you can trace the violation of universal natural law by the whole population, to all of the problems of individuals and society.

The Meaning of Violating Universal Natural Law

A universal law of nature is, by definition, inviolable in scientific usage. The term universal law of nature is reserved for those scientific principles that apply under all circumstances and for all times. For instance, upheld without exception throughout the universe is Einstein's discovery, in the context of his special theory of relativity, that the speed of light is the limiting velocity for the propagation of matter and information.

Therefore, when chemists refer to violations of universal natural law, they are simply referring to actions that not only fail to take maximum advantage of the universal laws of nature, but where the functioning of universal natural law results in negative consequences for the individual. For example, smoking cigarettes introduces known carcinogenic compounds that cause DNA mutation and cancers to form in the bronchi and lungs. Smoking is thus an example of an action that violates universal natural law. Furthermore, it's an action that stimulates certain universal laws of nature to produce additional undesirable consequences—like causing secondary smoke that people inhale who don't smoke, thus resulting in further deaths.

Between universal natural law and national law, there is a profound parallel. Both sets of laws define rules of behavior that are designed to be in the best interest of the individual and society. Both result in negative consequences of acting outside these boundaries of prescribed behavior. One obvious difference is that one can sometimes escape the consequences of violating national law, if one does not get caught. In contrast, one is always in the grip of universal natural law and its consequences.

Violation of Universal Natural Law Problems

So far, this chapter has described the all-pervasive character of universal natural law that everything that occurs is the direct result of the functioning of universal natural law. It is therefore not surprising that the failure to achieve one's natural aspirations (health, wealth, or success in life) results from the failure to apply natural law successfully, and that all significant achievements result from the skillful application of laws of nature.

Indeed, a careful examination of the problems that confront individual and national life is sufficient to confirm that all significant problems result from the violation of universal natural law. Such violations are directly responsible for the economic, health, social, and environmental problems facing governments today. These problems can be solved by bringing national life into accord with universal natural law. Now, let's review some of the specific mechanisms through which violation of universal natural law leads to society's most pervasive health and social problems.

Problems Caused by Violation of Universal Natural Law

This chapter has shown how the misapplication of universal natural law, even unintentionally, can cause unforeseen negative consequences. Certainly, the failure to achieve one's desired purpose through the misapplication of universal natural law is itself a problem. However, there are other ways in which violation of the universal natural law causes problems.

- If a violation takes the form of an action that is unhealthy or harmful to the body (e.g., smoking), certain violations of universal natural law cause physiological stress directly.
- The fulfillment of desires is dependent upon individual growth. Fulfillment of a desire brings happiness, and also raises the scope and the standard of future desires. Without such growth, which is essential to life, frustration is inevitable. Over time, this frustration causes physiological stress, which research shows impairs creativity and further restricts the ability to fulfill future desires. This can also lead to problems of health.

- Further violations of universal natural law are promoted by stress resulting from either mechanism. Stress causes a complex psychophysiological chain reaction in the human body. Chronic, acute stress leads to an out-of-balance biochemistry that is linked with anxiety, aggression, hostility, impulsive violent behavior, and substance abuse. In other words, behavior that violates natural law comes from accumulated stress, which compels an individual towards activities that are unhealthy. [3]

Now, let's briefly look at how microscopic atomic forces produce various macroscopic behaviors. A brief discussion will ensue that shows how hot objects cool and how this macroscopic behavior is related to the microscopic behavior of the constituents of macroscopic objects.

How Microscopic Atomic Forces Produce Various Macroscopic Behaviors

The goal here is to understand the properties of macroscopic systems, that is, systems of many electrons, atoms, molecules, photons, or other constituents. Examples of such macroscopic objects include familiar systems such as gases, liquids, solids, and polymers. Thus, macroscopic systems such as the neutrons and protons comprising the nucleus, the brain, and the galaxies, are less familiar to chemists' everyday experience.

You will find that the type of questions that are asked about macroscopic systems differ in important ways from the questions chemists ask about microscopic systems. An example of a question that they might ask about the macroscopic systems is, "What is the trajectory of the Earth in the solar system?"

However, have you ever wondered about the trajectory of a particular molecule in the air of your room? Why not?

According to Dr. Cooksy,

> There are several applications where chemists are concerned with the trajectories of individual molecules, including some impressive studies of how gases interact with solid surfaces in industrial catalytic reactions. One of the most remarkable examples of research depending on molecular trajectories is the work of Dudley Herschbach, Yuan T. Lee, and others in the field of crossed molecular beam reaction dynamics (for which Herschbach and Lee shared the 1986 Nobel prize in chemistry, together with John C. Polanyi, who worked on a different aspect of the problem). In these experiments, a chemical reaction between two molecules is studied by creating a narrow supersonic gas stream of each reactant in a vacuum chamber. The two streams intersect in the middle of the vacuum chamber. The distances separating the molecules in these streams is so large compared to the molecular diameters that most of the molecules fly through this intersection without noticing the other stream. But every now and then, a molecule from one stream hits a molecule from the other stream, and then a reaction between the two can occur. The result is that the product molecules from the reaction are deflected out of the paths of the two streams and fly towards the wall of the vacuum chamber at some other angle. By placing mass analyzers at various positions in the chamber, these researchers are able to map out how much of each product ends up a spot on the vacuum chamber wall. From this map, they can work backwards to determine the collision speeds and even the specific molecular orientations that best favor the chemical reaction. This is an elegant way to study a chemical reaction—which we traditionally investigate at the macroscopic scale—one molecular reaction at a time. [1]

So, how does the pressure of a gas depend on the temperature and the volume of its container? Well, according to Dr. Cooksy,

A traditional explanation of how gas pressures depend on other parameters of the gas is called the "kinetic theory of gases" and relies primarily on the microscopic perspective. Pressure is the average total force exerted by the gas on the walls of the container, divided by the total area of the walls. Looking at a particular region of the container wall, say a one-square-cm area, the molecules that will hit that region come from some small volume of space just above that one-square-cm patch of wall. They hit the wall and bounce back into the rest of the gas. The momentum that is transferred to the wall in those collisions accounts for the force the gas exerts on the container. If we double the volume of the container, keeping the number of molecules the same, then the density of molecules drops throughout the container, and the small volume of gas molecules that will hit our one-square-cm patch now holds only half as many molecules as it did originally. Therefore, only half as many molecules strike the wall in a given period of time, transferring only half the momentum, and the pressure drops in half. By increasing the temperature of the gas, we increase the average energy of the molecules, and therefore increase their individual speeds and momenta. The average speed and momentum increase only as the square root of the temperature, but this has two effects: (1) the momentum transfer from each collision increases, and (2) because the molecules are moving faster, more molecules hit the wall over the same period of time. Both terms grow by a factor of the square root of the temperature, so the pressure (which depends both on how many and how hard are the collisions) increases in direct proportion to the temperature. [1]

So, with this in mind, examples of questions that chemists might ask about macroscopic systems include the following:

1. How does a refrigerator work? What is its maximum efficiency?
2. What is the maximum energy that can obtain from a coal-fired power plant?

3. How much energy is needed to add to a kettle of water to change it to steam?
4. Why are the properties of water different from those of steam, even though water and steam consist of the same type of molecules?
5. How are the molecules arranged in a liquid?
6. How and why does water freeze into a particular crystalline structure?
7. Why does iron lose its magnetism above a certain temperature?
8. Why does helium condense into a superfluid phase at very low temperatures? Why do some materials exhibit zero resistance to electrical current at sufficiently low temperatures?
9. How fast does a river current have to be before the flow changes from laminar to turbulent?
10. What will the weather be like tomorrow? [3]

These questions can be roughly classified into three groups. Questions 1–3 are concerned with macroscopic properties such as pressure, volume, and temperature and questions related to heating and work. These questions are relevant to *thermodynamics,* which provides a framework for relating the macroscopic properties of a system to one another. Thermodynamics is concerned only with macroscopic quantities and ignores the microscopic variables that characterize individual molecules. For example, chemists will find that understanding of the efficiency of a refrigerator does not require a knowledge of the particular liquid used as the coolant. Many of the applications of thermodynamics are to thermal engines—for example, the internal combustion engine and steam turbines.

Questions 4–8 are based on understanding the behavior of macroscopic systems starting from the atomic nature of matter. For example, chemists know that water consists of molecules of hydrogen and oxygen which in turn consists of nuclei and electrons. Chemists also know that the laws of classical and quantum mechanics determine the behavior of molecules at the microscopic level. The goal of statistical mechanics is to begin with the microscopic laws of physics that govern the behavior of the constituents of the system and deduce the properties of the system as a whole. The bridge between the microscopic and macroscopic worlds is known as statistical mechanics.

Thermodynamics and statistical mechanics assume that the macroscopic properties of the system do not change with time on the average. The third set of questions concern macroscopic phenomena that change with time (see Questions 9 and 10).

NOTE

Thermodynamics describes the change of a macroscopic system from one equilibrium state to another.

Finally, related disciplines are nonequilibrium thermodynamics and fluid mechanics (from the macroscopic point of view) and nonequilibrium statistical mechanics (from the microscopic point of view). Thus, chemists' understanding of nonequilibrium phenomena is much less advanced than their understanding of equilibrium systems, although there has been progress in their understanding of nonequilibrium phenomena, such as turbulent flow and hurricanes.

Conclusion

This chapter concludes by restating that the violation of universal natural law leads to stress, and stress leads to further violations of universal natural law. This vicious cycle has taken an enormous toll on society. Due in part to the widespread violation of universal natural law by the whole population, society is experiencing an epidemic of stress. Statistics reveal an alarming rise in stress-related illness, such as stroke, hypertension, and heart disease. Indeed, 80% of disease today is caused or complicated by stress. This amounts to a trillion dollar drain on the national economy. An additional $500 billion is lost to U.S. industry annually due to job-related stress. High levels of stress among youth have eroded academic outcomes, increased drug usage, and raised juvenile crime and violence, particularly in the inner cities, to record levels. The prison population has grown to four million—the highest percentage of any country in the world.

There is a societal component to stress as well. Society is, ultimately, a collection of individuals. The quality of society is determined by the quality of its citizens. When individual citizens of society are stressed, one can say that the collective consciousness is characterized by stress. Just as individual stress has been scientifically linked to a wide range of diseases, similarly, epidemic levels of societal stress have been linked to the widespread rise of social disorders: crime, drug abuse, family disintegration, domestic violence, and the decline of moral and social values. This breakdown in social order and institutions has only exacerbated stress levels in society. In turn, this rising stress contributes to further violations of natural law by the individual and society.

Finally, while elected leaders universally decry this decline of moral and social order, there has been little understanding of its underlying cause in the widespread violation of natural law and resulting stress—and no practical knowledge of how to break this vicious cycle and reverse this slide in collective consciousness. Fortunately, together with scientifically proven procedures to bring individual and national life into accord with universal natural law, such practical knowledge is now available, thereby eliminating problems at their basis.

"One should refrain from making analogies between science and social behavior," explains Professor of Chemistry and Physics, Dr. Marcos Dantus (Department of Chemistry, Michigan State University). "Scientifically, we know that opposite-sign electrical charges attract and similar-sign electrical charges repel. What does this teach us about society and human behavior? It teaches us absolutely nothing. The human mind is very complex, and the behavior of individuals and groups can be extremely diverse. Every day humans do something that follows no logic, and we constantly question laws." [4]

References

[1] Dr. Andrew Cooksy, Associate Professor, Department of Chemistry, San Diego State University, San Diego, CA 92182-1030.

[2] Dr. James F. Harrison, Professor of Chemistry, Department of Chemistry, Michigan State University, East Lansing, Michigan 48824-1322.

[3] Al Globus, "Molecular Nanotechnology in Aerospace: 1999," Veridian MRJ Technology Solutions, Inc., 1999.

[4] Dr. Marcos Dantus, Professor of Chemistry and Physics, Dantus Research Team, Ultrafast Dynamics and Control of Chemical Reactions, Department of Chemistry, Michigan State University, E. Lansing, MI 48824-1322.

The Fabrication and Manipulation of Carbon-Based Structures (Fullerenes)

"Radio is a bag of mediocrity where little men with carbon minds wallow in sluice of their own making."

—Fred Allen (1894–1956)

One can design, fabricate, and manipulate various nanodevices and molecular machinery parts through the unique properties of carbon-based structures, or fullerenes, in dimension and topology. For example, carbon nanotubes (CNTs) are of significant importance to the scientific and industrial communities due to their astounding properties, which include ballistic electron transport, high mechanical strength and flexibility, and very high-aspect ratio. CNTs have numerous potential applications in nanoelectronics, biological probes, nanoscale structural materials, and field emission devices, among others. For many practical applications, deterministic placement of CNTs is required. The term deterministic implies precise control over the CNT location, size, and orientation, as well as the mechanically and electrically reliable contact to the substrate. Unfortunately, the structurally near-perfect CNTs prepared by laser ablation and arc discharge are produced in tangled mats and must be laboriously separated, purified, and cut to length before use. Also, to date all methods developed for placing these CNTs on a substrate in a deterministic way have been rather slow and quite difficult to perform, which makes them incompatible with commercial fabrication. CNTs produced by conventional thermal chemical vapor deposition (CVD) also are limited in terms of deterministic incorporation into potential devices.

Recently, chemists demonstrated the growth of randomly placed, but vertically aligned carbon nanotubules (VACNTs), which they refer to as nanofibers (NFs). They also independently demonstrated patterned growth of individual vertically aligned carbon nanofibers (VACNFs). Despite the morphological similarity with CNTs, carbon nanofibers (CNFs) have very different crystalline structure. While CNTs are composed of one or multiple concentric graphene cylinders (single- or multi-walled CNTs), CNFs' walls consist of disordered "funnels" (see Figure 18.1) [1], and a bamboo-like structure is often observed.

Therefore, the mechanical and electronic properties of CNFs are expected to be quite different from those of CNTs. To date, the crucial advantage of using VACNFs versus CNTs is the ability to grow them deterministically (the position, height, diameter, and, to some extent, orientation of VACNFs can all be controlled). A schematic representation of the VACNF growth technique utilized by chemists is shown in Figure 18.2. [1]

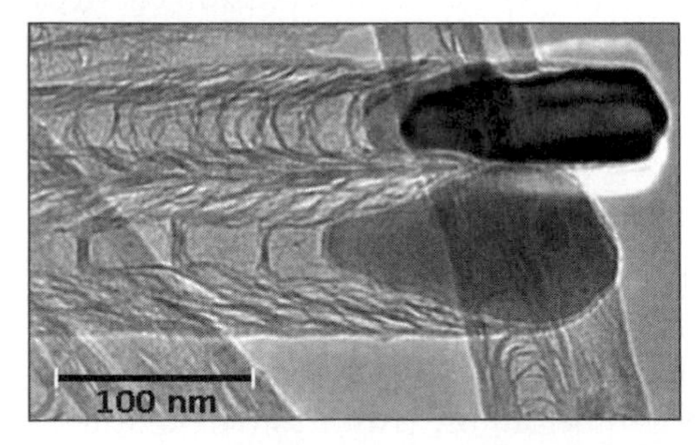

Figure 18.1 CNFs' walls.

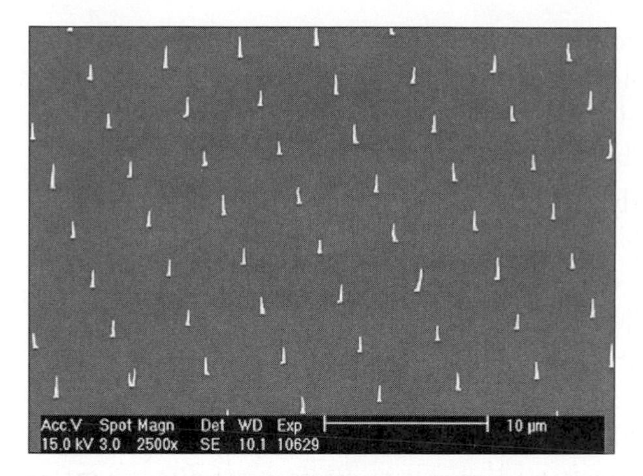

Figure 18.2 Schematic representation of the VACNF growth technique.

VACNFs are grown by direct current (dc) glow discharge plasma-enhanced CVD (PECVD). Continuous Nickel (Ni) thin films and Ni dots patterned using electron-beam lithography are used as a catalyst. In the case of Ni thin films, Tungsten (W) or W-Ti alloy is used as a buffer layer on the Silicon (Si) substrate to prevent the formation of Ni silicide. Titanium (Ti) is used as the buffer layer for the patterned Ni dots. Acetylene (C_2H_2) and ammonia (NH_3) are used as carbon source and etchant gases, and in some cases, helium has been used as a buffer gas. The etchant is needed to etch away graphitic carbon film that continuously forms during the growth from the glow discharge and would prevent the formation of VACNFs by completely covering the catalyst nanoparticles. The substrates are heated directly by placing them on a heater plate. This method has the advantage of enabling large-area deposition and good uniformity of VA-CNF films. Upon heating and ammonia plasma pre-etching the substrates at elevated temperatures (~700°C), the Ni layer breaks into nanoscale droplets that are the necessary precursors for the catalytic growth of CNFs. Following this, acetylene is introduced into the chamber and the VACNF growth begins. Each Ni droplet initiates the growth of a single VACNF, and then resides on top of the nanofiber and provides for its continued catalytic growth. A few examples of deterministic VACNF growth demonstrated by chemists are shown in Figure 18.3. [1] [2]

Most recently, chemists have developed a technique for fabrication of novel carbon nanoscale structures such as carbon nanocones (CNCs) and nanostructures with cylinder-on-cone shapes. By adjusting growth parameters such as the ratio of acetylene (C_2H_2) to ammonia (NH_3), vertically aligned CNCs, rather than CNFs, can be formed. Herein the prefix *nano* refers to the tip diameter of the CNCs; the CNC height and base diameter can be grown to micrometer dimensions, if desired. An example of a vertically aligned CNC array synthesized using PECVD is shown in Figure 18.4. [1] [2]

Dr. Benjamin L. Lawson (Materials Science and Engineering Department, North Carolina State University) states:

> Under the direction of Dr. Donald W. Brenner, our group performed molecular dynamic simulation of carbon nanocones, which have been found capping carbon nanotubes and as free-standing structures formed in carbon arcs, to determine their

Figure 18.3 Deterministic VACNF growth.

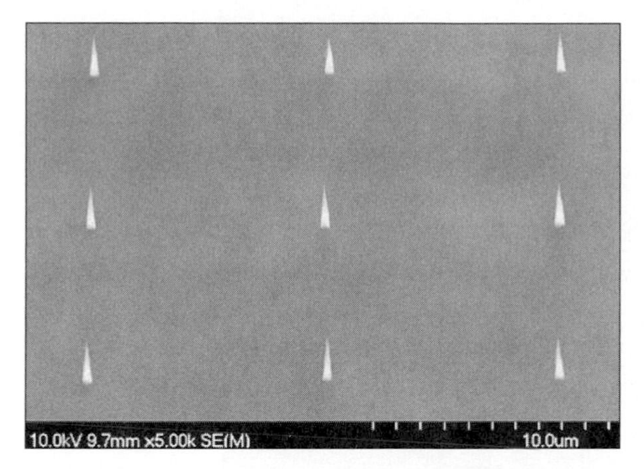

Figure 18.4 Vertically aligned CNC array synthesized using PECVD.

> predicted structural and electronic properties. My responsibility was to arrange the multiple nanocones into nanostructures and then relax these structures, via molecular dynamic simulation, to their lowest potential energy by utilizing an analytic many-body bond-order potential energy function for hydrocarbons. In order to use the theoretical knowledge of the properties of carbon nanocones in practical and relevant applications in industry, physical experiments need to be performed to confirm the properties we predicted. Refinements in nanomanipulation using SPMs or self-assembly are also necessary for assembling the nanocones into nanostructures. [3]

The remarkable aspect of the growth process is that the tip diameter of the cone does not increase during growth, and is determined by the size of the catalyst droplet. In contrast, the base diameter of the CNC, as well as its height, increases with growth time (conditions can be selected to permit both lateral and longitudinal growth). Consequently, very tall cones with sharp tips and large robust bases can be formed. Furthermore, by changing the growth parameters (in this case the relative acetylene content), the cone angle can be controlled. Examples of vertically aligned CNCs with different cone angles are shown in Figures 18.5 and 18.6.[1]

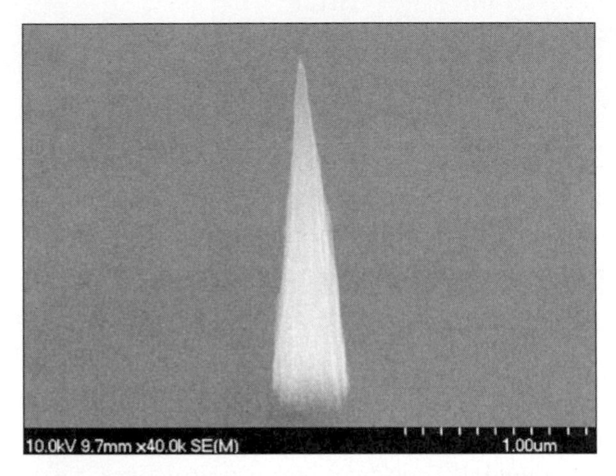

Figure 18.5 Vertically aligned CNCs with different cone angles.

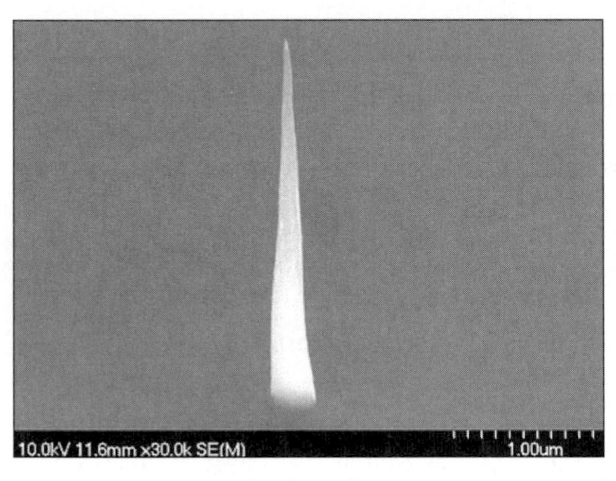

Figure 18.6 Vertically aligned CNC array synthesized using PECVD.

Higher acetylene content yields higher cone angles (increased lateral growth rate), and reducing the acetylene content can bring about a transition between the formation of vertically aligned cones and cylinders.

In addition, chemists have synthesized carbon nanostructures that consist of a nanocylinder (VACNF) grown directly on a nanocone. The growth process is as follows. The growth conditions initially are set for the synthesis of a cone. After a CNC of desired length and shape (cone angle) is obtained, the relative acetylene content in the chamber is reduced to suppress growth in the lateral dimension (C precipitation at the outer walls), thus forming a cylindrical VACNF on top of the CNC. Figure 18.7 [1] shows an example of a nanoscale cylinder-on-cone carbon tip fabricated using this technique.

Now, let's look at how chemists use molecular dynamics to investigate the properties and design space of molecular gears fashioned from carbon nanotubes with teeth added via a benzine reaction known to occur with Carbon-60 (C60, which is a naturally occurring molecule comprised of sixty carbon atoms). Because C60 molecules are naturally spherical, they tend to cluster together. They are ideal for use as building

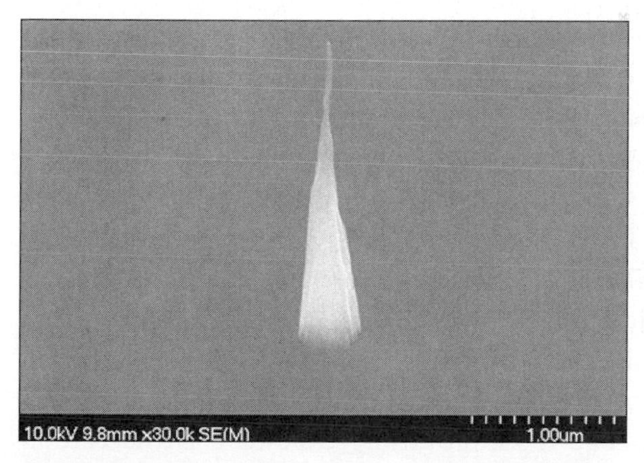

Figure 18.7 Nanoscale cylinder-on-cone carbon tip.

blocks in nanoscale fabrication because the open structure lends itself to the formation of new compounds.

For example, one gear can be powered by forcing the atoms near the end of the buckytube (a string of C60 atoms) to rotate, and a second gear is allowed to rotate by keeping the atoms near the end of its buckytube on a cylinder. Buckytubes may be enhanced during the manufacturing process with other compounds. Buckytubes, for example, are tangles of C60 strings mixed with a small amount of residual iron catalyst particles and coated with a layer of disorganized carbon. The buckytube is extremely strong mechanically and a very pure conductor of electric current. Recent discoveries have shown C60 strings cooled to 20K can carry electric current with no resistance, qualifying as a super-conductor.

Furthermore, the meshing aromatic hydrocarbon gear teeth transfer angular momentum from the powered gear to the driven gear. A number of gear and gear/shaft configurations have been simulated. Cases in vacuum and with an inert atmosphere have been examined. In an extension to molecular dynamics technology, some simulations used a thermostat on the atmosphere while the hydrocarbon gear's temperature is allowed to fluctuate. This models cooling the gears with an atmosphere. Results suggest that these gears can operate at up to 50–100 gigahertz in a vacuum or inert atmosphere at room temperature. Failed gears can be returned to operation by lowering temperature and/or rotation rate, because the failure mode involves tooth slip, not bond breaking. [2]

"Even though the molecular motors are very important and exciting," explains Dr. Oguz Gulseren (Professor of the Department of Physics, Bilkent University), "there is another very important application, especially for nanotechnology. Electronic devices are based on atoms, like transistors and integrated circuits. The most promising candidate is the carbon nanotube. A prototype of such a transistor was achieved." [4]

Fabrication and Manipulation of Simulations of Carbon Nanotube-Based Gears

As previously explained, the unique properties of carbon-based structures or fullerenes in dimension and topology allow one to design, fabricate, and manipulate various nano devices and molecular machinery parts. Shafts are multiwalled carbon nanotubes (see sidebar, "Building a Space Elevator"), and aromatic hydrocarbon gear teeth are benzine molecules bonded onto the nanotube. For example, benzole is a mixture of aromatic hydrocarbons with a high percentage of benzine, used as a solvent and as a fuel additive.

Building a Space Elevator

Science fiction writers and SciFi Channel TV shows like *Andromeda*, have long touted an elevator into the heavens, but pesky physics has always gotten in the way. There's simply no material that's light and strong enough to stretch to orbital heights without collapsing under its own weight.

Now, carbon nanotubes (nanoscale carbon structures 60 times stronger than steel), might make what was once science fiction a reality. According to NASA researchers, rockets would blast off to 22,000 miles, launching an anchor satellite that uncoils a ribbon made from carbon-nanotube composite fiber as it ascends to 62,000 miles. The ribbon flutters to the ground, where technicians attach it to a platform floating at the equator. The centripetal forces at the space end keep the ribbon taut and maintain it in a geosynchronous orbit. Electric elevators powered by ground-based lasers and carrying as much as five tons in payload would climb up and down.

The biggest challenge: The longest nanotube ever created is just microns in length. But by combining nanotubes with an epoxy resin and then extruding the mixture like monofilament fishing line, you can already make strands as long as you want.

Two other great obstacles remain: maintaining the elevator despite hurricanes, meteorites, and corrosive atomic oxygen; and building the laser system. It's a big engineering challenge, according to NASA researchers. There will be difficulties, but there's absolutely no physical reason why it can't be done, according to the researchers.

Most of a rocket's fuel is spent blasting through Earth's thick atmosphere and out of the planet's strong gravitational field. But here's an alternate strategy for getting payloads up to space: Construct a 62,000-mile-long cable jutting straight out from the equator, hold it in place with centripetal force, then lift satellites and spacecraft out of the atmosphere with a giant freight elevator. One major hang-up: Cable strong enough to support the system does not yet exist, though it could be made from carbon nanotubes. [5]

Nanotube-based gears are much simpler in structure and may be synthetically accessible. The idea of carbon nanotube-based gears comes from recent progress in fullerene science and technology. Multi-walled and single-walled nanotubes have been successfully prepared, and rapid advance has been made in controlling tube diameter, length, chirality, and the number of concentric shells. Carbon nanotubes have many attractive material properties. For example, experiment and theory have demonstrated that nanotubes have exceptionally high modulus, and nanotube electronic properties vary as a function of diameter and chirality. These properties have opened doors to electronic, optical, magnetic, and mechanical applications. It has been shown that nanotubes can be used as atomic scale field emitters, electronic switches, and pinning materials in high-Tc superconductors. In addition, the functionality of fullerene materials provides opportunities for fabrication of novel nanodevices. Filled nanotubes leading to improved catalysts and biosensors (see sidebar, "Carbon Nanotubes and Fullerenes Biosensors") are being developed and monoadducts and multiple adducts on C60 are finding applications. Therefore, it may be possible to make nanotube-based molecular gears by bonding rigid molecules onto nanotubes to form gears with molecular teeth. It may be practical to position molecular teeth in atomically precise positions required for gear design by, say, scanning tunnel microscopy (STM) techniques. IBM scientists have succeeded in positioning individual molecules at room temperature by purely mechanical means. They used the extremely fine tip of STM to position organic molecules having a total of 173 atoms and a diameter of 1.5 nm. Chemists are also investigating routes to chemical synthesis of these gears. [2]

"Organic chemistry has advanced to the point that we can now synthesize incredibly large and complex molecules," explains Professor of Chemistry and Physics, Dr. Marcos Dantus (Department of Chemistry, Michigan State University). "However, we are very far from being able to compete with nature. The virus, for example, is an unbelievably complex self-replicating machine. As we seek to build machinery in the nanometer scale, we should forget our world of nuts, bolts, and gears, and learn from the fabulous machinery in the cells of living organisms." [6]

Carbon Nanotubes and Fullerenes Biosensors

With fullerenes and carbon nanotubes, surface plasmon resonance (SPR) can be used for real-time protein-binding measurements. Attachment of one of the interacting partners to an SPR gold sensor surface allows for consistent and repeatable binding experiments. In a typical experiment, streptavidin was adsorbed to the gold sensor and a biotinylated C60-fullerene derivative was attached. Proteins with fullerene affinity, such as chemists' recently reported monoclonal anti-C60 antibody, could be flowed across the surface, and binding data obtained. Binding experiments for anti-fullerene antibodies with several different fullerene derivatives as well as carbon nanotubes are reported here. [2]

SPR

Through the use of gold, SPR can be used to monitor the real-lime interaction between molecules on a thin metal film surface. The surface plasmon phenomenon exists when polarized light reaches the interface between a thin metal film and a high-density medium in Kretschmann geometry. Ordinarily, it is difficult to couple light into surface plasmons for a non-zero wavenumber. In the Kretschmann geometry, though, it is possible to couple light into surface plasmons. In this geometry, light passes through a prism and is incident (with a certain angle of incidence) on a thin metal film. An evanescent wave propagates through the metal and excites surface plasmons on the other side of the film, which is immersed in some liquid. This situation is called surface plasmon resonance (SPR): Light is absorbed, and surface plasmons are created. Only transverse magnetic polarized light (the electric field polarized in the plane of incidence) may couple to surface plasmons.

The alternating electric field within the light causes oscillation of the firmly held electrons in the dielectric material. This oscillation produces evanescent waves that are non-propagating spatially decaying fields, in turn causing oscillations in the free delocalized electron density of the metal called surface plasmons. By influencing the effects of the electric fields, solutions and other material near or attached to the metal, on the opposite side of the incident light, alter the properties of the surface plasmons.

As it interacts with the surface plasmons, spectroscopy of surface interactions is made possible through detection of changes in the index of refraction of polarized light. Light incident on a reflective surface obeys the Law of Reflection, which states that the angle of incidence is equal to the angle of reflectance. Light that travels through the interface between two media with material indices of refraction (n_1 and n_2) follows Snell's Law, which gives the relationship between the incident and refracted angles. The limit of Snell's Law occurs when the incident light is totally reflected at the interface at a certain increased angle of incidence. The critical angle of incidence is reached when the angle of transmission increases toward 90° more quickly than the angle of incidence, resulting in light totally reflected into the high index medium. Although incident light can be totally reflected, the electromagnetic field components penetrate a short (tens of nanometers) distance into the medium of lower index of refraction, creating the exponentially attenuating evanescent wave that results in surface plasmon effects. Thus, reflection of light occurs in the apparatus being used for surface plasmon resonance spectroscopy.

Where some light interacts with the surface plasmons instead of being purely reflected, light incident on the metal-dielectric interface reaches its minimum in intensity of reflection at the plasmon angle (depending of course on the material). Resonance occurs when the intensity of the reflected light is reduced at the plasmon angle, due to the resonance energy transfer between the evanescent wave and surface plasmons when light is coupled to surface plasmons for a non-zero wave number. Light that has its electric field polarized in the plane of incidence may couple to surface plasmons and, in Kretschmann geometry, resonance occurs when wave vector components of light and surface plasmons match. The coupling of light to surface plasmons results in absorption of the light, rather than complete reflection at the thin metal interface. The changes in intensity of reflected light is detected by a photodiode array, which records the SPR signal around the plasmon angle.

The plasmon angle is extremely sensitive to the refractive index of the solution in contact with the metal. This makes SPR an excellent technique for accurately measuring conformational changes in the adsorbed molecules with high sensitivity, refractive indices, and adsorption of molecules on the metallic film. SPR is able to detect sub-microgram quantities of material at the metal surface. The percent intensity of the reflected incident light is higher at angles adjacent to the plasmon angle. Thus, within a few degrees of its known plasmon angle, spectroscopy surrounding this plasmon angle occurs with light incident on the metal-dielectric interface.

Influenced materials in contact with or adsorbed to the metal surface are known as resonance conditions. Also, the position of the SPR curve indicates the refractive index of the liquid on the sensor. Changes in detected reflected intensity of light in the incident angular region is related to changes in the index of refraction of the material or solutions on the opposite side of the metal surface, calibrated from a known index of refraction, such as 1.333 for pure water. The SPR curves can be compiled into a sensorgram that describes real-time binding interactions at the metal surface. Thus, surface plasmon resonance spectroscopy uses the interaction of polarized light on a metal-dielectric with surface plasmons to determine the index of refraction of the medium on the external metal surface. The measured changes in index of refraction are often defined as resonance units. Although biomolecules have been primarily used, SPR can be used to detect interactions of any two molecules. [2]

Finally, the data discussed here show that a biosensor surface can be prepared that detects the binding of fullerenes and carbon nanotubes. The qualitative analysis of these experiments demonstrates the extent of antibody-fullerene binding through surface and solution phase interactions. The data from these binding experiments will be rigorously analyzed to determine useful quantitative information, such as stoichiometry of interaction, association and dissociation binding constants, and equilibrium constants. Chemists' understanding of fullerene-protein interactions will provide new tools for nanotube analysis and will improve through data, techniques, and future SPR experimentation.

The chemical feasibility of nanotube-based gears is supported by chemists' extensive quantum chemical calculations and molecular simulations. A simple approach is to bond rigid planar benzine molecules onto a nanotube. The calculations are in agreement with experiments for napthylene and buckyballs (C60), while experimental verification using nanotubes has not been reported. This part of the chapter demonstrates how these processes form nanotube-based gears and evaluate the gears via molecular dynamics simulation. Chemists determined favorable gear working conditions and characterize gear performance by studying the rotational dynamics of gears under various conditions. [2]

Bonded Interactions Simulation Details

In order to describe bonded interactions, chemists have chosen a potential. This potential realistically describes bonding structure and properties in graphite, diamond materials, and small hydrocarbon molecules. In addition, its proper description of bond forming and breaking can be used to observe reactive collision to bond benzine onto fullerenes and, also, potential molecular gear failure modes. For intermolecular interactions between different gear atoms, chemists used and tested several force field potentials: with parameters derived from fitting graphite and C60 experimental data plus electrostatic interaction terms derived from benzene dimer energy and structure. Chemists carried out molecular mechanics calculations of C60, nanotubes, and molecular gears to test these force fields. Minimized energy and C–C bond length are the same as reported values for C60 and nanotubes. Energy minimization shows that the nanotube-based gears are in stress-relaxed structures, with a slightly higher energy than nanotubes (<5%, depending on length). Detailed comparison with quantum chemistry results should be given separately. [2]

Molecular Dynamics (MD) simulations are performed on the energy-optimized structures. The equations of motion have been integrated using a four-order predictor-corrector algorithm with a time step of 0.5 fs. The rotational velocity components in a total atomic velocity were subtracted for evaluation of the thermodynamic temperature. Angular velocity can be measured for each atom and averaged for each gear for each given input angular velocity. By using a parallel MD algorithm using replicated data, most MD runs are done on Numerical Aeronautics Simulation (NAS)'s IBM SP2. [2]

The gears must be kept correctly positioned relative to each other, in simulations of gear rotation. This is accomplished by constraining the atoms near the end of each tube (usually two rows of rings at each end). In the chemists' simulations, a spring model was used to connect the end atoms to a mount and model the interaction between them and massless mount atoms. Chemists call it the "hot-end model." In contrast, if end atoms are only allowed to rotate, with no thermal motion, chemists call this the "cold-end model." These two models can constrain tube end atoms to stay on a cylinder and therefore maintain the relative position of two gears near a constant value. Nevertheless, since it considers the effects of interatomic interaction on gear operation, the hot-end model is more realistic.

Also, to rotate, one gear must be powered. Some chemists suggest that lasers may be used to rotate the end atoms of a molecular gear. An MD simulation study also showed that the rotational motion of the end atoms of one nanotube could be induced by adding charges to the tube ends and applying one or two oscillating laser fields. Thus, chemists can model the powered gear by giving its tube end atoms an angular velocity. Once the end atoms are driven to rotate, strain between them and neighboring atoms will be induced. In order to release strain, neighboring atoms must rotate with the end atoms. Thus, rotational momentum is transferred to all other atoms and the gear is powered to rotate because of interatomic interactions. Since such interaction use strong bonded forces, one can expect very effective momentum transfer and very high rotation rate. On the other hand, weak van der Waals (VDW) interactions between two gears will cause the driven gear to rotate. MD simulations with atomistic interaction potential will rotate gears by following the motions of gear atoms.

The Results

The MD study can be started by chemists with a demonstration of synthesis of nanotube-based gears and functionalized C60. In simulations, a benzine molecule was given an initial velocity toward a nanotube. If the velocity was too small, a nonreactive collision was observed that appeared to be elastic. If the velocity was too large, dissociation of benzine and the nanotube was seen when the molecules collided. If an appropriate initial velocity was given, a reactive collision to bond benzine onto the nanotube occurs.

Nanotube-Based Gears Computational Synthesis

Two types of products from MD reactive collisions of benzine and nanotubes were observed by chemists: 1,2 and 1,4 adducts, for energy-minimized structures. From our calculations of quantum chemistry and molecular mechanics, both 1,4 and 1,2 adducts on the nanotube are stable while the 1,4 adduct is of slightly lower energy. In contrast, the 1,2 adduct of C60 was found to be much more stable than the 1,4 adduct, consistent with both the experiment and quantum chemistry calculations. More interestingly, the experiment and calculation of benzine adducts on planar napthylene molecules showed that the 1,4 adduct is much more stable than the 1,2 adduct. Considering that the diameter of the nanotube in the calculations is larger than that of C60 (12 vs. 9Å), stability of 1,2 or 1,4 adduct may be a function of nanotube diameter.

The 1,2 adduct is less stable for large diameters and more stable for smaller diameters (C60).

Some discrepancy existed for optimized four-membered ring geometry, but reasonable agreement in energy values of these adducts has been found for quantum chemistry and molecular mechanics. The valence angle in four-membered ring is close to 90° with quantum chemistry, but it deviates from normal by about 15° with the empirical potential. However, since it agrees well with most of the quantum chemistry results, chemists believe that the potential is fairly good. [2]

Where the [a,b] is the helicity notation for nanotubes, experiments suggest that the tube end is either closed (buckyball cap) for [a,0] semiconductive tubes or open (benzine end) for short [a,a] metallic tubes. Chemists' preliminary studies showed that the hydrogen-terminated [a,0] tube ends could exist in the energy minimized structure at 0K (absolute zero), but they were highly unstable at room temperature. Since the six-membered rings at the tube ends were treated by spring models, chemists' MD studies of gears were not involved in the termination of tube ends. [2]

According to Dr. Gulseren, "A notation (a,0), (a,a) and (a,b) is used to label zigzag, armchair, and chiral nanotubes. To be more correct, these labels should be integers, like (n,0), (n,n), and (n,m), since these indicate the rolling vector in terms of graphene (a single layer of graphite) lattice vectors. The way to produce the nanotubes, at least schematically, on a graphene sheet, is to go from origin to the end of rolling vector (for example, (n,m) = n*a1 + m*a2) and roll the sheet so the end of the rolling vector comes to the origin." [4]

Vacuum Rotation Dynamics of Gears

The first gear system is in a vacuum. The two gears were made of [14,0] nanotubes with a diameter of 11Å. Each nanotube has seven 1,2 benzine teeth where each pair is separated by two six-membered rings around the nanotube. The number of atoms in these simulations is about 1,000 (short tube) and 2,000 (long tube). The gears should work well in a vacuum since there is no drag to resist their rotation. The problem, however, is that heat generated by friction cannot be removed. The gears are eventually destroyed by the accumulated heat. [2]

Furthermore, the gears initially were at a minimized energy state at 200K. Powered atom input angular velocity was increased linearly from 0 to 0.2 rpps (revolutions per ps) in the first 10 ps and then stayed at 0.2 rpps to 50 ps. Consequently, the gears started rotation with a linearly increasing rate from 0 to 0.05 rpps. During this period, heat was accumulated and temperature increased from 200K to 600K. The gears worked very well. But heat resulting from gear atom friction accumulated as the temperature increased to >1000K after 30 ps. At higher temperatures, input energy could not be effectively converted into rotational motion; gears only wiggled, and the measured rotation rate decreased. If the input rate was again increased from 0.2 to 0.3 rpps from 50 to 60 ps, the gears resumed rotation and the measured rate also increased. However, increasing temperature (~2000K) forced the gears to return to the wiggling state with the measured rate decreasing. Continuing to increase the input rate failed to rotate gears at this high temperature. [2]

At high temperature, gears certainly cannot work. The critical temperature is about 600 to 1000K, as estimated from the current study. Well beyond this temperature, gears cannot work. [2]

NOTE

No bond or tooth breaking occurs up to at least 3000K.

Therefore, the input rate is not equal to the measured rate with hot-end conditions. This is a very interesting observation. When the end atoms of a nanotube are given an angular velocity, strain and stress are induced between them and neighboring atoms. These atoms must move to release the strain. In the cold-end conditions, end atoms are not allowed to move back; the neighboring atoms have to move forward and therefore input rate should equal the measured rate, which will be illustrated later in the chapter. In contrast, hot end atoms can move back as neighboring atoms go forward and input rate in this case is always larger than the measured rate. In either case, chemists have found that the critical temperature of 600–1000K for gear operation is similar. [2]

Vacuum Isothermal Rotation Dynamics of Gears

For gears to work, it is common sense to maintain a constant temperature. In MD simulations, constant temperature is often obtained by an artificial thermostat that controls temperature by a heat transfer equation, not by a real cooling medium. So, without any coolant-gear molecular interactions, such a model system can provide information on isothermal dynamics in a constant temperature system.

The measured rotation rate of the two gears increases with increased input rate in the first 120 ps, and the averaged values for the two gears are basically identical. The gears at constant room temperature work well until the measured rate is about 0.1 rpps (input rate of 0.4 rpps). This critical rotation rate was also found for cold end gears. Beyond this rate, slip occurs as the rotation rate continues increasing or stays at a higher value for the powered gear and the rotation rate of the driven gear decreases. When the gears slip, intergear energy and bonded internal energy jump to a higher value. Obviously, this jump does not signify bond breaking or tooth breaking because the energy change is small. Therefore, the gears should resume working if the input rate is reduced. Also, the measured rotation rate and energies return to reasonable values when the input rotation rate decreases. Nevertheless, a very interesting case can be made from the energy curve. When the gears are not rotating, they have a pronounced breathing mode and are quite flexible. However, as the rotation rate increases, centrifugal force stiffens the gears and the energy decreases. When gears fail, they slip rather than fragment (no bonds brake). When the conditions that led to failure are removed, intermolecular forces will cause the gear teeth to straighten out and mesh properly, and the gears go back to a low energy state and start working again. This offers an operational advantage. Without needing to worry about destroying them, a trial-and-error procedure can be used to establish operation conditions for physical gears. [2]

When the rotation rate approaches the critical value of 0.1 rpps, the benzine molecular teeth start tilting. This tilting allows the gear teeth to slip at the expense of increasing both intergear and intragear energy. A rough estimation is that tilting 20° will induce slip, but the gear still can work when tilting is up to about 10°. Relative to the energy of a stable tooth configuration, the tilting energy of 10 and 20° is 1.4 and 5.6 kcal/mol, respectively. Also, rotational kinetic energy of gear teeth is a function of the diameter of the gear and rotation rate. If the rotational kinetic energy is greater than tilting energy at 20°, major tilting and slip occur. Furthermore, if the kinetic energy is less than tilting energy of 10°, gears will rotate steadily and safely. [2]

In addition, the operation curve has been supported by MD runs for the gear systems using short [14,0] nanotubes. Chemists also did MD runs for longer tubes (about twice longer) and the critical rotation rate of 0.1 rpps was found again, although it took more input energy and time to reach this value. For cold-end gears, the 0.1 value also holds. [2]

For gear simulations, chemists have also tested several force fields. Slip occurs because the molecular teeth tilt. To make sure that no reactive slip occurs, chemists did comparative MD runs for two cases: with and without Brenner's reactive potential for intergear interactions. Also, for slip condition and energy values, no significant difference was observed.

Furthermore, electrostatic interactions also needed to be taken into account. When two benzine teeth come close, they try to maintain a displaced parallel configuration, even when slipping. This is similar to the favorable interactions observed for benzine dimers. It was found that the Buckingham (exp+6) force field plus electrostatic interaction terms was better at predicting the configuration and energy compared to 6-12 Lennard-Jones potential without the charge term. Therefore, chemists tested this case by placing partial charges on benzine atoms. It turns out that these different force fields do not make significant differences in rotational dynamics. This is perhaps because mechanical motion is not as sensitive to force fields as is local molecular configuration and conformation. As a matter of fact, with the difference being only in detailed local structures, most force fields for the same types of atoms predict almost identical structures and properties for molecular systems in condensed state.

Additional Operating Gear Systems Examples

From the preceding studies, chemists have drawn several useful conclusions. If gear temperature was less than 1000K, a trial-and-error procedure could be used to determine gear operation conditions without fear of destroying the gears. Furthermore, if tilting energy at several angles, say, 10° and 20°, was properly estimated and rotation rate was chosen so that rotation energy was less than the tilting energy, gear systems should work well without any slip. The operation would be easier if powered gears were driven by cold-end atoms since input rate was the same as rotation rate of teeth of powered gears. Following these points, chemists have also simulated operations of other types of nanotube-based gear systems. [2]

Gears with Multiple Teeth

Two types of gears with multiple rows of 1,2 teeth, have identical nanotubes [14,0] to the gears previously studied. Chemists call a left system the on-line tooth gears and a right one the off-line tooth gears. The critical slip rate of 0.1 rpps was observed again. In addition, the ratio of input to measured rate was also the same as that of single-row teeth. As expected, operation of the off-line multiple tooth gears, as indicated by the measured rotation rate, is smoother than on-line multiple tooth gears and single-row tooth gears previously discussed. This is because gear teeth sometimes do not mesh properly and will occasionally exert counter-rotational forces on each other. The teeth will interface face to face in proper working conditions. [2]

Shaft and Gear

The carbon nanotube is still [14,0] for the gear, but [9,9] for the shaft with diameter of 11 and 12.2Å, respectively. Two types of operations were performed in hot end conditions. One was powering the gear to drive the shaft, converting rotational motion into translational motion. The other was converting translation of the shaft into rotation of the gear. In the latter case, a constant linear velocity was given the atoms at one end of the shaft. Chemists can see that the gear-shaft system works well in either case. However, it takes more input energy to convert these two types of mechanical motions compared to the case of one gear driving another gear. The ratio of input rate to measured rate is about five for the shaft driving the gear, and around eleven for the gear driving the shaft. It takes more power for the gear to drive the shaft, since the mass of the shaft is almost twice that of the gear. [2]

Large and Small Gears

The nanotube is [18,0] for a large gear and [10,0] for a small gear. The ratio in diameter and number of atoms or mass is 1.8, with the smaller diameter being about 8Å. The chemists' simulations showed that this system could work, and similar operation curves to the previous ones were obtained. Since the tube diameters in this system are different from those previously studied, the critical slip conditions and the ratio of rotation rates are not the same as before. Operation of the large gear driving the small one is easier and smoother than the small driving the large because of the difference in mass. If the small gear is given a large acceleration, it does not drive the large one. Instead, like elastic collisions of a small ball between two boards, the small gear bounces back and forth several times. [2]

Gas Gear Rotation

So far, success in operating nanotube-based gears has been based on control of temperature, which was achieved by a software thermostat in a vacuum. To improve realism, chemists used a coolant gas instead of the thermostat. A new problem is that the gas will resist rotation. The higher the kinetic energy of the gas, the faster the cooling and the more drag, when the gas temperature is much lower than the gear temperature. Thus, there seems to be a conflict with the conditions to get both faster cooling and less drag.

The gears are placed in the center of a box of 2,000 gas atoms. The dimensions of the box can be adjusted to change gas pressure or gas density (in this example, it is 120 × 80 × 60Å). The gas atom type also can be changed from the input. No artificial thermostat is used for the gears. The heat in the gears is removed by gas. Here, the temperature is controlled by a software thermostat. [2]

Finally, let's look at the comparison of operations of the gears in gaseous Neon (Ne) at 200K and 0.25 atm with gears in a vacuum. The initial temperature of the gears is 150K. In both cases, the rotation rate is almost the same for the powered gears and the input since the end atoms in the powered gears are in cold end conditions. Angular velocity increases linearly from 0 to 0.05 rpps and then is constant. It can be seen, by comparing the rotation rate of secondary gears, that the operation of the gears after 100 ps is much better in the gas than in the vacuum. The fluctuation of the measured rate of the second gear around that of the first gear is very small for the gas-gear system, but it is very large for the gear-vacuum system. Obviously, the gears work much better with gas coolant than in a vacuum because the role of the gas in cooling gears. The gear temperature keeps increasing in the vacuum, but is controlled to some extent by gas. To gain confidence, more and longer MD runs are needed. Nevertheless, this preliminary work has demonstrated the possibilities of cooling the gears and making them work. [2]

Conclusion

The information presented in this chapter has suggested that nanotube-based gears can be made and operated. The MD simulations were checked by chemists using quantum chemistry calculation results of nanotube model molecules with the following results:

- The gear temperature can probably be controlled by a coolant gas. Additional work is in progress.

- The gears will work well if the temperature is lower than 600–1000K and rotational energy is less than the teeth-tilting energy at 20°.
- The predominant mechanism of gear failure is slipping due to teeth tilting. Gears will resume functioning if the slipped gears are slowed down. Therefore, a trial-and-error procedure can be used to find the optimized operation condition in physical gears without needing to worry about destroying gears. [2]

Also, the extraction of some information here for future work is useful. First, this consists of the force field as being not as sensitive to mechanical motion in nanodevices as to local molecular motion. This suggests that it may be possible to establish a set of force fields for all nanosystems, particularly for programmable molecular machines.

Second, simulation and analyses of nanosystems needs more computer tools and algorithms than other molecular simulations. (See sidebar, "Computing Chemistry.") They rely on visualization to test programmable nanodevices and nanomachines, object-oriented languages for assembling of both various components and various calculations, and efficient load-balancing parallel algorithms for large heterogeneous nanosystems.

Computing Chemistry

Computers made of molecule-sized parts could build themselves. Chemists and their notion of computer parts make today's technology seem as unwieldy as the radio tubes of days past. Indeed, if chemists have their way, the silicon components of today's machines will ultimately go the way of those electronic dinosaurs.

Consider this: There are a billion trillion water molecules in just one drop of water. That's hundreds of times the sum total of all the transistors in every computer ever made. If just a fraction of the molecules in a speck of matter could be made to act as electronic switches, able to control the electrical currents that are the basic language of computers, present-day computers with their paltry billions of transistors could quickly become obsolete. So would today's giant and expensive chip fabrication plants: A molecular computer might build itself in a set of chemical steps.

If this sounds like rank speculation, it's not. Those molecular switches are real. Recently, researchers have made and tested complex, artful molecules that work like switches you might buy at RadioShack, flipping from "off" to "on" and back when tripped by pulses of voltage. Other groups have made wires that are only atoms across, equipped with the molecular connectors that might soon hook up the tiny switches into computer circuits.

A droplet-sized supercomputer will become a reality only after scientists learn how to connect astronomical numbers of molecules into a working computer architecture and link the minuscule computer to the larger-scale world you live in. That could take decades. But simpler devices that use dense patches of molecular switches to soup up conventional chips are already in the works.

Computer switches need to do two jobs: open or close in a flash to process information, and stay open or closed for long enough to act as short-term memory. As recently as 1999, chemists at Yale University reported that they had met the first requirement by tailoring an electrically conductive molecule. When they positioned a few hundred of the molecules between tiny metal electrodes and increased the voltage across them, they found that at about two volts, the molecular switches turned on, allowing current to flow. Above or below that voltage, the switches were off. Now they have modified their switch molecule to create memory: After being flipped on or off, it stays that way for a few minutes. You can read, write, erase it. [6]

At the University of California-Los Angeles, chemists have built similar switches from molecules that actually change shape when they are zapped with a charge. Think of a dumbbell with a ring that slides back and forth along a shaft, turning the switch on or off.

Problems in Wiring

Connecting these tiny switches with wires as coarse as those on an ordinary microchip would squander much of the molecules' size advantage. So the UCLA group's collaborators at Hewlett-Packard are learning how to make metal or silicon wires just 10 or so atoms across. As a bonus, these wires don't need to be laid down one by one, like the connectors on current silicon chips; instead, they assemble themselves en masse, like long, thin crystals. At Pennsylvania State University, another group is making tiny wires that are predesigned to snap together into larger structures. The researchers coax platinum and gold atoms into the narrow pores in a membrane—then dissolve away the membrane to free breathtakingly thin wires that are striped with bands of the different metals, like a roll of Life-Savers. Molecular switches should stick readily to some of the bands, but not others, as do other molecules, including DNA, that serve as glue for linking the wires to one another. [6]

Chemists at Hewlett-Packard are putting together the first molecule-based devices. They hope to be able to deliver an entire 16-bit memory unit that could fit on top of the smallest wire in today's integrated circuits, with room to spare. Other groups may also be unveiling prototype molecular processors and memories by 2005, as well as taking the first steps to market. [6]

The design space of all potential gears is quite large. This space can be at least partially parameterized by:

- Coolant molecules, temperature, and pressure
- Diameter of gears

- Distance between gears
- Length of gears
- Molecular fragment used for teeth (chemists have used benzine and naphthalene, but others are possible)
- Nearby molecules (chemists hypothesize that long range forces will have some effect)
- Rotation rate
- Rotation rate acceleration
- Temperature [2]

Finally, throughout this space, the operating characteristics of the gears appear to change. Chemists have examined a few points in this design space and related slip conditions to input energy and tooth tilting energy. So designing and building complex machines will require characterizing some substantial portion of this multi-dimensional design space. [3]

References

[1] Oak Ridge National Laboratory, P.O. Box 2008, Oak Ridge, TN 37831

[2] Dr. Benjamin L. Lawson and Dr. Olga A. Shenderova. "Nanostructures Fabrication from Carbon Nanocones," Materials Science and Engineering Department, North Carolina State University, Raleigh, NC 27695, 2000.

[3] Dr. Benjamin L. Lawson, Materials Science and Engineering Department, North Carolina State University, Raleigh, NC 27695.

[4] Dr. Oguz Gulseren, Professor, Department of Physics, Bilkent University, Bilkent, Ankara 06800, Turkey.

[5] Steve Kettmann, "To the Moon in a Space Elevator?," *Wired News*, 660 3rd Street, 1st Floor, San Francisco, CA 94107, 2003.

[6] "Serious Computing," *Cybernest for Garbi & Torleif* [http://www.torleif.org/computing/], 2000.

Energy

Free Energy

"If we could produce electric effects of the required quality, this whole planet and the conditions of existence on it could be transformed. The sun raises the water of the oceans and winds drive it to distant regions where it remains in state of most delicate balance. If it were in our power to upset it when and wherever desired, this mighty life-sustaining stream could be at will controlled. We could irrigate arid deserts, create lakes and rivers and provide motive power in unlimited amount. This would be the most efficient way of harnessing the sun to the uses of man."

—Nikola Tesla (1856–1943)

Free energy! You either believe it's possible to create or harness or you don't. For many people, free energy is a buzzphrase that has no clear meaning. As such, it relates to a host of inventions that do something that is not understood, and therefore, is a mystery. For others, it means perpetual motion and therefore is dismissed by naysayers without consideration. This chapter is a mix of science and science fiction, which so often has come to pass, and may one day become reality. The chapter clarifies exactly what free energy is, how it hypothetically works and how it can be theoretically applied in everyday life for light, heat, and power.

The Ultimate Survivors: Have Advanced Civilizations Harnessed Free Energy?

Gazing up at the myriad stars visible to the unaided eye on a dark night, it's easy to wonder if any might sustain civilizations possessing capabilities eons beyond this one here on earth. But what would such advanced civilizations be like, and what sorts of feats could they be capable of performing? Can humans really hope to join their hypothetical ranks and one day control the weather, harness the power of the sun, or colonize the Milky Way Galaxy?[1] The answer of course, is only if humans can attain faster than light speed technology (as seen on *Star Trek)* or create stable worm holes that become highways to other worlds, as seen on SciFi Channel's *Stargate SG-1* or as discussed in Chapter 4.

It may seem foolish to describe a hypothetical advanced civilization millions of years ahead of the one here on earth. After all, only a century ago, who could have predicted the widespread use of jet aircraft for travel, multiple roundtrip journeys to the moon, and track-wheeled robots traversing the Martian landscape? So how could scientists ever hope to characterize civilizations one hundred years beyond this one, not to mention one thousand or one million years?

Dr. Patrick G. Bailey (Institute for New Energy, Los Altos, CA) states:

> It is very easy, if not commonplace, to assume that our society now has the most advanced science and technology that ever existed—in this country, on the Earth, or anywhere else. After all, if there was some other group with advanced physics, then where is the evidence? From one point of view, there isn't any evidence—as everything that we can see today has a simple explanation—by someone's definition! An example would be how the pyramids of Egypt were built by using piles of sand, with vast numbers of people rolling those huge blocks into place

over logs. From another point of view, there is evidence of the use of advanced technology all around us! How were those pyramids really built? A company from Japan tried to build a small stone pyramid in Egypt a few years ago, and failed miserably! Other examples include the statues on Easter Island, the lines seen from the air in Peru, and other exact geometrical patterns on the ground that can only be seen from the air, that cannot be seen while standing on the Earth. Who made those? When? And how did they get there?

In a society driven by money and profit, only those who seem to have all of the answers and all of the explanations tend to become the political leaders. Those who leave room for doubt and research are passed over and soon forgotten in the press and in the media. However, do not be fooled! There are many, many examples of unexplained science today, just as there were 50 or 100 years ago! Just because you have not seen something does not make it unreal!

Science will continue to advance and unravel as our ability to understand the new sciences grows and as we are able to incorporate the results into our society. As an example, in today's world of physics, antigravity effects are just now being demonstrated by a few scientists. So the science of that effect is real. However, due to the threat of terrorism and national security, the application of this technology will not be incorporated into our society for a long time. Thus we can say that such advanced science is real, but since it is not applicable in our daily life, we can say that it doesn't matter, or that for all practical purposes—it doesn't exist. The same can be said for any ancient Earth civilization, or off planet civilization, that may now be far ahead of the technology that we have in the U.S.: since we can't see it—it does not exist. Yet, I would strongly encourage you to consider that it is really there, and that if we can handle it, it may be available to us one day soon! [2]

Given that this civilization (modern humans) is barely 50 thousand years old, scientists can't help but let their imaginations run wild, conjuring up fantastic hypothetical possibilities. Still, humanity's quest to determine the features of advanced civilizations is not hopeless. No matter how many millions of years may separate them from us, one truth remains constant: Their actions must obey the fundamental laws of physics. Actually, beginning with the seminal work of Russian astrophysicist Nikolai Semenovich Kardashev (Russian astrophysicist and deputy director of the Space Research Institute of the Academy of Sciences, Moscow), scientists have, over the last few decades, employed knowledge of physical laws to analyze hypothetical advanced civilizations in great detail.

In 1964, Kardashev categorized extraterrestrial civilizations into hypothetical Types I, II, and III. They were ranked according to their capacity to harness a ubiquitous and measurable quantity—energy. In truth, the logic behind his classification is startlingly simple. Because energy, by definition, is the "ability to do work." For example, in simple universal units such as horsepower, civilizations could be quickly organized according to their output.

Given that the successive stages of human history can likewise be easily ranked according to energy assets, Kardashev's analysis is particularly useful. This stems from prehistoric times (when humans possessed only the power of their hands, about one-fifth of a horsepower); to slavery (when kings had hundreds of horsepower at their disposal); to feudalism (thousands of horsepower); to the industrial revolution (millions of horsepower); and, finally to modern times (billions of horsepower). But the real importance of Kardashev's work comes with the realization that, for the first time, a scientist had replaced wild unconstrained speculations with quantitative discussions of physical laws. And, for nearly four decades now, scientists have been working to determine the rate of evolution of such civilizations, to detail their mastery of various energy sources, and to place upper and lower bounds on their possible technological sophistication. In the meantime, they've generated a road map to immortality—hazards, obstacles, and pit stops included.

Scientists know the energy output of each type of civilization, because Kardashev's classification was based on universal, stable sources of energy, such as planets, stars, and galaxies. In turn, this makes it possible to compute the upper limit of energy available to each.

By possessing the ability to harness the energy reserve of an entire planet—up to one trillion (10^{12}) horsepower, Kardashev's Type I civilization is a planetary one. A Type II civilization is a stellar one, capable of controlling the energy output of a single star—up to 100 billion trillion (10^{23}) horsepower. Finally, a Type III civilization is one that has outstripped the energy output of a single planet or star, and has begun to construct a galactic civilization spanning many star systems. Type III civilizations possess an energy capacity of roughly 10 billion trillion trillion (10^{34}) horsepower. [1]

NOTE

Keep in mind that these are theoretical and/or hypothetical civilizations, and that some in the scientific community believe that the power output of these civilizations (if they exist) is impossible to achieve for many reasons.

At the dawn of the 21st century, humankind's energy capacity currently qualifies Earth's civilization for Type 0 status. Human's derive their energy not from harnessing global forces, but by burning tiny pockets of the fossil remains of dead plants (oil and coal). Scientists can barely forecast trends in the weather, let alone control them. [1]

NOTE

At present, this civilization's entire planetary energy production is a relatively puny 10 billion (10^{m}) horsepower.

However, it shouldn't take long for our current civilization to evolve into a Type I civilization, since economic activity and energy consumption are roughly proportional to each other, and since humans can safely assume that their planet's energy consumption grows at the modest rate of a few percent per year. Within 300 years or so, some scientists estimate that humans should attain Type I status.

Today, scientists can already see the seeds of a Type I civilization developing all around them. Like it or not, they are seeing the inexorable movement toward a planetary language (English), a planetary communication system (the Internet), a planetary economy (the forging of the European Union, NAFTA, etc.), and even the beginnings of a planetary culture (via mass media, TV, rock music, and Hollywood films). Because Earth is finite, but the rate of energy and telecommunications expansion is exponential, it is inevitable that, barring a catastrophe, humans will rise to Type I status. Nevertheless, in order to move more quickly to this end, looming natural catastrophes might generate the necessary impetus.

Dictated by the maximum energy output of planets, stars, and galaxies, in retrospect, scientists know that Kardashev's analysis only placed upper limits on the growth of advanced civilizations. Given the recent scientific interest in meteorite impacts, climatic changes (see sidebar, "Out in the Cold" later in this chapter), supernovae, and other natural catastrophes, astrophysicists have now tried to place lower limits on them as well. Indeed, any advanced civilization must grow fast enough to avoid life-threatening disasters—fall below the threshold and they will perish. It's a cosmic race against time!

On Earth, for example, ice ages take place on a timescale of tens of thousands of years (except for the mini ice age which lasted from 1300 to 1850 AD). Because Type I and Type II civilizations are likely to encounter various atmospheric or climatic disasters, they must learn to modify the weather within that time frame. A new ice age may commence within a span of thousands of years or less (see sidebar, "Out in the Cold"), which will bury Earth's Type 0 civilization under mile-high ice sheets. Therefore, a civilization will face extinction if it cannot progress to the next energy level.

There is also the danger of huge meteorite and comet impacts large enough to destroy most species of life on the planet (mass extinction that occurred both 250- and 65-million years ago). Earth's Type 0 civilization must evolve into a Type I civilization in order to prevent meteorite and comet impacts, or die. However, this should not pose too much of a problem for Earth's Type 0 civilization, since scientists expect civilization-killing meteorites and comets on a timescale of hundreds of thousands to millions of years.

NOTE

Some in the scientific community question the need for a Type I civilization. After all, current technologies may prevent these impacts. But, because humans as a general rule are complacent, the prevention of these impacts will in all probability, remain unlikely, until it's too late.

For Earth, the real immediate danger is internal catastrophes, such as terrorism, nuclear warfare, and global pollution, not external natural threats. Indeed, it is inevitable that any Type 0 civilization emerging in space will discover nuclear weapons and generate large quantities of chemical waste. Thus, the main obstacle preventing the transition from Type 0 to

Type I status is strictly self-inflicted, and perhaps the most dangerous transition of all is the one from Type 0 to Type I status.

Eventually, after several thousand years, a Type I civilization will exhaust the power of a planet (of course, solar and wind power will still be available) and will be forced to derive its energy by consuming the next available source of energy—their star. According to Kardashev, a Type I civilization should be able to make the transition to Type II within about 3,000 years, at a modest growth rate of one percent per year. [1]

In order to more efficiently capture its total energy output, some scientists have proposed that a Type II civilization may build a gigantic sphere around its star. Resembling the Oort Cloud of comets, perhaps a large collection of energy-gathering, or orbiting colonies, the sphere, speculate scientists, would not necessarily be solid.

NOTE

The preceding was dramatized in an episode on the TV series: *Star Trek: The Next Generation*, where the *Enterprise* came upon a civilization that had built a gigantic sphere around its star, but abandoned the sphere because the star became unstable. Some in the scientific community feel that the building of a sphere around a star is absurd and will never happen. Only time will tell, however.

However, a Type II civilization should be readily detectable from space, even if concealed by a sphere. By the Second Law of Thermodynamics, these protective spheres will rapidly heat up and emit large quantities of infrared radiation. In order to identify such advanced civilizations, some scientists, in fact, have long advocated looking specifically for infrared emission from distant spheres.

In all likelihood, the most worrisome natural threats to Type II civilizations come from nearby supernova explosions and gamma-ray bursters, whose sudden eruption could irradiate their planet with a withering blast of X-rays or gamma rays, killing all life. Such titanic explosions are rare, fortunately.

NOTE

Supernovae occur in the Milky Way at the rate of about one or two per century.

The Type III civilization is perhaps the most interesting of all civilizations. This type of civilization has reached for other star systems and has exhausted the power of a single star. A Type III civilization, harnessing the power of a galaxy, has a maximum energy output of about 10^{34} horsepower. According to Kardashev, it would take about 6,000 years or more for Earth to attain Type III status. That's at a growth rate of one percent per year. [1]

Because they possess the greatest capacity for achieving immortality, Type III civilizations are the most promising. Ice ages can be altered, meteorites can be deflected, and even supernovae or gamma-ray bursters will damage only a small part of a Type III civilization. A Type III civilization should last for millions or even billions of years, once it evolves in a galaxy.

Scores of Type I and II civilizations may have risen and perished long before humans arrived on the scene, since humans are fairly new to the universe—currently living about 15 billion years after the Big Bang. But, if a Type III developed in the Milky Way galaxy since the Big Bang, it could still be here. This led the late Carl Sagan and others to speculate that a Type III civilization may exist within the Milky Way Galaxy. In fact, it's the premise that intelligence searches like the Search for Extraterrestrial Intelligence (SETI) or optical SETI are based on.

Recently, progress in artificial intelligence, nanotechnology, and biotechnology has modified Kardashev's analysis. By far, the most mathematically efficient method to explore the galaxy is via artificial intelligence: Fleets of robotic Von Neumann probes are sent throughout the galaxy (named after mathematician John Von Neumann, who established the mathematical laws of self-replicating systems).

By definition, a small robot designed to reach distant star systems and create factories is known as a Von Neumann probe. Such a probe would reproduce copies of itself by the thousands. Amazingly, though, since these probes could easily land and take off from moons due to their low gravity, and also because the moons would not be eroded, a dead moon, rather than a planet, would make the ideal destination for Von Neumann probes.

NOTE

If these super-robots existed, landing on a planet would pose no problem, but self-assimilating robots that mine, refine, and machine raw materials on their trips are still very Sci-Fi, because Earth is still a Type 0 civilization.

By mining the raw ingredients to build a robot factory, Von Neumann probes could live off the land. Like the *Replicators* from various episodes on SciFi Channel's *Stargate SG-1*, they would create thousands of copies of themselves, which would then scatter and search for other star systems.

Just one self-replicating Von Neumann probe, like a virus, would exponentially generate a sphere of trillions of such probes, expanding in all directions, increasing in size at a fraction of the speed of light. Even a galaxy 100,000 light-years across, such as the Milky Way Galaxy, might be completely analyzed within a million years in this fashion. [1]

Some scientists have even raised the possibility of a Von Neumann probe resting on Earth's moon, left over from a previous visitation in this system eons ago. If this idea sounds a bit familiar, that's because it was the basis of the film *2001: A Space Odyssey,* as well as, *2010* and *2030.*

There is scientific speculation that a spacefaring civilization could use nanotechnology to build miniature probes, perhaps no bigger than your palm, to explore the galaxy. The tiny probes mentioned here will be so inconspicuous that it's no surprise that scientists haven't come across one. It's not the sort of thing that you're going to trip over in a backyard. So, it's possible that Earth could be surrounded by surveillance devices, if that is the way technology develops—smaller, faster, cheaper—and if other civilizations have gone this route. Of course, this is pure science fiction at this point. Or, is it?

By opening up entirely new possibilities, this analysis has also been modified by advances in biotechnology. In order to enhance their capabilities in Darwinian evolution, these probes may act as lifeforms, reproducing their genetic information in space, mutating and evolving at each stage of the reproduction process—and they may even have their own forms of artificial intelligence to accelerate the search for intelligent life.

Today, scientists realize that sufficiently powerful rockets may spare Earth from its inevitable death, or even the death of the sun. But how do humans escape the death of the universe itself? Either through a hot collapse or frigid runaway expansion, it seems inevitable that all intelligent life in the universe will die when the universe perishes.

Dr. Harold Aspden (Energy Science Ltd.) explains:

> Given our dependence upon the resources available to us in this 21st century, it is quite evident that it needs very little on the cosmic tragedy scale to eliminate mankind. The death of the universe is an unlikely event as our first, or rather next, encounter with what Nature has in store. What is not generally realized is that on a scale of every few hundred thousand years, our Earth

encounters something in space that causes tremendous upheaval, as if gravity is switched off for a few seconds, as body Earth moves across what I would call a "space domain wall." I say this, because these events are accompanied by a reversal of the Earth's magnetic field, and there is a pattern of such reversals recorded in the geological surveys of volcanic rocks which can be dated. The geological evidence also points to extinction of life forms on the more destructive of such transits.

Have you ever wondered why we are so fortunate as to exist during a small several-thousand year slot in the 4,000,000,000 years of Earth history? Can it be that 12,000 years ago, as is in evidence in geological records of the so-called "Stockholm Reversal" and in Australian geological data, that we traversed a "'space domain wall"?

What, you may well wonder, is a "space domain wall"? My answer is, that in my university research years at Cambridge in England (more than half a century ago), my interest was in magnetism and its energy anomalies. I then learned that in iron (even iron that is not polarized by magnetization), there are tiny domains of the order of 100 microns in width bounded by domain walls. Within each of those domains, the magnetism is enormously strong in one direction, but it can exist in a reverse direction in an adjacent domain. In later years, my research into the properties of iron and the internal cubic structure of its atoms gave me an insight into the analogous properties of space itself. Apart from thereby discovering the basis of quantum theory as a property of space, and the related gravitational action, I realized that space itself has a domain structure on a vast scale. I could see why a star might be created, one in each domain of space, because the action of gravity, as with magnetism, could not traverse a domain boundary. This was mere speculation, perhaps, or you might say wishful thinking, until I was further enlightened by a visit from J. Steiner, an Australian expert of geology, who opened

> my eyes to the link between my theory and the history of body Earth's geological record.
>
> The tragic phases in Earth history occur when our solar system, which moves at a cosmic speed close to 400 km/s (250 miles per second), crosses a space domain boundary at a very low angle of incidence. Dinosaurs no doubt failed to survive one such event. For our part, we need to pray that the evidence is correct and the last space domain boundary crossing really was some 12,000 years ago. Otherwise, to be sure, the next traversal is not too far away, even overdue—but, with luck, our next crossing will be in a few hundred thousand years. Future generations will need to prepare accordingly. Gravity will, in effect, be switched off or vastly reduced in strength for the half-minute or so of a normal traversal. So, if you happen to be in an aircraft, you will survive provided your pilot can find a runway on which to land that is not too uneven following the resulting earthquake.
>
> This is the horror we face, and not the death of the universe, although it may as well be the universe as far as we are concerned, given the way in which human life depends so much on what our modern world has provided. That, however, is really a secondary concern, given the impending catastrophe we confront, once our dependence upon oil and gas leaves us bereft of a viable energy resource. [3]

So, suppose that scientists extend the classification upward. Members of these hypothetical civilizations of Type IV, V, VI, and so on, would be able to manipulate the structures in the universe on larger and larger scales, encompassing groups of galaxies, clusters, and superclusters of galaxies. Via holes or portals in space, Civilizations beyond Type III may have enough energy to escape their dying universe.

There is renewed interest among physicists about energies so vast that quantum effects can rip apart the fabric of space and time to create wormholes. This has come about with recent advances in quantum gravity and superstring theory. Although it is by no means certain that quantum physics allows for large, stable wormholes, this does raise the remote possibility that a sufficiently advanced civilization may be able to migrate via holes in space. If one can successfully navigate through stable wormholes, then faster-than-light travel is no longer a problem. They merely take a short-cut through the galaxy. In effect, the transition time between a Type II and Type III civilization would be greatly cut down.

NOTE

Like Alice's Looking Glass, a wormhole is a hole in space, allowing one to instantly travel across vast distances in space and time.

When the environment becomes lethal, the evolutionary process forces lifeforms to leave their environment. Perhaps, by obeying this evolutionary imperative, an advanced civilization, facing the death of its universe, will create another.

One of the originators of the inflationary universe theory, Physicist Alan Guth of MIT, has computed the energy necessary to create a baby universe in the basement. The required temperature is 1,000 trillion (10^{15}) degrees. This temperature would be within range of civilizations beyond Type III.

Dr. Aspden states:

> All I can say is that, though I am myself a well-qualified physicist, I see no merit in inflationary universe theory. I think, if one seeks to create babies, one should do it the conventional way at room temperature, rather than engage in fantasy linked to a belief in the Big Bang. The latter belief has no secure foundation; it relies on the unproven assumption that an electromagnetic wave can travel at the speed of light for billions of years without loss of frequency. One day, I hope, physicists will

> ask how an electromagnetic wave, with its ongoing lateral displacement of something electrical in space, can exist without something moving dynamically in space that provides a lateral counterbalance. There is something missing from Maxwell's equations. Find what it is and you will discover that waves can preserve their dispersionless properties and decrease in frequency as they propagate. If you, the reader, are a physicist, remember the saying: "Action and reaction are equal and opposite" and ask yourself how this is accounted for in Maxwell's equations. So much depends on your answer, because once you understand how light from a remote star can suffer loss of frequency in transit, you will see that the notion of interpreting frequency loss as evidence of an expanding universe is ill-founded. You will then need to question belief in the Big Bang event and related hypotheses including those 1,000 trillion degrees of temperature. [3]

Scientists will use satellite arrays to identify hundreds of Earth-like planets in space in the next ten years, and perhaps one day, amateur astronomers will experience something of an existential shock as they gaze toward the night sky, holding in their hands an encyclopedia listing the coordinates of hundreds of small, Earth-like planets, wondering if any of them harbor life, perhaps some with civilizations more advanced than Earth's.

So, with this in mind, how do Earth's human inhabitants really transition from a Type 0 to a Type I civilization? As previously explained, the secret of this transition is a planetary one, possessing the ability to harness the free energy reserves of an entire planet. Is this the final secret of free energy?

"Oh, no!" exclaims Dr. Aspden. "This is not the secret of free energy! The secret resides in understanding the underlying cause of gravitation and a radially directed electric field produced when a star is formed and how that imports energy from the omnipresent quantum underworld of space itself—our aether, but see my comments later in the chapter." [3]

The rest of this chapter is dedicated to the detailed explanation of different alternative energy sources to achieve self-sufficiency through free energy. The naysayers claim that free energy has never been seen or demonstrated (as described next), because it violates several laws of physics. But isn't this what science is all about? To rewrite itself and its laws through exploration and testing—always wondering what's beyond the next hill or threshold of the imagination? Let's take a look.

Free Energy

Free energy, in the simplest terms, is any energy that is provided by the natural world. In science, energy is defined as "the ability to do work." Free energy is called by many names, such as renewable energy, alternative energy, or nonconventional energy, to list a few. Examples of free energy technologies include a wind generator on a remote homestead, or a solar panel on the International Space Station. But this is only the tip of the iceberg. Free energy also includes amazing technologies like a car powered by a water fuel cell, a battery charger powered by the earth, or a home furnace powered by permanent magnets. Without detrimental effects to the environment, and at extremely low cost for the maintenance of the equipment, the best free energy systems deliver energy at no ongoing cost to the user.

Current Free Energy Technology

In the late 1880s, trade journals in the electrical sciences were predicting free electricity in the near future. Incredible discoveries about the nature of electricity were becoming commonplace. Nikola Tesla (see sidebar, "Who Was Nikola Tesla?") was demonstrating "wireless lighting" and other wonders associated with high frequency currents. Like never before, there was an excitement about the future.

Who Was Nikola Tesla?

In the beginning of the 20th century, a Croatian* engineer, emigrant to America, Nikola Tesla, measured the electrical charge of the planet Earth and found it of a very high potential. He made his observation during thunderstorms.

NOTE

Tesla was in fact a Serb who was born in the Croatian village of Smiljane in the Lika region, which at the time was part of Austrian monarchy. His father was an Orthodox priest.

Tesla's instruments were affected by stronger discharges that were taking place at great distances, than by those nearby. This puzzled him very much. No doubt, whatever remained, he was observing stationary waves. As the source of the disturbances (thunderstorms) moved away, the receiving circuit came successively upon their nodes and loops. Impossible as it seemed to Tesla, this planet, despite its vast extent, behaved like a conductor of limited dimensions. The tremendous significance of this fact in the transmission of energy by his system had already become quite clear to him. Not only was it practicable to send telegraphic messages to any distance without wires (as he recognized long ago), but also to impress upon the entire globe the faint modulations of the human voice—and more still, to transmit power, in unlimited amounts, to any terrestrial distance almost without loss (attenuation).

In fact, in the late 1880's, Dr. Nikola Tesla accidentally discovered an electrostatic super-charging effect while trying to verify Hertz' discovery of electromagnetic waves. After hundreds of experiments, he learned how to control and maximize this phenomenon. This led him to the discovery that electricity is made up of different components that can be separated from each other, and that a pure, gaseous, aetheric energy can be fractionated away from the flow of electrons in a circuit designed to produce short duration, unidirectional impulses. When all of the conditions were right, this aetheric energy would manifest itself as a spatially distributed voltage that would radiate away from the electrical circuit as a light-like ray that could charge other surfaces within the field.

NOTE
This is also called radiation light or eletromagnetic waves.

Tesla found that this effect was greatly magnified when these impulse currents were produced by the discharge of a capacitor. This huge explosion of electrostatic energy, which radiates away at right angles from the capacitor discharge pathway, is the primary operating principle of his Magnifying Transmitter. With this amazing device, Tesla planned to broadcast energy to the whole world from his facility at Wardenclyffe, New York.

After Tesla was prevented from bringing his World Broadcast System into full manifestation, he worked for years to develop a smaller version of the device that harnessed the same principles. By the 1920s he had succeeded. This specialized electronic circuit (aetheric energy) is what powered his infamous Pierce-Arrow automobile.

It should now be quite obvious to the reader that Nikola Tesla was a pioneer in many fields of electrical theory and technology. He was, in fact, the first to utilize alternating current, conceiving an effective system for its generation, transmission, and utilization. Edison appealed to the public, warning that the alternating current of Tesla would cause great harm to its users, being dangerous, and that only direct current can be harmlessly used. Tesla referred to Edison as an inventor, to himself as a discoverer. Today everyone knows that alternating current, with the help of the polyphase induction motor, can be converted into mechanical energy more effectively and economically than direct current. He invented new forms of dynamos, transformers, condensers, and induction coils. He discovered the principle of the rotary magnetic field, upon which the transmission of power from the Niagara Falls and other waterfalls and dams is carried. A regal recluse, he despised the short-seeing men of science. Many of his pioneer inventions he carried with him to his grave, but he believed in the destiny of man who searches, discovers and invents, designs and constructs, and covers with monuments of beauty, grandeur, and awe, the star of his birth.

NOTE

When Nikola Tesla died, January 7, 1943, in a world at war, the FBI showed up within hours to open his safe—though Tesla had become an American citizen in 1891, his many boxes and crates were put under seal and unaccountably turned over to the Office of Alien Property (OAP). Many were released in 1952 to the Tesla Museum in Belgrade; some have not resurfaced. His is a legacy of brilliance and enigma.

Not only have the contemporaries of revolutionary ideas in science repeatedly rejected the idea of free energy, but they have also rejected other such ideas from generations to generations throughout history. For example, Archimedes rejected the heliocentric system of Aristarchus; Brahe rejected the system of Copernicus; and Galileo was deaf and blind to the discoveries of Kepler, just as Edison warned against the alternating current developed by Tesla. And, who was more competent to judge than Archimedes, in his time, Brahe in his, Galileo in his, and Edison in his? [4]

Automobiles, airplanes, movies, recorded music, telephones, radio, and practical cameras would all be here in 20 years. The Victorian Age was giving way to something totally new. For the first time in history, common people were encouraged to envision a Utopian future, filled with abundant modern transportation and communication, as well as jobs, housing, and food for everyone. Disease would be conquered, and so would poverty. Life was getting better, and this time, everyone was going to get a "piece of the pie." So what happened? In the midst of this technological explosion, where did the energy breakthroughs go? Was all of this excitement about free electricity (which happened just before the beginning of the last century) just wishful thinking that real science eventually disproved?

"In our society," explains Dr. Bailey, "science is ruled by corporations and profit. Nikola Tesla is said to have discovered how to create a certain type of alternating electricity, and knew how to transmit it through the air to any receiver, on land or in the air. This would have replaced transmission lines, batteries, and the need for gasoline for cars and planes. However, the banker who was promoting Tesla's work found that he could make more money by investing in copper mines and then using the copper from his mines to make wire that the electricity could travel over. He therefore terminated Tesla's work."

He continues:

> The age of enlightenment has given way to the age of profit. The age of helping one another appears to be giving way to the age of making a buck just to survive. While this condition exists, more and more attention will be focused on making money from easily obtainable resources—regardless of their long-term consequences—while holding back any other technologies that could replace them. Examples of this abound in our society. Our society is run by oil. As long as gasoline can be made to cost about the same as water, then no other energy conversion technologies are deemed to be needed. International banks, including the Federal Reserve Bank, look to the ownership of oil reserves as assets to back their bank and the value of their currency. Other or newer technologies that could replace the use of oil are usually perceived as a threat by these bankers and large corporations. On

one hand, the countries and corporations that own the oil are actively looking for alternate technologies to help reduce the rate of oil extraction from their resources. On the other hand, the companies and corporations that import oil into countries like the United States, make their profit on the flow of the oil, and they do not want to see that flow reduced. The politics of oil and the greed for profit seems to be causing a race for the corporate control of all of the world's oil reserves. What will happen then is anyone's guess, and, it looks like the end of that race is in sight.

Another factor that now suppresses scientific advancement today is "national security." Any invention or discovery that is sent to the U.S. Patent Office automatically undergoes a review by a panel of military representatives, representing each of the armed services. If anyone on that panel thinks that any patent can be applicable to their branch of the military, then that patent is removed from the normal patent processing flow, and is "classified." The individual is then sent a "Secrecy Order," which by federal law requires them to send in all of their research notes and data, and keep silent about the patent application that he has submitted—under the threat of heavy fines and imprisonment. Examples of these "Secrecy Orders" can be found on the Internet, using "SECRECY ORDER (Title 35, United States Code (1952), sections 181–188)." On the one hand, keeping secret exactly how to double the gas mileage for a tank may be very beneficial for Army operations, and may be a strategic advantage over a foreign army. But, on the other hand, your car still gets the same miles per gallon as it did in 1950. That's why. There must be hundreds of really technically advanced inventions and devices that have been developed by inspired inventors and researchers since the 1950s. You can imagine where the ones that were submitted to the U.S. Patent Office have gone. Is this the way things should be done? It is your country. [2]

Actually, the answer to the preceding question is a flat no! In fact, the opposite is true. Spectacular energy technologies were developed right along with the other breakthroughs. Since that time, multiple methods for producing vast amounts of energy at extremely low cost have been developed. None of these technologies have made it to the open consumer market as an article of commerce, however. Exactly why this is true will be discussed shortly. But first, let's look at a short list of "free energy" technologies that scientists are currently aware of, and that are proven beyond all reasonable doubt. The common feature connecting all of these discoveries is that they use a small amount of one form of energy to control or release a large amount of a different kind of energy. Many of them tap the underlying aether field in some way, a source of energy conveniently ignored by "modern" science. For example, there are:

- Radiant energy
- Permanent magnets
- Mechanical heaters
- Super-efficient electrolysis
- Implosion/Vortex
- Cold fusion (so far, it has been disproven)
- Solar assisted Heat Pumps
- Wind power
- Sound

NOTE

The naysayers claim that Michaelson/Morely showed that aether energy does not exist. Therefore, the claim that it was conveniently ignored by modern science, must be a paranoid accusation.

According to Dr. Aspden:

> There is much in evidence in the free energy field, but it lies on the fringe of orthodox science and is shunned by those we trust to take our power industry forward. The reason is the belief that energy is conserved and the inability to see the necessary source that can be tapped to provide the energy input. Physicists heeded Einstein's doctrines and so abandoned their earlier belief in the existence of the aether. The aether did not match up to their assumptions as to how it regulated the speed of light and so they thought it could not exist. It was replaced by an imaginary mathematical world of four dimensions in which time is interwoven with space. Yet the more dominant assumption, the one that is justified experimentally in our real three-dimensional world of space, is that it can store electrical energy and so must exist. It is only when one portrays the aether in a form that can explain quantum physics (the physical nature of its fundamental unit of quantum angular momentum) that the effects of storing electrical energy in the vacuum medium can be understood. We can then see how it is conditioned to shed some of its own hidden store of energy. Keep in mind that the pioneers making anomalous energy discoveries in free energy technology did not think in terms of quantum physics and, though lucky in stumbling on a path that gives a glimpse of excess energy, they could not see the picture underlying Nature's mechanism for releasing that energy. Lacking such understanding, they could not move forward and find support in contending with the formidable barriers that exist in our technological world when confronted by what seems to be a crazy idea. [3]

Radiant Energy

Aetheric power is the power source of the future. Until recently, only a handful of researchers have understood how to tap it. For over one hundred years, there have been persistent rumors that a number of brilliant inventors had succeeded. These included Dr. Nikola Tesla, Dr. T. Henry Moray, Edwin

Gray, Sr., and more recently, Paul Baumann in Switzerland. Each reported and showed hundreds of eyewitnesses a cold form of electric power that could be produced with relatively simple equipment. This natural energy form can be gathered directly from the environment (mistakenly called static electricity) or extracted from ordinary electricity by the method called fractionation. Radiant Energy can perform the same wonders as ordinary electricity, at less than 1% of the cost. It does not behave exactly like electricity, however, which has contributed to the scientific community's misunderstanding of it.

NOTE

Here, the naysayers and skeptics of free energy claim that all of this is pure fantasy and an absurd assertion. But not everyone agrees with them.

Dr. Aspden explains:

I have put my case elsewhere, as in a talk I gave in Berlin on June 14, 2002 to a group of scientists interested in free energy. This is on record in Chapter 9 of my book *The Physics of Creation,* which is published on my Web site (*www.aspden.org*). In that book, I presented the theory that should enlighten quantum physicists on this question of how to harness the vast energy resource locked in the quantum properties of the aether. I show how Mother Nature created the sun and the planets, including body Earth, although, as previously pointed out in this chapter [this volume], physicist Alan Guth may have "computed the energy necessary to create a baby universe in the basement." But it needs just a little more know-how of the kind I present in my book just referenced, before you can lay your hands on the necessary influx of energy. Set to work in a basement if you wish and try to create a star if you can, by throwing a switch that controls gravity. You will set up a radial electric field, only transiently, but long enough to import energy from the aether. Alternatively, spin a magnet, one that is of alloy composition such as Alnico and so electrically conductive. That will induce a radial electric field inside the magnet and import aether energy. But, for a steady energy gain, the magnet strength has to pulsate. That is why machines, such as that of Lee Borman (1954), exhibited perpetual motion properties. He used interacting rotating systems of Alnico magnets, which brought pairs of magnets into line along a common axis, while sharing the rotor spin, which meant that their juxtaposition caused fluctuation in magnetic field strength. Instead, do something radically different and produce that pulsating radial electric field in a different way, as by using powerfully electrically charged capacitors of concentric construction. Then you are really on target for a solution of our future energy problems. You will surely be tracking the early 20th century discoveries of Nikola Tesla, as well as Dr. Henry Moray and Paul Baumann in a Swiss community of our present era. Here is the secret to free energy, but though the next question might set your minds at work it needs that insight I have just provided—the reference to the radially directed electric field—to guide you to the underlying secret. Think about it and read on! [3]

Magnets That Are Permanent

The world has already changed, through the harnessing of the invisible force called *magnetism*. It has given humans electricity, radio, television, computers, and thousands of other things. But its greatest gift to mankind is yet to be realized. Magnetism can provide a source of inexhaustible, pollution-free energy. In the last 120 years, dozens of inventors have reported success in harnessing magnetism to produce excess mechanical energy, electricity, and heat.

NOTE

Again, here, the naysayers and skeptics of free energy claim that all of this is pure fantasy and an absurd assertion.

Recently, there have even been reports of using magnets to produce refrigeration effects. More and more researchers are probing the unknown properties of magnetism, since permanent magnets are getting stronger and cheaper all the time.

Magnets That Are Used in Machines

The way magnets push and pull and twist each other with no apparent connection between them is quite fascinating. It's tempting to suppose that there must be some way of arranging them to extract energy from them. For example, one of the earliest perpetual motion machines proposed to use a lodestone (a lump of naturally magnetic iron ore) to pull a ball up a slope towards a hole through which it would drop to cycle back to the start. This didn't work. Neither do its modern derivatives, and for the same reason. So, to get back to the starting position, any work that a magnet does on an object has to be undone.

NOTE

This is also why an unbalanced wheel won't work.

Moving something in a closed loop in either a magnetic or a gravitational field causes it neither to gain nor lose net energy. Since this applies to all objects, it applies to every part of a machine, no matter how complicated it is. So, to get a positive result, there's no way of combining many zeros.

This doesn't stop people from proposing motors that are driven only by magnetic fields. These motors have rotors that are pushed or pulled most of the way around its circle by some arrangement of magnets. There's nothing impossible about this, but the designers then expect the rotor to suddenly ignore the magnetic field and to complete the cycle, thus getting the rotor back to its starting point after delivering a net output of energy. This is impossible.

Magnets and Wires That Are Used in Machines

So forget magnets acting alone. Let's mix in some wires and electric currents. Earth's whole civilization depends on devices that move wires in magnetic fields to generate electrical power and other devices that pass an electric current through a wire to generate motion—that is, on electrical generators and electric motors.

People are always asking scientists whether their odd configuration of wires and magnets will generate a power output from lesser input power. The answer is no, but it is not always clear from elementary physics textbooks why this is so. Without explaining what the implications are, textbooks are inclined to describe the forces involved.

NOTE

Here, the naysayers and skeptics of free energy claim that the preceding violates the second law of thermodynamics, but they never explain why.

Wires Voltage Generation

Let's start with generators. If you have a uniform magnetic field (say between the poles of a horseshoe magnet), and you move a straight wire through the field so that the field, the length of the wire, and the direction of motion are all at right angles to each other, then voltage will appear between the ends of the wire. In this simple case, it is easy to calculate the voltage; it is equal to BvL volts. *L* is the length of the wire in meters, *B* is the strength of the field in webers/square-meter and *V* is the velocity of motion in meters/second. To put this effect in perspective, 1 weber/square meter (10,000 Gauss) is a much stronger field than you can get from a small permanent magnet. To generate a field this strong and a meter across, would require an electromagnet weighing several tons. Wiggling a few inches of wire in the field of a small magnet will produce a few millivolts. Only the relative motion counts, because it makes no difference whether it is the conductor or the field that is moving. [5]

Looking at this effect in isolation, scientists have a length of wire moving continuously in a uniform field. No work is needed to keep the wire moving since, once scientists have started it off, it is generating no electrical power output. When scientists started the wire moving and added some more energy to establish the initial field between the ends of the wire, it took some more energy to overcome the wire's inertia, but scientists will still overlook them.

NOTE

When the wire stops moving, scientists can, in principle, recover that energy.

Loop Completion

Suppose scientists want either to do something useful with the voltage generated or measure it. To attach a voltmeter or a load resistance, scientists must complete a loop that contains the bit of wire they are looking at. Suppose scientists truly have a uniform field and that it is very large, at least in the direction they are moving. Imagine, for example, that they are traveling north on rails between the poles of a magnet that is several miles wide. The scientists' test wire extends from east to west, and the field goes from south to north. So, between the ends of the wire, they are traveling fast enough to generate one volt.

So that scientists can hook their voltmeter up to the east end without any trouble, they are sitting at the east end of the wire in order to make it easy. How do scientists connect the voltmeter to the west end of the wire? With another wire, of course. But that wire is also going from east to west and is traveling at the same velocity as the first one. It also has a voltage of one volt induced in it. The west end of both wires have the same voltage on them; when scientists connect them together, the net voltage around the circuit is zero, and that is what their voltmeter will show. No matter what route the second wire takes from east to west, it will have exactly the same voltage across it as the test wire. If you connect a load resistor to the loop of wire, no current will flow, and no output power will be generated. So, in order to generate voltage in a loop of wire, one of two conditions must hold. First, the field in one part of the loop must be different from the field in another part; and second, the velocity of one part of the loop must be different from the velocity of another part of the loop. [5]

Loop Types

For example, by using a small magnet and a large loop so that only a part of the loop is in the strong field, the first condition can be met. So obviously, at any given velocity, this limits the length of time scientists can generate a voltage.

Therefore, sliding the wire through the field on rails is one way of meeting the second condition. Since the rails would not be moving they would not contribute to the voltage.

These illustrate a general principle that says that no voltage is generated around the loop, so long as the product of a magnetic field and the area of a loop remains constant. Either the area of the loop or the strength of the field must change to generate voltage. As previously discussed, you can vary the strength of the field within the loop by moving the loop from a weak field to a strong field or by changing the effective area of the loop.

So, if you rotate loop in the field, you'll find another way to vary the effective area of a loop. This makes its effective area go from +A to –A and back to +A every complete turn. Or, look at it this way: twice per turn, one side of the loop will, briefly, be moving in the opposite direction to the other. It will generate voltage, which adds to the first side's voltage rather than subtracting. Half a turn later, the loop will be generating a voltage in the opposite sense. Thus, if scientists hook up the two ends of the loop to slip rings, they will see an output voltage that alternates as the loop turns. Not only can scientists measure the voltage, but they can get a useful output. The scientists have just invented the alternator. [5]

A brush on each side of the ring will pick up pulses of voltage that all have the same polarity, especially if the two ends of the loop are connected to a split ring around the shaft. By adding loops, each at a slightly different angle, and connecting each to a pair of segments on a commutator, at any moment the brushes will connect only to the loop that is moving the fastest in the field. This results in a DC generator, because it gives an almost steady output voltage.

Outside the Loop

Now, let's look at that original piece of wire. If it is not at right angles to the field or if it is not moving at right angles to its length, the voltage generated will be less than in the right angle case. It will be proportional to the sines of the angle between the motion and the field and the sine of the angle between the wire and its direction of motion. No voltage will be generated, if the motion is along the field or the wire is moving lengthwise. [5]

By adding up the contributions from each little piece of wire, scientists can use this relationship, at least in theory, to calculate the voltage around any loop of wire moving at any velocity in any magnetic field. Some quite ingenious ideas fall flat when this is done. The inventors have looked only at the interesting part of their machine and have ignored the fact that all magnetic fields loop back on themselves somewhere. The voltage generated in that part of the loop also has to be taken into account. In the rest of the system, it often cancels out the voltage.

Energy: Where Does it Come In?

As previously mentioned, moving a wire through a field requires no energy. That rotor will spin until friction stops it.

NOTE

The eddy current losses are being ignored here.

You can hang a voltmeter on the output and measure the voltage without any significant effect. By giving the loop of wire many turns, you can generate as big an output voltage as you please. Using a stronger field, a faster rotor speed, or a longer rotor also increases the output voltage. No energy is needed to keep things moving unless some current flows. [5]

Current Flows: What Happens?

When you play with wires and magnets, another phenomenon occurs. This time let's hold that wire in the uniform field in a fixed position and pass a current through it. What happens is that the wire tries to move in a direction at right angles to both the field and the current. The force needed to keep the wire still is given by ilB where i is the current in amps, l is the length of the wire in meters, and B is the field strength in webers per square meter. The force is measured in Newtons. [5]

The wire will start moving across the field, if scientists let it go. Its acceleration will be proportional to the current times the field strength. There is no limit to how fast the wire will go in Newtonian physics.

NOTE

This principle has been proposed for firing payloads into space.

However, when the wire moves, voltage is generated across its ends. This is where scientists come in. The faster the wire moves, the higher the voltage generated. This voltage acts in the opposite direction to the current scientists are feeding in, making it harder and harder to force that current through the wire. For a given current, low speed equals low back voltage and hence low electrical power input. A high speed generates a high back voltage and thus requires a high electrical input power. Now, the mechanical energy out is simply equal to the force exerted, times the speed. Low speed equals low output power and high speed equals high output power. Are you beginning to see a pattern? High mechanical power out equals high electrical power in, low mechanical power out equals low electrical power in. [5]

Input and Output Computing Powers

The product of the current and the voltage is the power in watts that scientists are feeding into the wire. If the wire is moving at a constant velocity and is lifting a weight at so many meters per second, then the mechanical power being generated is just the product of the velocity times the force on the wire. The force is proportional to the current, and the back voltage is proportional to the velocity, so the input power is proportional to the output power. It turns out that the mechanical power out is exactly equal to the electrical power in, when you take the time to multiply out all the units.

NOTE

This is done with the usual provisos about friction and resistive losses being negligible.

Turning Things Around

When you apply mechanical power to a wire, exactly the same thing applies the other way around. As the wire moves, it generates a voltage proportional to its velocity. If a current flows, the electrical output power is proportional to the product of the voltage and the current; that is, it is proportional to the product of the current and the velocity. Now, when a current flows in the wire, it generates a force on the wire. Surprise, surprise! This force acts to oppose the motion of the wire. To keep it moving, you have to push it harder. The mechanical energy you must apply is proportional to the velocity and the reverse force caused by the output current. So, the output electrical power is equal to the input mechanical power in this case.

This is rather wonderful in a way. It means that scientists can convert mechanical power into electrical power or electrical power into mechanical power with practically no losses. This is also practical with thermal power (heat engines). You can go to any coal/oil/gas/nuclear power plant or look under the hood of your car to see how it works.

You can never come out ahead, unfortunately, since the equality of power in and power out applies to each little piece of wire, no matter how it is moving and in whatever magnetic field. Any device, no matter how ingenious, that generates an output current also generates a force opposing its motion. Any device that generates motion from a current, also generates a back voltage that opposes the input current.

A Case That's Practical

Until you connect something to its output terminals, the alternator described here will spin happily with almost no input power. Then a current will flow that is given by the output voltage, divided by the total resistance of the wire loop plus the resistance of the load you apply. If the loop resistance is not zero, the output voltage will drop. The useful output power is this lower voltage times the current, so it pays to use low-resistance wire.

You are applying a mechanical load to the alternator by allowing a current to pass. It thus requires more input power to keep it turning. The more output power you take, the more difficult it becomes to turn the rotor. If you have a DC motor with a built-in reduction gearbox, it is easy to turn its output shaft by hand when its leads are open circuit, but almost impossible to turn it if they are shorted together. Power in, at best, equals power out.

The Homopolar Generator

It can be quite difficult to see just where the current loop is sometimes. Michael Faraday discovered that if you turn a conducting disk in a magnetic field, you can measure the voltage between the shaft of the disk and its rim. Thus, both of these devices have been baffling to people ever since: the homopolar generator and its equivalent, the homopolar motor.

Dr. Aspden states:

> Although I agree that the induction of electricity by a homopolar motor has posed problems, these have arisen from thinking that the magnetic field directed along the axis of the rotor actually spins with the rotor. This is not the case. Just consider a free conduction electron and its counterpart positive ion sitting

> inside a conductive disk. They are transported by the disk at right angles to the radial direction centered on the spin axis. They are moving through a magnetic field directed parallel with that axis and so are subject to radial electromagnetic force, the electron being urged along such a radial path in a direction opposite to that of the ion. The ion, being constituted by an atom of that disk, cannot be displaced, but the electron is free and so can move and set up current flow. Hence one has an EMF acting in a radial sense, set up by the disk rotating in a magnetic field. Of course, here again, there is interplay with the aether and, just as the aether can store electrical field energy, so it can also store magnetic field energy too, but it needs a physical picture and not just some mathematical symbolism to understand how it does that. Take away the aether and, yes, you do have problems understanding the homopolar motor.[3]

Consider a disk that is rotating in a uniform magnetic field. The field passes through the disk at right angles to its surface. Any radius of that disk is moving through the field. The parts of the radius near the shaft are moving slowly, and the parts near the rim are moving quickly, but they are all moving in the same direction, so the voltages that generate by each little bit of the radius all add up. The result is a voltage all around the rim of the disk that is higher than the voltage at the center.

You can measure this voltage if you mount a sliding contact at any point on the rim. It won't be a large voltage, since only a single conductor is moving through the field and parts of it are not moving at the full speed. Suppose you had a copper disk 20 cm (8") in diameter, turning at 10,000 revolutions per minute, and that you could put it in a uniform 0.1 weber field.

CAUTION

This would be a very dangerous thing to try!

The rim will be moving at 105 m/s, so the mean velocity of a radius is half that. The length of the radius is 0.1 meters and the field is 0.1 webers, so the voltage will be 0.52 volts, about a third of what you'd get from an alkaline cell [5]!

Quite a substantial current would flow from the shaft to the rim if your disk had a low resistance, and if the brush contact and the external circuit also had a low resistance. Because the voltage is generated on all radii of the disk, you could put brushes all around the rim and reduce the effective generator resistance. Unfortunately, the frictional forces on the edge of the rim chew up a lot of input power. As in any other generator, the motion of the disk is resisted by the force which the magnetic field exerts on the output current. Thus, homopolar generators are only useful if you don't care how much energy it takes to turn the disk and if you need a high current at a low voltage.

TIP

You should use a uniform magnetic field.

You might think that having a field just between the shaft and the point on the rim where the brush is placed would be more efficient. Unfortunately, when the field is not uniform over the whole disk, local currents are generated in the disk, heating it up and wasting input power. This "eddy-current" effect is the basis of the magnetic brake. Electricity meters are one common application, and some car speedometers also depend on this effect.

Homopolar Motor

The magnetic field will generate enough force on it to make it rotate despite the brush friction, if you pump a huge current into the disk. Unfortunately, a notoriously inefficient process is the generation of large direct currents at low voltages.

NOTE

Early experimenters made contact to the rim of the disk by making it pass through pools of mercury. Since mercury vapor is rather poisonous, this is not done any more.

AC can be transformed down to get a low voltage, but the rectifying device needed to convert the output to DC tends to drop about half a volt, making the efficiency less than 50% before you start moving anything. You guessed it! About the only device that can generate large currents at low voltages at all efficiently is a homopolar generator.

According to Dr. Aspden:

> The homopolar motor claims for free energy reveal anomalies. But, as pointed out earlier, taking current from the rim of a rotating disk in a magnetic field directed along the spin axis or from a rotating magnet (an alloy conductor such as Alnico) allows low-voltage induction and high current needed for real power output measurement difficult and uncertain. The answer I see is to avoid drawing current from the rim. The small voltage induced will displace electric charge radially within the conductor, setting up a charged system that induced aether spin coextensive with the conductor. Aether spin is a quantum-coupled reaction between matter and the space medium owing to the radial electric field displacing the synchronously moving charges that exhibit the quantized motion of that aether underworld. To keep synchrony, those aether charges themselves must rotate about the same axis of spin as the conductor. All the machine input supplies is the energy needed to displace charge, but that spin has its own equal measure of kinetic energy tapped from the aether itself owing to the phase-lock action and external constraint. It is kinetic energy that knows no way of returning to its original source except by being dissipated and eventually melding with radiant energy to be absorbed into the quantum underworld somewhere in far-off regions of space—that is, unless it

> melds with the spin kinetic energy of the conductor, which certain homopolar motor evidence suggests to be a possibility. The only way such devices can ever draw power steadily from the aether is to provide within them the means by which the magnetic field pulsates so as to harness the kinetic energy extracted from the aether and keep renewing the input of such energy. That is a formidable task, given that heat produced by eddy-current induction can become a formidable problem. Unless that problem can be overcome, I see homopolar motor research as pioneer research giving insight into aether properties, but only thereby leading us to better techniques exploiting the same underlying principles. Such techniques were used by Nikola Tesla and Dr. Henry Moray, but these techniques did not require fast rotation of heavy rotors and their consequent heat losses, owing to very high internal current flow. [3]

Defeating Your Own Object

By attaching the magnets generating the field to the rotating disk, one hopeful inventor thought that you could generate an output voltage without generating a reverse force. Unfortunately, this doesn't work. Only relative motion of the field and the current-carrying disk generates voltage. What misled this inventor was that he did measure voltage when he connected a meter between the shaft and the rim of the disk. He hadn't realized that the ring magnets were generating two fields, one in the disk between the magnets and another toroidal field in the space around the disk. The disk wasn't generating voltage, but the wire leading to the meter was being cut by this rotating field and was generating a small voltage. Connecting a current meter would have shown an output current. However, this current would have reacted on the rotating field to slow the disk down. As previously mentioned, you have to consider what is happening everywhere in the system, not just focus on one part of it and ignore the rest.

The effect was the same as if the disk and magnets were stationary and a contact had been spun around the edge of the disk. The wire going to the contact would have been moving in a magnetic field, and all the usual rules would apply.

No arrangement of wires and magnetic fields and moving parts is going to generate more electrical power than the input mechanical power or generate more mechanical power than the input electrical power. If you want to experiment, by all means have fun, but please don't think you are going to bring about an energy revolution. Above all, unless you want to spend the rest of your life dodging irate investors, spend only your own money, not other peoples'.

Mechanical Heaters

There are two classes of machines that transform a small amount of mechanical energy into a large amount of heat. The best of these purely mechanical designs are the rotating cylinder systems. In these machines, one cylinder is rotated within another cylinder with about an eighth of an inch of clearance between them. The space between the cylinders is filled with a liquid such as water or oil, and it is this "working fluid" that heats up as the inner cylinder spins. Another method uses magnets mounted on a wheel to produce large eddy currents in a plate of aluminum, causing the aluminum to heat up rapidly. By using the same energy input, all of these systems can produce more heat than standard methods. Each of these has a Coefficient of Performance (COP) of less than one. Current heat pumps have COPs on the order of five to six.

Super-Efficient Electrolysis

By using electricity, water can be broken into hydrogen and oxygen. Standard chemistry books claim that this process requires more energy than can be recovered when the gases are recombined, but this is true only under the worst-case scenario. When water is hit with its own molecular resonant frequency, it collapses into hydrogen and oxygen gas with very little electrical input.

NOTE

Here, the naysayers and skeptics of free energy claim that this is an absolutely false statement. But again, they never explain why.

Also, using different electrolytes (additives that make the water conduct electricity better) changes the efficiency of the process dramatically. It is also known that certain geometric structures and surface textures work better than others do. The implication is that unlimited amounts of hydrogen fuel can be made to drive engines (like in your car) for the cost of water. Even more amazing is the fact that a special metal alloy that was patented by Freedman (USA) in 1957 spontaneously breaks water into hydrogen and oxygen with no outside electrical input and without causing any chemical changes in the metal itself. This means that this special metal alloy can make hydrogen from water for free, forever.

NOTE

Here, the naysayers and skeptics of free energy claim that this is absolutely false. Catalysts work, but require higher temperatures for some processes. Nothing catalyzes water into hydrogen and oxygen at room temperature without power being supplied.

Now, let's look at some of the research that is going on in this area—especially in the area of hydrogen fuel cells. The future of hydrogen is simple, according to scientists.

The Future of Hydrogen

The future of hydrogen is nonpolluting, inexhaustible, nontoxic, and so very basic in its chemical structure (H, one proton, one electron). The first, lightest, and most common element in the universe. The stuff that turns oil into margarine. The stuff that made the Hindenburg float. The stuff that combines with oxygen to make water and with carbon to make methane. The stuff that sends the space shuttle skyward and could someday power your car, office building, house, cell phone, even your hearing aid. This is the stuff that could clean up the planet. [6]

Think of a world in which cars are whisper quiet, they emit only water vapor, and OPEC is out of business because the price of oil has fallen to five dollars a barrel. Global warming (see sidebar, "Out in the Cold"), smog, California-style blackouts, a whole host of ills will be solved by hydrogen, according to scientists.

Out in the Cold

Much of North America and Europe were in the throes of a little ice age, roughly from 1300 to 1850. And now, there is mounting evidence that the chill could return. A growing number of scientists believe conditions are ripe for another prolonged cooldown, or small ice age. The next cooling trend could drop average temperatures 5 degrees Fahrenheit over much of the United States and 10 degrees in the Northeast, northern Europe, and northern Asia. No one is predicting a brutal ice sheet like the one that covered the Northern Hemisphere with glaciers about 12,000 years ago, except perhaps 20th Century Fox's *The Day After Tomorrow,* which opened in theaters on May 28, 2004.

According to scientists, it could happen in 10 years. Once it does, it can take hundreds of years to reverse. Scientists are alarmed that Americans have yet to take the threat seriously.

You'd have an idea of what this would be like, if you recall the coldest winters in the Northeast, like those of 1936 and 1978 (and in the Northern and Central Plains in 1977, 1978, and 1979), and then imagine recurring winters that are even colder.

A drop of 5 to 10 degrees entails much more than simply bumping up the thermostat and carrying on. Both economically and ecologically, such quick, persistent chilling could have devastating consequences. An abrupt climate change could generate agricultural losses alone at $200 billion to $360 billion, while damage to ecologies could be vast and incalculable. A grim sampler: accelerated species extinctions, disappearing forests, increased housing expenses, dwindling freshwater, and lower crop yields.

The reason for such huge effects is simple. A quick climate change wreaks far more disruption than a slow one. People, animals, plants, and the economies that depend on them are like rivers. For example, high water in a river will pose few problems until the water runs over the bank, after which levees can be breached and massive flooding can occur. At particular thresholds of temperature and precipitation, many biological processes undergo shifts.

Demographic changes since the last ice age could make survival far more difficult for the world's poor. During previous cooling periods, whole tribes simply picked up and moved south, but that option doesn't work in the modern, tense world of closed borders, to the extent that abrupt climate change may cause rapid and extensive changes of fortune for those who live off the land, removal of one of the major safety nets for distressed people would be the inability to migrate.

Still, climate science is devilishly complex and the onslaught of a little ice age is not certain, at least at this stage of research. Scientists the world over are weighing the potential for a rapid North Atlantic cooling.

But first things first. Isn't the Earth actually warming?

Indeed it is. So how could such warming actually be the surprising culprit of the next mini-ice age? The paradox is a result of the appearance over the past 30 years in the North Atlantic of huge rivers of freshwater (the equivalent of a 10-foot-thick layer) mixed into the salty sea. No one is certain where the fresh torrents are coming from, but a prime suspect is melting Arctic ice, caused by a buildup of carbon dioxide in the atmosphere that traps solar energy.

Today, the story that's making major news in ocean science circles is the freshwater trend. Arguably the largest full-depth changes observed in the modern instrumental oceanographic record is the drop in salinity and temperature in the Labrador Sea (a body of water between northeastern Canada and Greenland that adjoins the Atlantic).

By subverting the northern penetration of Gulf Stream waters, the trend could cause a little ice age. Normally, the Gulf Stream, laden with heat soaked up in the tropics, meanders up the east coast of the United States and Canada. As it flows northward, the stream surrenders heat to the air. Because the prevailing North Atlantic winds blow eastward, a lot of the heat wafts to Europe. That's why many scientists believe winter temperatures on the Continent are as much as 36 degrees Fahrenheit warmer than those in North America at the same latitude. Frigid Boston, for example, lies at almost precisely the same latitude as balmy Rome. And some scientists say the heat also warms Americans and Canadians. It's a real mistake to think of this solely as a European phenomenon.

Having given up its heat to the air (in a process oceanographers call thermohaline circulation), the now-cooler water becomes denser and sinks into the North Atlantic by a mile or more. This massive column of cascading cold is the main engine powering a deepwater current called the Great Ocean Conveyor that snakes through all the world's oceans. But as the North Atlantic fills with freshwater, it grows less dense, making the waters carried northward by the Gulf Stream less able to sink. The new mass of relatively fresh water sits on top of the ocean like a big thermal blanket, threatening the thermohaline circulation. That in turn could make the Gulf Stream slow or veer southward. At some point, the whole system could simply shut down, and do so quickly. There is increasing evidence that the North Atlantic is getting closer to a transition point from which it can jump to a new state. What could yield a big response are small changes, such as a couple of years of heavy precipitation or melting ice at high latitudes.

It's not just in the Labrador Sea. This cold, freshening area is now invading the deep waters of the entire subtropical Atlantic.

You have all of this freshwater sitting at high latitudes, which could literally take hundreds of years to get rid of. So while the globe as a whole gets warmer by tiny fractions of one degree Fahrenheit annually, the North Atlantic region could, in a decade, get up to 10 degrees colder. What worries scientists is that history is on the side of rapid shutdown. They know it has happened before.

Clearly, the little ice age from 1300 to 1850 wasn't kicked off by humans releasing greenhouse gases into the atmosphere. But natural climate cycles that melted Arctic ice could have caused thermohaline circulation to shut down abruptly. Scientists are almost certain that this was the cause of the last little ice age.

Because of a thermohaline shutdown, a more recent event is perhaps better evidence that a climate can cool quickly. In the late 1960s, a huge blob of near-surface fresher water appeared off the east coast of Greenland, probably the result of a big discharge of ice into the Atlantic in 1967. Known as the Great Salinity Anomaly, it drifted southward, settling into the North Atlantic in the early 1970s. There it interfered with the thermohaline circulation by quickly arresting deepwater formation in the Labrador Sea. It continued to drift in a counterclockwise direction around the North Atlantic, re-entering the Norwegian Sea in the late 1970s and vanishing soon after.

Scientists believe it shut the system down for just a few years. Particularly in Europe, the result was very cold winters.

Fortunately, that fresher-water mass was small enough to disperse in a short period of time. However, the one accumulating up there now is just too big.

Because it is dependent upon the gathering and interpretation of millions of data points, climate science is extraordinarily complex. If the National Weather Service has trouble predicting tomorrow's weather, how can anyone forecast a change in global climate a few years hence? One answer is even more data. At the moment, there are about 450 floating sensors bobbing around in the Atlantic monitoring temperature and salinity changes, and that is not enough. The models don't have enough resolution to capture all the physics yet. Prediction is tough.

Should a little ice age arrive, its impact will be told in human suffering, not scientific terminology. Tales of woe, depicting the plight of European peasants during the 1300 to 1850 chill, include famines, hypothermia, and bread riots. In the late 17th century, agriculture had dropped off so dramatically that Alpine villagers lived on bread made from ground nutshells mixed with barley and oat flour. Finland lost perhaps a third of its population to starvation and disease.

Finally, life was particularly difficult for those who lived under the constant threat of advancing glaciers in the French Alps. One, the Des Bois glacier on the slopes of Mont Blanc, was said to have moved forward over a musket shot each day, even in the month of August. When the Des Bois threatened to dam up the Arve River in 1644, residents of the town of Chamonix begged the bishop of Geneva to petition God for help. In early June, the bishop, with 300 villagers gathered around him, blessed the threatening glacier and another near the village of Largentire. For a while, salvation seemed at hand. The glaciers retreated for about 20 years, until 1663, but they had left the land so barren that new crops would not grow [7][8].

Whether or not hydrogen will become the fuel of choice in the foreseeable future is controversial: There's no shortage of energy pundits who proclaim that scientists and other boosters grossly underestimate the expense of making hydrogen, not to mention the technical hurdles that must be overcome. These naysayers argue that hydrogen's role in the world's fuel mix is likely to remain marginal for decades. Immense practical barriers exist.

Still, deployed according to a unique plan, some scientists believe that hydrogen's virtues make its widespread use in the near future virtually inevitable. Unlike other environmentalists who claim clean technologies will thrive only with government insistence, some scientists contend that the entire fossil-fuel economy will give way to hydrogen because of simple obsolescence and efficiencies, in much the same way that vinyl records gave way to CDs. Scientists' plan for switching to a hydrogen-based economy stresses the money to be made. By starting now, the transition can be profitable at every step.

Can scientists usher in the hydrogen age? Should they? The answer, ultimately, depends not upon scientists, but upon the nature of hydrogen itself. Despite its ubiquity, the hydrogen molecule lies outside your everyday experience. Because it bonds readily to other elements, scientists don't run into it in its elemental state. Before the hydrogen-based energy economy becomes a reality, scientists need to know: Is hydrogen plentiful, cheap, safe, and powerful enough to run the world?

"Is the use of hydrogen gas safe?" asks Dr. Bailey. "Yes!"

> Most people remember the movie footage of the crashing and burning of the Hindenburg, an airship (blimp) that was filled and lifted by hydrogen gas. The movie shows huge flames as the entire airship caught fire and fell to the ground. Many people were burned and burned to death—by diesel fuel—not by the hydrogen gas. As hydrogen gas is lighter than air, like the helium in balloons, it rises very fast, and tends to burn upwards.

The use of hydrogen gas is very safe, and many universities and research corporations have hydrogen gas piping and outlet valves in their facilities. Can it explode? Yes. Just like using natural gas in gas stoves.

Is hydrogen plentiful? Yes! Hydrogen gas can be easily and cheaply made from the electrolysis of water, releasing both hydrogen and oxygen.

Can we use hydrogen gas for transportation and other uses? Yes! Several organizations exist today that are actively promoting the use of hydrogen gas for transportation and heating. Hydrogen gas can even be stored in blocks of certain metals and released when heated.

Is the burning of hydrogen gas clean? Yes! Combining with oxygen, the burning of hydrogen creates water—pure water. The only pollution in the air or in the water will result from the impurities or additives in the hydrogen or in the oxygen, or air, that is used.

Is liquid hydrogen safe? No! Liquefied Natural Gas (LNG) is very unsafe as well. A gas is made to become liquid by lowering its temperature. Liquid natural gas and liquid hydrogen both can freeze flowers or an arm in seconds. However, the danger is not in the cold temperature; the danger is in the explosiveness of the liquid when it evaporates and is mixed with the oxygen in the air. A rule of thumb that I recall from my working at the Bevatron high energy laboratory in Berkeley, CA, was that a small milk container of liquid hydrogen, when dropped, could mix with the air and provide an explosion large enough to blow up the heavy concrete Bevatron building. A study was done in the 1980s that compared the safety of an English petrochemical plant with the safety of a U.S. commercial nuclear reactor facility, using the same set of standards that are used in the nuclear safety area (the WASH1400 report at that time). The results showed that, because of the use of LNG, the probability of serious danger to a person per year near that chemical plant was

> about the same as that of dying in a car accident, while the probability of danger from the nuclear plant was 100 to 1,000 times less. Hydrogen energy has a future, just like nuclear energy, if it is handled safely, and if it can compete economically. [2]

"Hydrogen has the potential to become an abundant, benign, portable energy source," explains Dr. Robert E. Jervis (University of Toronto).

> Whether it will fill this role in the near future depends on its source(s). Much hydrogen today is extracted from fossil fuel sources, but this is a waste of a nonrenewable resource and is marginal in cost. Additional hydroelectric power is limited in future availability for electrolysis of water. A practical solution is to produce hydrogen from uranium by using off-peak power from nuclear reactors. Nuclear plants are most effective when operated continuously on a 24-hour-a-day basis. The technology for producing, storing, and dispensing hydrogen and using it in fuel cells for transportation is already at hand. Surely, bringing hydrogen technology to everyday, economic use would have taken place more rapidly were it not for the overwhelming infrastructure and universal oversupply and use of internal combustion vehicles. [9]

According to Dr. William J. Garland (Department of Engineering Physics, McMaster University):

> Hydrogen is just a currency, like electrons. As you said, it exists only in combined forms in nature (i.e., low energy state). The game is to energize it by breaking the chemical bonds and later re-bonding to release energy. But as you indicate, the real nuts to crack are the issues surrounding technology insertion. Look at Beta versus VHS, Linux versus Microsoft Windows, steam engines versus piston engines. The better technology is no measure of success. Steam engines were killed, for instance, by an outbreak of hoof and mouth disease around 1918 or so. Maybe if we start with small fuel cells for phones, laptops, and the like,

> then people would get used to hydrogen and would be more likely to accept it before it went mainstream. [10]

A surprisingly old idea is hydrogen as a fuel. In Jules Verne's novel *The Mysterious Island,* published in 1874, a shipwrecked engineer suggests that when fossil fuels run out, water will one day be employed as fuel, that hydrogen and oxygen which constitute it, used singly or together, will furnish an inexhaustible source of heat and light of an intensity of which coal is not capable. Verne knew his physics: Pound for pound, hydrogen packs more chemical energy than any other known fuel. Hydrogen also fits the arc of history: From firewood to coal to oil to gasoline to methane, the world's fuels of choice have become increasingly decarbonized. Carbon adds bulk and smoke without adding energy. Hydrogen seems to be the logical omega point. It is the only carbon-free combustible fuel.

Verne's visionary engineer imagined burning the hydrogen, but most modern schemes revolve around the fuel cell. This is a device that combines hydrogen with oxygen to generate electricity. This idea is hoary too: In 1839 Oxford-educated barrister Sir William Robert Grove figured out that if electricity could split water into hydrogen and oxygen—a process known as electrolysis—then combining the atoms would make electricity. Though Grove built a working hydrogen fuel cell, the advent of cheap fossil fuels relegated his invention to the sidelines. There it languished until the 1960s, when NASA began using fuel cells to power space missions. A fuel cell cranked out power at an attractive weight-to-voltage ratio. Therefore, astronauts could drink its principal by-product: pure water.

Fuel cells exist in many incarnations. Today, the proton exchange membrane (PEM) version is one of the most popular, because it is the lightest and easiest to manufacture. (See sidebar, "Building a Practical Hydrogen Fuel Cell.") The thin proton exchange membrane is coated with a catalyst, usually platinum. When pressurized hydrogen gas (H_2) is forced through that catalyst, it is stripped of its two electrons. The membrane allows the hydrogen's protons to flow through but stops the electrons, which zap through an external circuit as electricity. On the other side of the membrane, the protons combine with both oxygen and the electrons that have flowed through the circuit (and powered electrical devices in the process) to form water.

According to Dr. Jen-Shih Chang (Department of Engineering Physics, McMaster University), "Any fuel-cell application must be for cogeneration of applications for electricity and heat or hot water; otherwise, efficiency is too low. The current target in industry is 80%, but it needs to improve." [11]

Building a Practical Hydrogen Fuel Cell

Hydrogen fuel cells will provide the power to propel all vehicles, converting the universe's most abundant element into electricity. The by-products are heat and water. The hydrogen, which does not occur freely on Earth, can be separated from methane, natural gas, or petroleum. It can be separated from water using electricity, preferably generated by wind, solar, or geothermal energy. A hydrogen-powered world might have a number of interesting benefits, such as slowing temperature increases around the globe, ending North America's dependence on foreign oil and even vulnerability to terrorists who might blow up a nuclear plant. In the latter case, if you just plugged enough hydrogen-powered cars into wall sockets while they were parked so that they could make extra power (calculations show four percent of those in any given city would do the job), they could energize that city. Too small to be of interest to terrorists, the plugged-in cars could serve as tiny generators.

Hydrogen-fuel-cell-powered buses are already slated to hit the streets in Germany, Iceland, and Mexico. The vision of some scientists (a far cleaner and more efficient technology for producing electric power) will likely revolutionize vast reaches of the economy and society. Indeed, the cornerstones of sustainable development are the fuel cells and hydrogen.

The engine of change will not be a battery (see sidebar, "Fuel Cell Options and Achievements"); it will be a hydrogen fuel cell. Fuel cells weren't new. Sir William Grove discovered the principle in 1839, long before the invention of the internal combustion engine. NASA's Gemini project had used fuel cells built by General Electric to power onboard electrical systems. But fuel cells, which simply strip the electrons off hydrogen atoms to produce electricity, seemed impractical at the time: They were expensive to produce. Thus, the idea of some scientists was to take this esoteric, space-oriented, million-dollar-a-kilowatt item and engineer it with commercially viable materials, so you could mass-produce it at the same price as gasoline. It didn't bother scientists that no one thought it could be done.

By the mid-1990s, a hydrogen-fueled car began to seem feasible. Scientists had developed fuel cells so powerful and compact that a stack of them could fit in the same space as an internal combustion engine.

Then, scientists began to develop fuel cells and drivetrains for electric cars powered by hydrogen extracted from methanol onboard. Daimler rolled out a prototype Ballard-cell-powered minivan in 1996. Then California passed a law requiring 10 percent of cars sold there to be zero-emission vehicles by 2003, creating a new market for vehicles run by fuel cells. Now, fuel-cell-powered vehicles are being developed by just about every major auto manufacturer.

Still, there is no free lunch, according to critics. Hydrogen may be ubiquitous, but splitting it off from other molecules, such as methane or water, can require more energy than the hydrogen can produce. Most scientists find this objection stupid and shortsighted. All systems have energy losses: Four-fifths of the energy generated by a gasoline engine is dissipated as heat. The critical question is whether the technology that generates the hydrogen makes less pollution, and that will depend on whether society has the will to develop technologies such as wind and solar to make hydrogen. A hundred years ago there was oil in the ground, but it had no value until someone developed the internal combustion engine. Now, where you can use hydrogen as fuel is the endgame.

Who will buy a fuel-cell car when they can't buy hydrogen to run it, and who will build the corner pump when no one's driving hydrogen-powered cars? Between here and there stands a catch-22. To address that thorny problem, scientists need to create a hydrogen-fueling infrastructure.

"On-site generation of hydrogen is critical for hydro carbon fuel," states Dr. Chang. "And this technology may soon generate free hydro-carbon fuel, as well as minimize CO poison for fuel cell materials." [11]

Fuel Cell Options and Achievements

Even though cars powered by fuel cells have been built and tested, the aerospace fuel cells could not deliver high power quickly when the driver wanted to accelerate. Today's hybrid electric cars (see sidebar, "The Premature Death of Electric Cars") carry a battery that supplies the acceleration power, and the prime power source, whether an engine or fuel cell, is not stressed with sudden load peaks. Zero air pollution becomes attainable when fuel cells supply the prime power on a hybrid vehicle.

There are hundreds of chemistries that combine a fuel and oxidant to produce electric power. In describing the coming age of personal electricity, a fuel cell's electrochemical energy conversion process does not have to observe the Carnot-cycle efficiency limit that applies to heat engines. However, in order to attract the development funding needed for entering the commercial world, only a few fuel cell systems have been able to raise the attention needed.

In order to develop a fuel cell that had electrodes made from platinum, Sir William Robert Grove used his own money between 1839 and 1849. His first cell electrolyzed water into hydrogen and oxygen, and then combined these gases into water, producing electricity. Made by reacting zinc with acid, his later fuel cell consumed oxygen from the air and hydrogen.

The first fuel-cell-powered farm tractor, which ran on hydrogen and oxygen, was introduced by Allis Chalmers in 1959. Ammonia fuel and cryogenic oxygen were once used to power a forklift. To provide the counterweight function, which had previously been supplied by the battery, cast steel had to be added to the rear of the vehicle!

There have been many successes in powering manned spacecraft with fuel cells. However, Earth-surface use of fuel cells has been limited by the cost, which has been around $8,000 per kW. To introduce the century of personal electricity, there are also many electrochemical options that could be developed to overcome this cost disadvantage.

In order to respond to rapid load changes, today's fuel cells have limited capability. However, there are fast-response processes for combining oxygen and fuel to deliver power. An example is the animal muscle. It converts fuel energy into mechanical power with a 50% efficiency. A baseball batter, after evaluating the path of a ball arriving at 100 miles per hour, can command his or her muscles to swing his or her bat in a manner that gives the batted ball the optimum velocity and direction. [12]

Pumping Electric Fuel into the Vehicle

The zinc-air fuel cell consumes zinc pellets, and its exhaust is zincate, which is a zinc oxide. A fuel pump extracts the zincate from a utility vehicle and refills the vehicle's fuel cell with zinc pellets.

The zinc-air fuel cell has a lower compartment in which oxygen from the air combines with the zinc pellets to produce electric power. The product zincate dissolves into the electrolyte and is carried out into the vehicle's electrolyte management tank. The consumed zinc gets replaced from the zinc hopper. At the "filling station," the electrolyte and zincate are pumped out and replaced with fresh electrolyte. Zinc pellets are delivered into the zinc-pellet hopper. The vehicle departs and the zinc-refueling station then proceeds to electrolyze the zincate and purify the potassium-hydroxide electrolyte.

"Zinc is toxic!" exclaims Dr. Chang. "We must minimize leaks and emissions." [11]

This fuel-cell system was tested by mounting two 17-cell stacks in a Cushman cart, which was driven around a 0.6 km track. The fuel-cell stack delivered 175 ampere-hours in five hours of operation. The maximum power was 4 kW. A second cart was equipped with two 20-cell stacks for further testing. [12]

This technology is to be applied to lift trucks, industrial sweepers, and commercial lawn and garden equipment. It could also be practical in standby power sources for small businesses and residences.

The Premature Death of the Electric Cars

A growing fleet of General Motors electric cars await an uncertain fate. The celebrated ride of the car that spawned the nation's toughest emissions regulation ends at a parking lot in Southern California. Dozens of the green, metallic blue, and bright red futuristic autos are lined up behind a chain-link fence at the edge of a freight rail line in Van Nuys. This is a sure sign that the world's largest automaker has pulled the plug on a vehicle it heralded, as recently as 2001, as the car of the future.

GM is taking the cars off the road when leases expire because it can no longer supply parts to repair them, as California retreats from its strict pollution regulation. The automaker is scrapping the cars, shipping them to museums and universities for preservation, sending them to a research lab in New York, and cannibalizing them for parts for the few still on the road.

Once touted as the company's clean air centerpiece, it's a long way from a program, and it comes as fans in California fight to keep some electric cars on the road, as the state rewrites its so-called zero-emissions vehicle rule. GM's effort to get the cars off the road, is a heartbreaking prelude to the imminent death of the battery-powered vehicle, and to the scores of drivers who embraced the technology. This effort continues, as state air regulators continue to weaken rules that would have required ten percent of cars for sale in 2003 to be nonpolluting.

They've gone from being regulators to just asking politely. Would you do this? To those driving battery electric vehicles, they feel as if they've been left hanging out to dry. [13]

Plans Anew

California launched its ambitious zero emission vehicle program in 1990 to help clean up America's smoggiest skies, only after seeing the promise of the first GM electric car in the late 1980s. New York and Massachusetts have followed suit, and other states are mulling similar regulations, but they are watching to see how California's rule-making plays out.

As car makers vigorously fought at hearings and in court to halt the regulation, over the past decade, state regulators have caved to pressure. Because the vehicles were limited to a range of about 100 miles required lengthy recharges and their high cost made them unappealing to a wide group of drivers, major automakers have stopped production.

The California Air Resources Board is poised to make changes that reflect that the cars are a commercial failure and to promote more promising technologies that have emerged. With a combination of low-polluting gas-powered vehicles, gas-electric hybrids, and a couple hundred fuel-cell cars down the road, the board's staff has suggested a new plan, letting auto companies reach the ten percent quota.

Automakers would also be able to apply credits for electric cars it once put on the road and electric golf-cart style vehicles that zip through neighborhoods, office parks, and campuses. Honda concluded that the limited popularity of the electric car wouldn't effectively contribute to cleaner air.

Honda is now focused on its hybrid models, natural gas-powered vehicles, and fuel-cell program. It plans to have five fuel cell models, which run on the electricity from a chemical reaction between oxygen and hydrogen, in the Los Angeles city fleet by 2005. The range of electric cars has already been doubled by the fuel cell cars.

The auto companies never seriously gave the cars a chance and didn't do enough to improve the technology or promote the cars to the public, according to supporters of battery-powered vehicles. These are claims automakers dispute.

Staying the Course

There were long waiting lists of people who wanted the cars. Before it ever hit the road, automakers predicted the demise of the vehicle.

Even while they built the cars, they were singing this tune. Back in 1991, when the California Air Resources Board laid down the zero-emission rule, there were no electric cars, it was a dream. They're prepared to abandon it, now that the dream is a reality.

Honda extended leases for some drivers, unlike GM. About 100 of Honda's original 300 or more EV Plus cars are still on the road. Of the more than 1,000 two-seater sporty EVI cars built by GM, only about 375 are on the road. By the end of 2005, the plan is to have them off the road.

Drivers who embraced the technology are not counting on a new lease on the life of their aging electric car. If they just scrapped the vehicles, it wouldn't be too all surprising. [13]

Anything but hydrogen will foul proton exchange membrane fuel cells. So, a gadget called a reformer is the second crucial technology in hydrogen-energy schemes. This splits hydrogen from the molecules to which it clings. Most hydrogen is made by "reforming" methane with high-pressure steam; the steam interacts with the methane to separate the hydrogen from the carbon. Reformers can also wring hydrogen from paper-mill waste, coal, sewage, and garbage.

Making hydrogen is already a large, mature industry. It consumes some five percent of total methane production, with about 100 billion cubic feet of hydrogen devoted each year to such industrial tasks as refining petroleum and making hydrogenated oil for food. There are already a lot of real experts out there who understand making, handling, and storing hydrogen safely. According to Dr. Chang, "hydrogen transport technology and more efficient production are critical to industrial needs." [11]

So why do automobiles (the grand prize in any alternative-fuel scheme) still use gasoline, if hydrogen is so great? It's classic catch-22 economics: No one will set up a nationwide hydrogen production and distribution infrastructure until there are cars that demand it. But until they can get a sure source of hydrogen, no one will build hydrogen-powered cars in bulk.

"A hydrogen car must solve heat or hot water pollution emitted from its principle operation," explains Dr. Chang. "We need more technological development for pollution control for a fuel-cell car." [11]

Before you can sell the first hydrogen-powered car, many people think you need a $400 billion hydrogen production and distribution infrastructure. That argument stopped a lot of people from thinking about it. How do you create demand? How, in other words, before even one corner gas station is ready to offer a hydrogen fill-up, could it make economic sense to build thousands of hydrogen-powered cars? [6]

Buildings are the answer! Buildings use 65% of America's total electricity. Imagine a high-tech, computer-dependent operation that typically might fork over $1 million annually to keep standby generators humming to ensure constant power. Far better for that plant to install an on-site methane reformer and a fuel cell.

Company employees would become customers for the first fuel-cell-powered vehicles, with an on-site reformer merrily extracting hydrogen from methane. Private cars are parked 96% of the time. If you lease hydrogen fuel-cell cars first to the people who work in or around buildings where fuel cells have been installed, then when you drive to work, you can plug a supply hose into your car to feed it hydrogen from the building's reformer. The car would use that hydrogen while it sits in the garage to generate electricity for sale. You plug your car into the electric grid. Now, your second biggest household asset has just become a profit center, making enough electricity to return to you a third of the cost of owning the car, while you sit at your desk. [6]

Gas stations are expected to install their own methane reformers and hydrogen pumps, once a critical mass of hydrogen-powered cars cruise the roads. That, in turn, will force the revamping of the national natural gas pipeline system to handle hydrogen as well as methane.

NOTE

Natural gas is mostly methane with a foul smell added to it to help detect gas leaks.

Hydrogen tends to embrittle typical methane pipes, but those that are relined or built from scratch to accommodate hydrogen can transport it safely. Ultimately, most homes will have a hydrogen-powered fuel cell in the cellar, heating, cooling, and producing power.

But this scenario won't unfold by simply trying to stick fuel cells into today's heavy SUVs. For hydrogen-powered cars to make sense, their gas tanks must be sufficiently small to allow room for people and groceries to fit inside and travel reasonable distances—at least 300 miles. That requires making cars vastly lighter and more slippery, or aerodynamic, than today's models. With the same amount of power it takes to run an SUV's air conditioner, some researchers have promoted the so-called hypercars—ultralight, ultrastreamlined vehicles that can achieve highway speeds.

Researchers have recently designed a vehicle made of lightweight carbon fiber, a stronger version of the material used to make tennis rackets and skis. At half the weight of a comparably sized vehicle, such as the Lexus RX300 SUV, the car could not only travel 330 miles on 7.5 pounds of compressed hydrogen, but also meet federal standards to protect occupants in a head-on collision at 30 miles per hour with a steel-bodied SUV moving at the same speed. [6]

According to detractors, hydrogen has one huge, basic flaw: It's an energy storage medium, not an energy source. Like a battery, more energy must be expended in its production than can be provided by its use, so while hydrogen is clean and efficient at the point of use, it just pushes the pollution and waste upstream to the point of production. Hydrogen is the most abundant element on Earth, but free hydrogen just isn't around. Scientific analysis has shown that technology has a better shot at being economical. With hydrogen, you have huge losses in production and distribution. The economics just aren't there.

You have to get the hydrogen from somewhere. You can have a sort of free lunch, that you can get more energy out of it than you put into it.

There are costs associated with making hydrogen, but fuel-cell cars could use hydrogen at least 2.5 to 3.5 times more efficiently than today's cars use gasoline. This means that hydroelectric dams could make big profits by using off-peak power to crack hydrogen from water. Those utilities could get five to seven times more for hydrogen than they can charge for electricity, which makes the economics attractive. This argument is even more compelling in Europe and Japan, where taxed gasoline prices are commonly three to four times U.S. levels.

"Hydroelectric is limited by finding new construction sites to build dams and also generates more pollution (greenhouse) and safety problems," explains Dr. Chang. "It may need a balance of nuclear and hydroelectric for off-peak power that is used."[11]

Ultimately, reforming methane into hydrogen will be just a bridge to a pollution-free, renewable-based-energy future. The endgame will be using solar cells or wind farms to electrolyze water. These intermittent power producers would be able to store the energy they gather on sunny or windy days as hydrogen and use it to power both automobiles and fuel cells in buildings, as well as to feed the electric grid.

But far from a sure thing, is the profitability of such an arrangement. Today, it is doubtful that any all-hydrogen scheme can unfold without massive government regulation, not to mention far more private investment and much more time.

The cost would be relatively modest, since converting 21,000 gas stations to hydrogen nationwide could cost about $7.4 billion. It is absolutely possible, and it can be done profitably. Analysts calculate a ten percent return on investment at every step.[6]

NOTE

Ten percent is the standard industry threshold for deciding whether to invest.

There are also safety concerns. What will happen if, in the first year of distributed hydrogen generation, an entire building blows up? To make sure this stuff is safe, the scale of investment is pretty high.

Escaped hydrogen likes nothing better than to dissipate—it's very buoyant and diffuses rapidly, unlike spilled gasoline. It does ignite easily, but this requires a fourfold richer mixture in air than gasoline fumes do, or an 18-fold richer mixture, plus an unusual geometry, to detonate. Moreover, a hydrogen fire can't burn you unless you are practically inside it, in contrast with burning gasoline and other hydrocarbons, which emit searing heat that can cause critical burns at a distance. "A hydrogen car needs better control of misfire and noise pollution control technology," explains Dr. Chang. [11]

Fuel cells are indeed coming, but the solid-oxide version will lead the charge over hydrogen-dependent proton exchange membranes for the foreseeable future. Of the roughly 500 fuel cells now cranking away worldwide, nearly all are stationary models capable of handling a mix of fuels, not just hydrogen, which gives crucial flexibility in a world of shifting fuel prices.

NOTE

There are a lot more than 500 fuel cells working at this time—a laptop will soon come on the market that runs on a fuel cell (you must supply methanol for its fuel).

These make economic sense right now—they are far more reliable than diesel generators for remote applications such as radar sites, cellular towers, and the like. But solid-oxide fuel cells, which are heavy and run at temperatures as high as 1,800 degrees Fahrenheit, tend to be impractical for vehicles because they require longer warm-up times. Cars might be the last place you'll find solid-oxide fuel cells. According to Dr. Chang, "A fuel-cell power generator is good if it can be a hybrid with solar and wind power for remote site application if on-site hydrogen generation can be done for hydrocarbon fuel or water with efficient reforming." [11]

Hopefully, the public will come around on the hydrogen safety issue. Hydrogen is inherently safer than gasoline, but scientific fact isn't always sufficient to sway the average consumer. In the 1980s, researchers built a pair of hydrogen-powered buses. People began referring to them as Hindenbuses. That kind of comment does not help.

Why are DaimlerChrysler, Ford, General Motors, Toyota, Nissan, Honda, and Mazda running fuel-cell research programs if such obstacles are real? Well, you are going to hedge your bets if you are a prudent car company, facing regulatory pressure from the government. [6]

A hydrogen economy is possible and desirable. As any student of economics knows, when some desirable object or behavior can be had more cheaply, people get or do more of it. If, despite doubts, hydrogen-powered hypercars become the automobile of choice, a vast irony looms.

Frankly, this is going to be a fun car to drive. It will be a kick. The more people like it, the more they will drive. People might simply dream up more and more things to do with this relatively benign power until they stress the system to the limit all over again, if the hydrogen economy does take off.

That scenario is a long way down the road. For the hydrogen economy to become a victim of its own success, it must first succeed. Amid charges and countercharges, one fact is clear: The transition to hydrogen will not be nearly as simple as the molecule itself.

Implosion/Vortex

In order to cause expansion and pressure to produce work, like in your car engine, all major industrial engines use the release of heat. Nature uses the opposite process of cooling to cause suction and vacuum to produce work, like in a tornado. Viktor Schauberger (Austria) was the first to build working models of implosion engines in the 1930s and 1940s. Since that time, a number of researchers have built working models of implosion turbine engines. These are fuelless engines that produce mechanical work from energy accessed from a vacuum. There are also much simpler designs that use vortex motions to tap a combination of gravity and centrifugal force to produce a continuous motion in fluids.

NOTE

Here, the naysayers and skeptics of free energy claim that this is absurd. A vacuum is a region of lower pressure. Work must be done to create a vacuum. For nuclear and subatomic regimes, you can talk about the energy of the vacuum and Heisenberg Uncertainty Principle, but not for macroscopic systems.

Cold Fusion

Two chemists from the University of Utah announced in March 1989 that they had produced atomic fusion reactions in a simple tabletop device. The claims were debunked within six months and the public lost interest. Nevertheless, cold fusion is very real. Not only has excess heat production been repeatedly documented, but also low-energy atomic element transmutation has been catalogued, involving dozens of different reactions! This technology definitely can produce low-cost energy and scores of other important industrial processes, such as the transmutation of radioactive elements. So, even though Cold Fusion researchers have been thoroughly debunked, research continues based on the promise of the technology. See Chapter 20 for a detailed discussion of fusion.

Solar Assisted Heat Pumps

The only free energy machine you currently own is the refrigerator in your kitchen. It's an electrically operated heat pump. It uses one amount of energy (electricity) to move three amounts of energy (heat). This gives it a COP of about 3. Your refrigerator uses one amount of electricity to pump three amounts of heat from the inside of the refrigerator to the outside of the refrigerator. This is its typical use, but it is the worst possible way to use the technology. Here's why. A heat pump pumps heat from the source of heat to the sink or place that absorbs the heat. The source of heat should obviously be hot and the sink for heat should obviously be cold for this process to work the best. In your refrigerator, it's exactly the opposite. The source of heat is inside the box, which is cold, and the sink for heat is the room temperature air of your kitchen, which is warmer than the source. This is why the COP remains low for your kitchen refrigerator. But this is not true for all heat pumps. COPs of 8 to 10 are easily attained with solar-assisted heat pumps. In such a device, a heat pump draws heat from a solar collector and dumps the heat into a large underground absorber, which remains at 55°F, and mechanical energy is extracted in the transfer. This process is equivalent to a steam engine that extracts mechanical energy between the boiler and the condenser, except that it uses a fluid that boils at a much lower temperature than water. One such system that was tested in the 1970s produced 350 hp, measured on a Dynamometer, in a specially designed engine from just 100 square feet of solar collector. The amount of energy it took to run the compressor (input) was less than 20 hp, so this system produced more than 17 times more energy than it took to keep it going! It could power a small neighborhood from the roof of a hot tub gazebo, using exactly the same technology that keeps the food cold in your kitchen. Currently, there is an industrial scale heat pump system just north of Kona, Hawaii that generates electricity from temperature differences in ocean water.

"The heat pump principle does indeed warrant special consideration as an ultimate power generating source based on tapping the heat energy of our atmosphere as sustained by the sun," claims Dr. Aspden:

> I draw attention to the energy loss anomaly found in electrical power transformers, where research shows that eddy-current loss can be several times that predicted from standard electrical theory. I believe this is due to the fact that heat flowing from the laminated transformer core moves through those magnetic domains in iron that I mentioned earlier in the chapter. In physics, there is a phenomenon known as the Nernst Effect by which heat flow transverse to a magnetic field will set up a voltage in the mutually orthogonal direction. So, you see, here is heat, heat going to waste, that becomes electrical energy, only to augment the induced EMF that accounts for the eddy-currents that cause much of the heat in the first place. This is a closed cycle action, which makes the actual loss much greater than expected. Surely, there should be research aimed at exploiting the 100% conversion of heat to electricity involved here by diverting the current from the core laminations before it is allowed to revert to heat. One can imagine a future in which large laminated magnetic cores can be used with an internal cooling system to initiate inflow of heat from our surroundings and convert much of it into useful electricity before it reaches the cooling system. Yes, I think there is something to consider that links with heat pumps in our search for free energy. I do not see the temperature difference in ocean water in my vision of the future, but rather heat pumps that provide heat flow into an iron-cored structure, heat energy of 10 times that needed to operate the heat pump, but heat energy with a near to 100% conversion rate into useful electricity by exploitation of the Nernst Effect. [3]

Dr. Chang concurs, "Heat pump technology is essential to the future and requires more efficient heat exchanger and less pollution refrigerant (non-toxic, non-ozone depletion, and less greenhouse effect, etc.) technologies." [11]

Wind

The autumn wind gusts with new promise this year on 200 farms across central South Dakota. Meteorological instruments rise above the cornfields, capturing precise measurements of air current speed and direction. The numbers will be crucial for the partners putting together financing for one of the nation's largest energy projects.

The plan, called Rolling Thunder, could be the breakthrough that lets the public in on a secret that is well known in the electricity business. Wind power, once a costly proposition on the flaky fringe of the energy debate, is going mainstream fast. Technology advances have made it possible—given suitable wind conditions and scale—to produce wind energy more cheaply than any other form, and the market is responding. U.S. wind-power capacity will increase by 200% by 2005; indeed, wind is now the world's fastest-growing energy source. Rolling Thunder, ten times as large as any previous wind project, aims to show that wind can be a big-scale contributor to a major energy market, the Midwest.

"Wind power is not continuous (20 to 30% operation days)," states Dr. Chang, "and needs a hybrid power generator with diesel or a micro-gas turbine power generator, as well as a 20-year life cycle. The technology also requires cutting tall trees for wide spaces. Also, 8% to 15% of wind generators are struck by lightning every year, with an average $2.6 billion in incidental damages. We need more fundamental technology development; otherwise we cannot become the dominant power generation (more then 10% of electricity in this area). We need a balanced power generation plan." [11]

Promising Winds

But wind enthusiasts (which now include business people as well as environmentalists) think it deserves more attention from policymakers, especially at this time of national crisis. Wind still provides less than one percent of the nation's energy, even though the federal government's own researchers have shown that 20% is feasible with current technology—the same share that nuclear energy enjoys, but without the terrorist target potential. Wind is also immune to wild price fluctuations, unlike natural gas, which roiled energy markets in 2001. And, of course, wind doesn't pollute.

The most far-reaching thinkers even see in wind the potential to manufacture fuel for a new generation of cars that would break the nation's dependence on Middle East oil. You can run an economy on wind. A future can be imagined where the blustery Great Plains become the Saudi Arabia of wind power, where land values skyrocket on windy terrain and meteorologists play the role of petroleum geologists today. This optimism is based on early 1990s federal research concluding that 12 central states had wind potential to produce four times the amount of electricity consumed nationwide. North Dakota alone could then have met 36% of U.S. energy needs. [14]

Of course, deploying that wind force nationally would require leaps in the technology for storing and distributing energy—science being worked on by automakers, oil, and new-line energy companies. In the meantime, the business of simply capturing wind power and pouring it onto regional electrical grids is thriving. No state better exemplifies the boom than Texas. The nation's old oil capital will see 900% growth in wind power in 2005, thanks to deregulation, state renewable energy policies, and gusty weather. The King Mountain project, near Odessa, will serve more than 120,000 customers with 659 wind turbines. It is economical to do this, and to do it on a large scale. [14]

Turbine Technology

Today's windmills are a far cry from the gently turning lattice wheels that riled Don Quixote. Sleek, white rotor blades take advantage of the wind just as jet propellers do. The generators, with the help of power electronics and computer modeling, adjust readily to variations in air speed. The turbine's top, with three blades, rotates to capture wind from any direction. Most important, today's wind turbines are large: On thin shafts more than 200 feet tall, the rotors span 231 feet. Average cost: $5.8 million, about 24 times as much in today's dollars as a 1981 model, while generating 120 times the energy. [14]

As a result, the cost of generating wind power has dropped from about 38 cents per kilowatt hour in the 1980s, to about three cents today, though initial financing costs would add to that. By comparison, the average U.S. retail electricity price is about seven cents per kwh. The wind-power boom has had an important side benefit, providing income to farmers and ranchers who lease their land for the turbines. [14]

Development is happening even in regions not known for breeze. A new 16-turbine wind farm opened in 2001 on a ridge in Pennsylvania's Allegheny Mountains, where wind averages more than the minimum required 11 miles per hour.

Farmers in four South Dakota counties have agreed to lease land for the 2,000 turbines for the Rolling Thunder project. It's kind of a new class of wind-power project. The 3,000-megawatt project, on a par with the largest U.S. nuclear power plants, would feed the grid serving Chicago. Until now, most developers have located wind projects as close as possible to electricity transmission grids. But when you start looking for a really great wind resource, you find it's in fairly remote areas. It will take at least five years to obtain rights of way and build transmission lines to central Illinois.

"We need careful ecological assessments," claims Dr. Chang. "Normally, a remote island out in the middle of an ocean has better efficiency, although it would require 40% more maintenance costs. Energy generated by land use, which is cost, is required as well as for a balanced choice of technology for power generations." [11]

Dead Air

Can a modern electricity delivery system rely on something as changeable as the wind? In fact, no purchasers of wind power are left in the dark when the air turns calm. Wind power is fed into the grid via power lines, where it mixes with electricity from other sources. The mix contains more wind-generated power when it's windy and none when the turbines are still (either because of no wind or extreme winds). Grid operators discount wind power accordingly; they count on a megawatt of wind energy to power 350 homes, compared with 1,000 homes for other types of generation.

The most far-reaching vision for wind energy involves breaking down the limitations of geography and weather. If enough wind infrastructure is built, the spare energy produced at night could be used to run an electric current through water (a rather simple process) to produce hydrogen. Hydrogen is what automakers and indeed, most oil companies, see as the viable alternative to the petroleum that now fuels the U.S. economy. Scientists already know how to make the "engine" of hydrogen-powered vehicles, the fuel cell, which has been used in the space program for decades. But they believe it will be the end of the decade before they can mass-produce cells on a scale that would make them competitive in cost with the internal combustion engine.

The oil industry also is investing heavily in hydrogen research, although it still sees a role for gasoline. Don Hubefts, chief of Shell Hydrogen, points out that an infrastructure needs to be built to allow consummers to fuel up with hydrogen as they now tank up at gas stations. This scenario is logistically attractive and would cut emissions. But others argue it makes more sense to invest in eliminating pollution entirely, as would be possible with vehicles that use 100% hydrogen, generated by cheap wind power. The transition to a new energy economy has begun. But it is not moving fast enough.

The war on terrorism has given new fuel to renewable-energy advocates. Although gas prices have risen and the fighting does involve oil producing nations like Iraq, the precarious political situation in Arab countries, particularly Saudi Arabia, has heightened concern about U.S. dependence on imported oil. The Bush administration has urged the Senate to adopt his energy plan to increase domestic development of fossil fuels. But House Minority Leader Dick Gephardt recently said that fuel-cell commercialization is so important for energy security that a Manhattan Project-type national program is called for. Others talk in similarly sweeping terms. The vision they're talking about is the kind Kennedy had to put a man on the moon by the end of the decade. Energy security is national security, and there's no question the U.S. could have it based on renewables.

Sound

The Thermoacoustic Stirling Hybrid Engine (TASHE) performs the same basic job as an ordinary car engine or gas-fired turbine: It converts heat into motion. But the similarity ends there. TASHE operates entirely on pressure waves, using high-intensity sound to do the work of steel. As a result, it has no moving parts, can be constructed from cheap, basic materials, and yet it is just as efficient as a typical modern internal combustion engine. Ultimately, sound engines could take dozens of forms, from big ones that liquefy plumes of natural gas to little ones laboring in the cellar that would provide supplemental home electricity. What sound allows scientists to do is build invisible machinery. It's the next level of mechanical engineering.

TASHE relies on a mechanical blueprint that dates back to the era of steam power, even though the idea of using sound to drive an engine is new. In 1816, Robert Stirling, a multi-talented minister of the Church of Scotland, patented a simple design for an external combustion engine; unfortunately, it proved too costly to mass produce. Stirling's engine consists of a sealed chamber filled with gas that shuttles back and forth between a "cold" end, often at room temperature, and a "hot" end, which can be heated by any energy source. A displacer piston within the chamber moves the gas between the two ends, while a power piston oscillates in response to the movement of the gas as it expands when heated and cools when it contracts. To do the work, the power piston can be attached to a crankshaft.

The Stirling engine was passed by—by time and the internal combustion engine, but scientists and engineers continued to be intrigued by it. Then, in 1979, Peter Ceperley, a physics professor at George Mason University in Fairfax, Virginia, showed that the work done by heat in a Stirling engine could also be carried out by a sound wave. After all, sound is nothing but motion—you hear because pressure waves traveling through the air vibrate your eardrums at varying frequencies. Just as heat moves a piston back and forth, those waves, Ceperley realized, could bat a slug of gas back and forth in a Stirling-like cycle.

With limited success, lots of people tried to put flesh on that idea. To test Ceperley's ideas, researchers built their own test engine, starting with a baseball-bat-shaped resonator made from inexpensive steel pipe. The resonator determines the operational frequency of the engine, in the same way that the length of an organ's pipe determines its pitch. At the "handle" end of the bat, the researchers bolted on a doughnut-shaped metal chamber to hold the hot (about 1,300°F, or 700°C) and cold (7°F or 20°C) heat exchangers. The researchers then filled the device with compressed helium.

Until it becomes inconceivably powerful, the heat exchangers in TASHE act like a huge stereo speaker—creating sound, sending it down the resonator, and amplifying the feedback repeatedly. If you were in that wave, permanent hearing loss would be the least of your problems. It's loud enough to set your hair on fire. The operating engine is remarkably muted, however, quieter than an idling car. Quarter-inch-thick steel walls, needed to contain the highly compressed helium, maintain the silence. The cavity walls are extremely stiff. They don't flex, so the sound wave hardly escapes.

Researchers are just beginning to sort out what their sound contraption can do. Soon it may provide a better way to recover natural gas. In the course of drilling, offshore oil rigs can liberate natural gas, which is often just burned as a waste product. The sound engine could provide a cost-effective way to capture and ship the gas to the mainland. Engineers are building a huge model of the engine (40 feet tall and four feet in diameter) that can cool and liquefy 500 gallons of natural gas per day. The heat needed to run it will come from burning a little of the cast-off fuel. Engineers are conserving a resource and cutting the pollution caused by flaring off that gas.

Sound engines could perform a similar conservation coup in the home. Gas-fired hot-water heaters dump unused warmth into millions of basements around the country. The sound engine could tap that thermal waste and use it to move a spring-mounted piston driven by acoustic waves. The piston, in turn, could run a household generator. You burn natural gas, and instead of putting the heat directly into the water, you'll use that heat to run an acoustic engine to make electricity.

There's more than one way to tap into the power of sound. At Los Alamos laboratories in the late 1980s, researchers worked on a resonator capable of creating far more intense sound waves than those generated by devices like TASHE. By vibrating the resonator with an electric motor, researchers generated sound waves having energy densities thousands of times greater than had ever before been achieved. Researchers began exploring ways to use extreme sound to perform a variety of jobs that normally require complex machinery, such as manufacturing pharmaceuticals, grinding up materials, mixing chemicals rapidly, compressing gas, turbocharging engines, and recycling plastics. It's a factory in a bottle. There's a level of control there that has never existed before.

Finally, chemicals can be heated and cooled 600 times per second over temperature swings as large as hundreds of degrees Celsius, or turbulently mixed 1,200 times per second. Researchers can create a wide range of physical effects that were simply impossible to attain before. Researchers call their sound-generating process resonant macrosonic synthesis, and they think it will someday find applications as diverse as the laser. But first the technology will show up in more conventional applications such as acoustic compressors, which can be used in refrigerators, air conditioners, and cooling systems for microprocessors. The researchers' little experiment at Los Alamos could soon become the thrum heard around the world, especially given the way this work is going. [14]

Conclusion

There are dozens of other free energy systems that have not been mentioned here; many of them are as viable and well tested as the ones that have just been discussed. But this short list is sufficient to make a point: Free energy technology is here, now. It offers the world pollution-free, abundant energy for everyone, everywhere. It is now possible to stop the production of greenhouse gases and shut down all of the nuclear power plants. Research engineers can now desalinate unlimited amounts of seawater at an affordable price and bring adequate fresh water to even the most remote habitats. Transportation costs and the production costs for just about everything can drop dramatically. Food can even be grown in heated greenhouses in the winter, anywhere. All of these wonderful benefits that can make life on this planet so much easier and better for everyone have been postponed for decades.

Finally, the source of free energy is inside of you. It is that excitement of expressing yourselves freely. It is your spiritually guided intuition expressing itself without distraction, intimidation, or manipulation. It is your open-heartedness. Ideally, the free energy technologies underpin a just society where everyone has enough food, clothing, shelter, self-worth, and the leisure time to contemplate the higher spiritual meanings of life. Do you not owe it to one another, to face down your fears, and take action to create this future for your children's children? Perhaps you are not the only one waiting for you to act on a greater truth!

Dr. Aspden concludes:

> The beginning of the preceding paragraph opened with the words: "The source of free energy is inside of you." The intention was to arouse your interest and stir you into action, but there is something you must be prepared for: The assertion you may face that free energy means perpetual motion—an impossible dream not warranting serious attention by those who seek advice from the scientific community. Do, therefore, keep in mind that perpetual motion is a reality, a reality evident in every atom of your body, by the sustained motion of electrons in their interaction with the quantum underworld of the aether, a reality

that scientists cannot deny. All they can argue is that you are using outmoded terminology. In other words, the name "aether" is what they prefer to refer to as "space–time" and describe in mathematical jargon that has no meaning in free energy terms. Even when you die, those atoms and that aether live on—ever displaying their intrinsic perpetual motion. Our task in confronting the problem of free energy is really two-fold. First, we must find a way of replicating that coupling with the quantum underworld, by setting up that radial electric field that I mentioned earlier. And second, locking into the resulting aether spin. But also, and possibly the harder task, we must somehow influence the minds of physicists committed to Einstein's doctrines and get them to accept the reality of an aether composed of electrical charge and having a physical structure of three-space dimensional form. [3]

References

[1] Nikolai Semenovich Kardashev, "Transmission of Information by Extraterrestrial Civilizations," *Journal Of Soviet Astronomy, 8*, 217 (1964).

[2] Dr. Patrick Bailey, President, Institute for New Energy (INE), Los Altos, CA.

[3] Dr. Harold Aspden, Energy Science Ltd., c/o PO Box 35 Southampton, SO16 7RB, England.

[4] Tesla, Nikola, "Short Biography," Copyright (c) 1996 by Encyclopaedia Britannica, Inc., hosted by University of Pittsburgh, Pittsburgh, PA 15260

[5] Geoff Egel, *Encyclopedia Of Free Energy, Volume 3,* 18 Sturt Street, Loxton 5333, Australia.

[6] Peter Hoffmann, "Energy Department Releases Integrated Long-Term Hydrogen Posture Plan," *Hydrogen & Fuel Cell Letter,* Grinnell Street, Post Office Box 14, Rhinecliff, NY 12574-0014, April, 2004.

[7] Sharon Waxman, "Global Warming Ignites Tempers, Even in a Movie," *The New York Times*, May 12, 2004.

[8] Andrew C. Revkin, "NASA Curbs Comments on Ice Age Disaster Movie," *The New York Times*, April 25, 2004.

[9] Dr. Robert.E. Jervis, P. Eng., Professor Emeritus, Nuclear Science and Engineering, Dept. of Chem. & Environ., Eng. and Applied Chem., University of Toronto, TORONTO, ON, Canada.
[10] Dr. William J. Garland, Department of Engineering Physics, McMaster University, Office: NRB 117, 1280 Main Street West, Hamilton, Ontario, Canada L8S 4L7.
[11] Dr. Jen-Shih Chang, Department of Engineering Physics, McMaster University, Office: NRB 118, 1280 Main Street West, Hamilton, Ontario, Canada L8S 4L7.
[12] "Fuel Cell Technologies Reports First Quarter 2004 Results," *Fuel Cell Today*, Canada News Wire, Kingston, On., May 13, 2004.
[13] "California Gives Up on Electric Car Goal: Air Board Now Backs Fuel Cells and Hybrids," Detroit Free Press Inc. (Associated Press), 600 Fort, Detroit, Michigan, 48226, March 6, 2003.
[14] Stuard Baird, M. Eng., M.A., "Wind Energy," Energy Fact Sheet, ICLEI, USA Office, 15 Shattuck Square, Suite 215, Berkeley, California, USA 94704, 2004.

Nuclear Fusion and Waste

"Ours is a world of nuclear giants and ethical infants. If we continue to develop our technology without wisdom or prudence, our servant may prove to be our executioner."

—Omar Bradley (1893–1981)

Arthur Eddington, in 1920, suggested that the energy of the sun and stars was a product of the fusion of hydrogen atoms into helium. This was his dream of harvesting energy from the same reaction that powers our sun. Since the 1950s, great progress has been made in nuclear fusion research. However, the hydrogen, or thermonuclear, bomb is the only practical application of fusion technology to date.

To supply electricity, researchers stress that nuclear fusion has an almost unlimited potential. The hydrogen isotopes in one gallon of water have the fusion energy equivalent of 300 gallons of gasoline. A nuclear fusion power plant would also have no greenhouse gas emissions, and would generate none of the long lived, high level radioactive waste associated with conventional nuclear fission power plants.

Dr. Harold Aspden (Energy Science Ltd.) says:

> I first heard of research on nuclear fusion reactors some 46 years ago when I worked for a major power plant producer in the UK. Being highly qualified legally and professionally by having to secure patent protection for the company's inventions, and by regular contact with research management, I well remember the day when I was informed that research scientists in this work had

> run out of ideas for containing the plasma discharge in their reactor, and were no longer keeping their problem secret. My university research having been on electromagnetism and associated energy anomalies, I even suggested a reactor design that my company then patented. But that was long ago and, watching onward developments, it seems that the future promise of nuclear power from hot fusion has always been some twenty years ahead, just about the expected retirement age of the average senior researcher involved. [1]

"Nuclear fusion generates radioactive wastes from a fast neutron, such as 14.7 Mev, from the reactor wall," states Dr. Jen-Shih Chang (Department of Engineering Physics, McMaster University). "However, this is much less than a fission reactor. In order to minimize radioactive waste, you need a good engineering design." [2]

"Not every fusion reaction produces helium," explains Dr. Dave Jackson (Department of Engineering Physics, McMaster University). "For the D-T reaction, the mass decrement is about 0.4%." [3]

Leading experts predict that the world is still at least 40 years and trillions of research dollars away from having electricity generated from nuclear fusion, despite its theoretical potential. The complexity of a reactor that would be capable of sustaining nuclear fusion is largely due to the enormous size. According to Dr. Chang, "The only thing that we are missing in current fusion power is more engineering work, but not too much for plasma physics." [2]

NOTE

Nuclear fusion involves the binding together of hydrogen atoms, creating helium. The total mass of the final products is slightly less, one percent, than the original mass, with the difference being given off as energy. If this energy can be captured, it could be used to generate electricity. On the other hand, nuclear fission

NOTE

is the splitting of an atom into two or more parts. When such an occurrence takes place, a very large amount of energy is released. This can occur very quickly as in an atomic bomb, or in a more controlled manner, allowing the energy to be captured for useful purposes. Only a few naturally occurring substances are easily fissionable. These include uranium-235 and plutonium-239, two isotopes of uranium and plutonium. Isotopes are forms of the same chemical element that have the same number of protons in their nuclei, but a different number of neutrons.[4]

Fusion Reaction

It has been found that it is easier to promote fusion by using two isotopes of hydrogen, deuterium and tritium, rather than using normal hydrogen atoms. Isotopes are forms of the same chemical element that have the same number of protons in their nuclei, but a different number of neutrons. Deuterium is a naturally occurring isotope of hydrogen that has one extra neutron. One hydrogen atom in 6,700 occurs as deuterium and can be separated from the rest. Tritium has two extra neutrons and is very rare, because it is naturally radioactive and decays quickly. Tritium can be manufactured by bombarding the naturally occurring element lithium with neutrons from either a fission or fusion reactor. Current thinking is that tritium would be created by having a blanket made of lithium surrounding a containment vessel. A reactor such as this, which breeds its own fuel, is called a breeder reactor.

"I would rather believe that there are ongoing processes at work in water," explains Dr. Aspden, "which determine the relative abundances of hydrogen and deuterium, and can account also for the existence of tritium. Such processes involve interaction with an energetic vacuum medium—the quantum underworld medium of space itself. This involves 'cold fusion'—not ultra-high temperatures. I have even recorded my account of this in an Energy Science Report dating from 1994 and entitled *Power from Water: Cold Fusion* which is on record on the Internet. (See my Web site: *www.aspden.org*.) The deuterium/hydrogen ratio 6700 is then derived theoretically as is the lifetime of tritium, but the real significance of this is that nuclear fusion can occur at room temperature, and nuclear fusion research ought to be guided accordingly, without trying to generate trigger temperatures of 100,000,000°C." [1]

"The term 'breeder reactor' is applied to fast fission reactors and not a term applied to fusion reactors," explains Dr. Jackson. "The lithium surrounding a fusion reactor is called the 'breeder blanket.'" [3]

There is no convenient method of starting a nuclear fusion reaction, unlike nuclear fission, which is used in conventional nuclear power plants. Fusion can only be accomplished at temperatures typical of the center of stars, about 100 million degrees Celsius. At such temperatures, the fusion components exist in the form of a plasma, where atoms are broken down into electrons and nuclei. No known solid material could withstand the temperatures involved in nuclear fusion. Therefore, to keep the plasma away from the walls of the vessel in which it is contained, a powerful confinement system (magnetic fields) is required.

Dr. Aspden states:

> "Fusion can only be accomplished at temperatures typical of the center of stars," is a statement I question because scientists have no way of measuring the temperature at the center of a star and, indeed, have no sure knowledge that the energy radiated by a star is the product of a nuclear fusion reaction. It is all an assumption. Let me explain my reasons. First, the hydrogen atoms that form the sun are held closely together by the force of gravity—so close that the electrons of adjacent atoms can crash into one another. My physics education tells me that under such conditions, some

electrons will come free. There will be ionization, and that means free protons as well. It also means a temperature of some 6,000 degrees Celsius, as measured for the sun's surface. More than this, however, because two free protons experience a mutual acceleration rate owing to gravity. In other words, that is 1,836 times applicable to interacting electrons. There will also be a surplus of protons within the body of the sun, there being a balance of gravitational attraction against the mutual electrostatic repulsion of the resulting positive charge density, and there will be a preponderance of electrons in surface regions of the sun. In short, our sun will have a uniform mass density within its radiating surface. It should also have a temperature much the same as that at its surface, subject to there being no nuclear reactions in its core. The decisive factor is then its mass density, which we measure as 1.41 gm/cc on average. Now, if you calculate the mass density of a system of hydrogen atoms with their electrons in near contact, given our knowledge of the atomic dimensions of the hydrogen atom, you will obtain 1.41 gm/cc. Just look up the mass of the hydrogen atom and divide it by the cube of the $1.058 \times 10^{(-8)}$ cm diameter of the electron orbit. Surely, astrophysicists need to explain why they have ignored these facts. I have tried to tell them, but what I say has been ignored. See for example my contribution to *Physics Education, v. 34(5), 263 (1999)* as published by the UK Institute of Physics of which I am a member. It was entitled: "The Imaginary Sun" and challenged the claim that "the very central core of the sun is extremely hot and dense with a temperature of about 150 million kelvin and a density around 160,000 kg per cubic meter," said to be conditions that favor nuclear fusion. So, those astrophysicists may then well ask how deuterium gets into the make-up of the sun, if not by hot nuclear fusion at 100,000,000°C. I have explained that in my previous comment, but equally I could give another possible reason. As I have explained elsewhere, the sun will traverse a space domain wall every few hundred thousand years or so. You may have heard of antimatter comprising systems of particles for which charge polarity is reversed in relation to normal matter. So it is for the electrical make-up of the aether, that such a reversal occurs, as between opposite sides of a space domain wall. Protons in transit can become antiprotons. The deuteron is a proton united with a neutron to form the nucleus of a deuterium atom, but such a neutron could well be an antiproton that has adopted a site in the aether normally occupied by an aether particle having a negative charge. It would then appear to be electrically neutral. Apart from the obvious implication this has for understanding neutron stars, I infer from this that the atomic nuclei of atoms other than hydrogen could well have to be created in the turmoil of the space domain boundary crossings. It need not be the fusion of atomic nuclei that has created the spectrum of atomic matter from hydrogen as a source, but rather the transient existence of antiprotons owing to polarity reversal of the proton at such times. After all, we do not know what it is that determines whether a fundamental particle has a positive charge or a negative charge, but I believe it to be a question of phase of oscillation as charges of the same family share energy in their interplay—half expanding as half contract. So, the timing can easily go adrift during transit through a space domain wall. [1]

The Current Research

When it was reported in 1989 that scientists had achieved fusion at room temperatures with simple equipment, fusion research was big news. (See sidebar, "Cold Fusion.") Unfortunately, the scientists involved could not prove their claims. Therefore, present-day fusion researchers still cannot avoid their greatest barrier, the ultra-hot temperatures required for sustaining nuclear fusion. Magnetic confinement and inertial confinement are currently two methods of confining the hot plasma, that are being studied around the world.

Cold Fusion

Fusion refers to the ability of small atomic nuclei to fuse together to form a larger nucleus. The small decrease in mass (due to the increased binding energy per nucleon) gives rise to a large release of energy. Fusion of hydrogen nuclei to form helium is the source of the energy output of the sun (4 × 10^26 W [Tungsten]) and other stars. The fusion of tritium (detonated by the fission of uranium) is the basis of the hydrogen bomb. [5]

For several decades, physicists have worked on schemes to control the fusion of hydrogen (or its isotopes) in order to generate energy, either as heat or electrical power. The task is not easy, since it involves the generation of very high temperatures. Hydrogen fusion is possible in the sun's core at a temperature of a few million kelvin because of the large pressure and density, but hot fusion capable of generating electrical power in a fusion reactor requires temperatures of over 100 million kelvin. "The large temperatures and densities are induced by immense gravitational fields, which can't be duplicated on Earth," states Dr. Jackson. [3]

At these temperatures, the thermal energy greatly exceeds the ionization energy of an atom, so all matter exists as a plasma containing positively charged ions and negative electrons. Magnetic fields can be used to contain the plasma (to prevent it from cooling by contact with the walls of a containment vessel) in a machine known as a tokomak, but there are numerous modes of instability that have made it difficult to achieve the necessary temperatures for a sufficient period of time to create useful energy from fusion reactions.

An alternative approach to hot fusion is inertial confinement, in which a large number of powerful laser beams converge to bombard a small plastic pellet containing a few milligrams of deuterium and tritium fuel. During the laser flash (lasting a few nanoseconds and containing a power level several times that of the combined electric power plants in the world), the pressure would rise to 10^12 atmospheres, generating a fuel density 10 times higher than that of lead and 10^10 times higher than that of a magnetically confined plasma. "High power laser technology needs a longer lifetime and improved efficiency technological development for fusion applications," explains Dr. Chang." [2]

According to Dr. Jackson, "'Magnetic' instead of 'hot'—inertial confinement is also very hot indeed. The United States is no longer a world leader in Magnetic Confinement Fusion (MCF), having been surpassed about a decade ago by Europe and Japan. However, the U.S. does lead the world in Inertial Confinement Fusion (ICF) with the National Ignition Facility (NIF) at Lawrence Livermore National Laboratory now at an advanced stage of construction. This $5 billion project will use the most powerful and advanced laser systems ever constructed." [3]

Although both types of machine have generated fusion reactions, none has exceeded the break-even point at which more power is liberated than used to heat the fuel. Since deuterium is a constituent of seawater, the supply of fuel is almost inexhaustible; a fusion reactor would generate radioactive by-products, but less than a fission reactor. Although many experts believe that power generation by hot fusion will eventually be feasible (and necessary), the required development cost will likely be in the tens of billions of dollars, and each machine is likely to cost several billion dollars to build.

Given these costs and technical difficulties, it is not surprising that intense interest (among scientists and in the public media) was generated by University of Utah researchers. These researchers claimed to have observed fusion reactions at room temperature in a small electrolytic cell containing heavy water (deuterium oxide, D20) and connected to a DC source of a few volts. The negative electrode of the cell was made of palladium, a metallic element that can absorb large quantities of hydrogen or deuterium (many times its volume at room temperature), and it was suggested that deuterium nuclei within palladium metal get squeezed together close enough to initiate fusion.

The main evidence for fusion was that the heat output of the electrolytic cell (measured by calorimetry) exceeded the resistance heating by a factor between 1.05 and 2.1. However, the researchers claimed also to have detected neutrons coming from the cell (a normal result of fusion reactions), although only at a low level (three times the natural-radiation background). They also claimed to have detected tritium in the cell, in very small amounts, but above the natural background. [5]

Within days, many researchers around the world were trying to duplicate the experiments of the University of Utah researchers. Some initially reported positive results, but major laboratories at Caltech, MIT, Princeton, and Harwell (UK Atomic Energy Authority) came up with negative findings.

One of the criticisms concerned the relative lack of neutrons. At the power output claimed by University of Utah researchers (up to 26 W/cm^2 Tungsten/centimeters^2), the number of fusion events would exceed 10^12 per second, and since each fusion ion should produce an energetic neutron, the resulting radiation dose should have been lethal. Experiments at Caltech showed the neutron output from an operating palladium electrode to be less than two per minute (a factor of 10^5 lower than reported by the University of Utah researchers). "In my opinion," states Dr. Jackson, "cold fusion was a hoax, a fraud, or just plain bad science." [3]

Research on cold fusion continues, but on a small scale and privately funded. Because of the bad feeling created by the original discoveries, it is impossible to obtain government funding for cold-fusion research.

Most of the recent research on fusion reactors has been based on the *tokomak* system, and most experts feel that magnetic confinement has the greatest potential. Tokomak is an acronym for the Russian words meaning torroidal magnetic chamber. The tokomak system was developed in the former U.S.S.R. and is under study in the U.S., Japan, and Europe. A doughnut-shaped steel structure in which the fusion plasma is confined by means of powerful coils of super-conducting material which create a strong magnetic field, is known as a torroidal magnetic chamber.

Inertial confinement is the other method of confining fusion plasma. This is where small amounts of a deuterium–tritium mixture are rapidly heated to extremely high temperatures with a high-powered laser beam or a beam of charged particles. Very high-power lasers are needed, and work on inertial confinement is not as far advanced as that on magnetic confinement.

The Tokomak Fusion Test Reactor (TFTR) in the U.S. and the Joint European Toms (JET) are the most advanced test reactors. They use the tokomak design and have come close to break even conditions. In fact, in November, 1991, the British-based tokomak reported break-even conditions. This occurs when the energy given off by the fusion reaction is equal to the energy input required to sustain the reaction. In order for a fusion reaction to generate useful amounts of electricity, the energy given off must be many times greater than that required to sustain the reaction. Before this stage is reached, even the most optimistic researchers feel that it will be well into the 21st century.

Fusion Power Plants

A full-scale fusion reactor capable of generating 1000 MW (1 MW = 1 million watts) of electricity, comparable to conventional nuclear power plants, would be a very large and complex machine. (See sidebar, "Star Power Reactors.") The minimum size and output of a fusion reactor would be similar to that of today's largest nuclear plants, while fission reactors can be made small enough to be used in submarines or satellites.

According to Dr. Jackson, "The size of the fusion reactor would be about the same, although the complexity would be greater." [3] On the other hand, according to Dr. Chang, "Fusion power generation needs better technology for energy conversion technology, and directly from plasmas and protection of the first t wall. We need more research funding on these engineering technology developments." [2]

Star Power Reactors

For a moment, in a small campus of anonymous white buildings in New Jersey, a miniature sun lights up inside the Princeton University plasma physics Laboratory. Within a tangled merry-go-round of wires and red copper coils, a ball of plasma—a charged gas of hydrogen nuclei and electrons—heats up to 70 million degrees Fahrenheit. The plasma blazes supernova-bright, then instantly goes black. Although the whole episode lasts less than 500 milliseconds, physicists seem pleased. For plasma that's a long time. The particles do a lot in a hundred milliseconds.

"Gee-whiz!" exclaims Dr. Jackson. "You've been sucked in by the Princeton PR machine. The glory that was Princeton has greatly faded in recent years. General Atomics in San Diego now houses the most advanced magnetic fusion experiment in the U.S.—Doublet IIID. Much larger tokamak facilities are in Europe (JET) and Japan (JT-60U), and the U.S. is in a distant third place in terms of MCF fusion expertise with NIF it is the world leader in ICF." [3]

Physicists are searching for a practical way to harness nuclear fusion, the process that causes stars to burn. If they succeed, the Earth's energy worries are over. Fusion reactions are incredibly potent and produce almost no pollution. The fuel for a fusion reactor is hydrogen, which can be extracted from a cup of ordinary seawater. There's one big catch: Despite a 50-year effort to make fusion commercially viable, no one has been able to start a reaction, keep it going, and keep it contained. Yet now, fusion researchers insist they are getting closer to success, just as the United States casts about for alternatives to imported oil.

Physicists around the world largely agree about how to tap fusion power. Trap a hydrogen plasma in a magnetic field—most fusion experiments use a doughnut-shaped magnetic bottle called a tokamak. Drive the temperatures up to about 180 million degrees Fahrenheit. Then sit back and let nature take its course. Under those conditions, by releasing an enormous amount of energy in the form of fast-moving neutrons, hydrogen nuclei hit each other with so much force that they sometimes stick together and fuse into helium. [5]

To get out more energy than you put in is the whole point of fusion, but no one has succeeded in doing that yet. In 1994, the Princeton Plasma Physics Laboratory's $300 million gymnasium-sized Tokamak Fusion Test Reactor generated a record 10.7 million watts of power, yet it operated at a loss. The Joint European Torus reactor near Oxford, England, has done better but still hasn't passed the magic break-even point. Physicists have turned theory into reality, but at a deficit in energy and only for a few seconds. If they're going to make cheap electricity, they need to do it 24/7.

"TFTR was not particularly successful as a fusion experiment compared to JET and JT-60U," states Dr. Jackson. "Both have produced plasma conditions that would yield more fusion energy (if tritium were present) than the heating energy required to bring the plasma to that condition—that is, break-even. JT-60U holds the record of about 1.3 compared to break-even defined as 1." [3]

In 1985, Princeton physicists thought about these problems and theorized that a magnetic bottle in the form of a spherical torus, resembling a cored apple, could produce the same power output using less energy than a tokamak. The apple-shaped reactor would require smaller magnets that cost less to build. Best of all, making fusion easier to achieve, the torus could keep plasma in a tighter configuration.

The Princeton physicists' vision took shape as the $25 million National Spherical Torus Experiment, or NSTX, which began operating in 1999. The machine has 12 long, curved magnets running top to bottom around the outside and a cylindrical array of magnets through the core. The magnets confine a sphere of hydrogen plasma while blasts of microwaves heat the particles to millions of degrees. The reaction chamber is lined with 2,700 graphite tiles and a multitude of sensors that measure the density and temperature of the plasma. Current test runs with the reactor are not hot enough to attain significant amounts of fusion, but they reveal a great deal about the behavior of plasma in a spherical torus.

According to Dr. Jackson, "there are scores of smaller fusion experiments throughout the world, employing many different magnetic field configurations for plasma confinement. For example, Japan started up the world's largest stellerator LHD (cost approaching $1 billion) in 1998, and Germany will complete another of similar scale in 2006. NSTX will no doubt be an interesting and worthy machine, but it's small potatoes in terms of its scale and impact on the world fusion scene." [3]

The news is good so far. NSTX is twice as efficient as earlier reactors, achieving a beta value (the ratio of plasma pressure to the pressure of the machine-generated magnetic field) of 25%. Physicists are getting more bang for the buck. Furthermore, the spherical magnetic bottle has an unexpected benefit: Under certain circumstances, the ability of the system to contain energy improves with increased plasma pressure. With a conventional tokamak it's usually the reverse. The plasma has a self-healing property that is working against instability. [5]

Additional support for the spherical design is being provided by a complementary machine in England, the $15 million Mega Amp Spherical Tokamak, or MAST. Now MAST is where physicists sort out the details. In order to scale down the size of the reactor, which would significantly reduce construction costs, the primary goal would be to reduce the size of the column of magnets at the center of the torus.

An international fusion collaboration is going the other way, while the British and American groups think smaller and cheaper. The planned International Thermonuclear Experimental Reactor could be the first machine to generate significant amounts of surplus fusion power—if it is built. The reactor's staggering price tag, estimated to be $6 billion, prompted the United States to back out of the project a few years ago. Even if the fuel is free, a $6 billion power plant would not be a commercially viable venture. MAST and NSTX point to a less expensive approach, but neither one is close to breaking even, much less generating any usable surplus.

Many in the scientific community wonder whether fusion will ever transcend its history as an expensive curiosity. "The post-fossil fuel age will be defined before fusion comes along. Fusion is a centralized business, meaning that the government is the only one spending money on it. Many see the world heading instead toward decentralized, renewable energies such as solar, wind, biomass, and fuel cells. Renewables now provide 8% of the total energy in the United States, but they could climb to 50% by the middle of the century, leaving little place for fusion. However, many physicists wouldn't rule it out, although it's a long shot.

The federal government, so far, is standing on its bet. The Department of Energy spends a steady $250 million a year on fusion research. The largest chunk, about $70 million, goes to Princeton. However, commercial fusion reactors will not be a reality until 2050. Many scientists are preparing for a time far into the future when fossil fuels are gone or no longer acceptable due to global warming, and renewables alone cannot provide enough juice.

Reactors could supply about half of America's electricity by 2100. This is a long-term issue.

The difficult part is creating a sustainable fusion reaction, although a fusion reactor capable of generating electricity has never been built. Capturing the energy given off by the reaction in the form of heat and transforming the heat to electricity is very similar to generating electricity from a conventional fission reactor.

For instance, where doughnut shaped atoms, tritium, and heat are extracted from the walls of the containment vessel is the basic components of a fusion power plant using a magnetic confinement system. The only inputs are a constant flow of deuterium and a periodic input of lithium to the "blanket," to allow for the breeding of the tritium needed for the reaction to continue. A 1000 MW fusion generator would have a yearly fuel consumption of only 150 kg of deuterium and 400 kg of lithium. [5]

Fusion-Powered Spacecraft

As previously explained, fusion has been seen as the power source of the future for the last five decades. Controlled fusion (joining two lightweight nuclei to get a slightly heavier nucleus and a lot of energy) has been challenging. In their quest to exceed $Q = 1$, the break-even point, scientists have moved from low energy yields of $Q = 0.0000000000001$ in the late 1950s to $Q = 0.3$ today, and have developed a large body of engineering and scientific knowledge showing that it can be made practical. [6]

From the NASA perspective, the challenge is to adapt fusion for space propulsion. Magnetized Target Fusion (MTF) is one of the major approaches that scientists are studying. NASA/Marshall is working with Los Alamos National Laboratory and the Air Force Research Laboratory to adapt MTF for propulsion. MTF tries to operate in an intermediate regime between the conventional magnetic fusion and inertial confinement using a laser. The problem with conventional magnetic confinement is it operates at very low density. To achieve sufficient power, the fusion reactor must be large, which translates to a high cost.

On the other hand, inertial confinement fusion uses a tiny plasma, 1,000 trillion times denser than in a magnetic confinement scheme. But that requires a driver (usually banks of intense, short-pulse lasers) that heat and compress the target in a short time. That also drives the cost up. [6]

MTF tries to operate at not too low or too high a density and to achieve a reasonable rate of fusion activity with a density 10,000 to 100,000 times higher than magnetic confinement and 10,000 to 100,000 times lower than laser fusion. It's more economical and uses pulse-power drivers (powerful capacitor banks that drive electromagnetic implosion) that are available today at low cost. It does not have the implosion speed generated by a laser beam, but a magnetic field confines the target plasma and insulates the inertial wall that implodes to cause the fusion. [6]

The Compact Model

Even if fusion is achieved, current methods are too cumbersome to use in rockets. The mass is quite prohibitive. Scientists want to make the physics work without using very large magnets. The mirror magnets for a fusion rocket would weigh about 401 tonnes (metric tons), about 16 times a single Space Shuttle payload. The heat radiators would add 240 tonnes. [6]

Scientists are now experimenting with a droplet radiator design that, using liquid lithium as a coolant, could reduce the radiator mass to 57 tonnes. They recently flew a test model aboard NASA's KC-135 low-gravity aircraft to test a model radiator. [6]

A rotating magnetic field could induce a magnetic field and electrical currents, a clever way of fooling the plasma into

Figure 20.1 Peering into the heart of a star. What looks like a 1950s model of an atom is a hollow cathode with a tiny plasma cloud contained inside an IEC fusion chamber small enough to sit atop a lab bench (UIUC).

behaving as if it were in a conventional magnetic mirror system. In turn, the mass of the spacecraft would come down from 720 to 230 tonnes, and the 44-meter (144-ft) long engine would have a specific impulse of 130,000 seconds. It's quite impressive. [6]

One of the most intriguing possibilities raised actually dates back to the 1950s and a concept developed by Philo Farnsworth (see Figure 20.1) [7], who pioneered most of the fundamental technologies for television in the 1920s and '30s. This is a really neat concept, something you can literally put your hands around. You can use the power it would generate to power electric propulsion, or use the plasma for thrust.

A Bottled Star

The technique is called inertial electrostatic confinement (IEC) (see Figure 20.1), a technique that avoids the use of

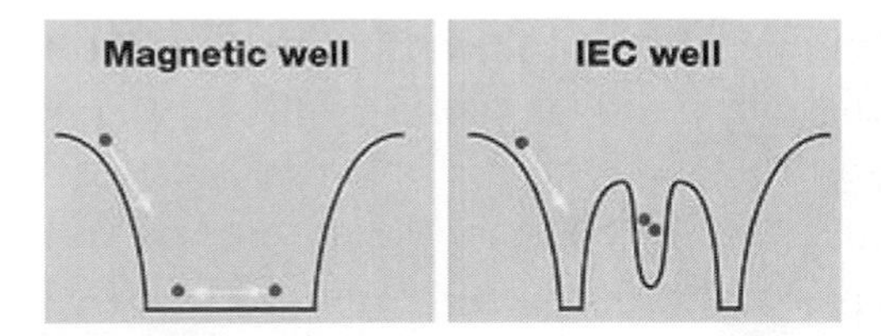

Figure 20.2 A schematic of the energy well in the middle of a conventional magnetic field, and in the IEC chamber where fusion is induced (UIUC).

massive magnets and laser systems used in other fusion-power techniques. Instead, the IEC device uses a hollow cathode, and the natural charges of electrons and ions, to form virtual electrodes that confine ions in a spherical region at the center of the 61 cm (2 ft) diameter IEC vacuum chamber (see Figure 20.2).[7]

Using a pulsed megawatt power supply, the IEC achieved its highest pulsed current yet—17 amps at 40,000 volts. The IEC has also gone from producing one neutron (released by deuterium–deuterium fusion) in every 10 cycles to more than 100 neutrons per cycle.[7]

IEC fusion would work best with a couple of unusual fusion cycles. One uses deuterium (heavy hydrogen), easily refined from water on Earth, and helium 3 (helium lacking one neutron), which is quite rare here, but possibly abundant in lunar soil exposed to 4 billion years of solar wind. The other fires protons into boron 11.[7]

Fusion missions would need just micrograms to reach the Oort cloud, the deep freeze of comets beyond the orbit of Pluto. Reaching the stars would require metric tons.

Advantages and Disadvantages of Fusion

If it can be developed, nuclear fusion would have several advantages over conventional fossil fuel and nuclear fission power plants. The fuels required for fusion reactors, deuterium and lithium, are so abundant that the potential for fusion is virtually unlimited. Oil- and gas-fired power plants as well as nuclear plants relying on uranium will eventually run into fuel shortages as these nonrenewable resources are consumed. Like conventional nuclear plants, fusion reactors have no emission of carbon dioxide, the major contributor to global warming, or sulfur dioxide, the main cause of acid rain. The large contributors to global warming and acid rain are fossil fuel power plants burning coal, oil, and natural gas.

Public concern over operational safety and the disposal of radioactive waste is one of the barriers to the widespread use of conventional nuclear power plants. (See sidebar, "Nuclear Waste Disposal.") Major accidents, such as Chernobyl, are virtually impossible with a fusion reactor because only a small amount of fuel is in the reactor at any time. It is also so extremely difficult to sustain a fusion reaction that should anything go wrong, the reaction would invariably stop. Long-lived highly radioactive wastes are generated by conventional nuclear plants; these must be safely disposed of and represent a hazard to living things for thousands of years. The radioactive wastes generated by a fusion reactor are simply the walls of the containment vessel that have been exposed to neutrons. Although the quantity of radioactive waste produced by a fusion reactor might be slightly greater than that from a conventional nuclear plant, the wastes would have low levels of short-lived radiation, decaying almost completely within 100 years.

Nuclear Waste Disposal

The high-level waste that everyone is concerned about that comes from nuclear reactors is from the fuel used to power the reactor. This comes in the form of little pellets of uranium oxide. Each pellet is about the diameter of a pencil and about half an inch long. Hundreds are stacked one on top of each other in tubes about 10 to 14 feet tall. Then the tubes are put into a special grid with spaces between the tubes. There are usually between 81 and 196 tubes in a fuel bundle (depends on who makes it and which reactor it goes into) and a little over a hundred bundles that make up a reactor core. Every two years or so, it is necessary to replace some of the fuel bundles in a reactor core to keep things running properly. The bundles that are pulled out (called spent fuel) are what eventually becomes nuclear waste. [8]

When a bundle is first removed from the reactor, it is very hot. "Hot" in this case, has two meanings. It is hot from the heat coming off of the tubes, *and* scientists gave things that have a lot of radioactivity in them the nickname "hot." The interesting part here is that the radioactivity is what causes the bundles to give off heat in the first place. When you split a uranium atom, the two parts it breaks into are usually radioactive. There are lots of different ways it can break in two (several thousand in fact), all of which behave slightly differently. Some become stable (non-radioactive) very quickly; some take thousands of years or more to become stable. (See sidebar, "The 10,000 Year Curse.") The ones that take a while are the ones that matter to people trying to deal with the waste. Since the bundle needs to cool off for a little while (usually a year), nuclear power plants and the like need a temporary place to store it. This is called the spent fuel pool. In most nuclear reactors, this pool was designed to hold about one to two times the total number of bundles in the core of the reactor. After this cooling-off period, the waste should be transferred to someplace else for permanent storage. This is where things have gotten messed up. [8]

U.S. Options

Since there isn't a permanent place to store it yet (more on that in a bit), nuclear power plants have had to come up with other solutions. Some reactors that had enough space in the main building built another pool to store the used bundles in. Some others have built really big cement and steel canisters that can hold several of the bundles at a time (the canister provides a shield from the radiation inside the bundles), and these big canisters can be set outside on a big concrete slab to wait until a permanent place has been built. [8]

NOTE
The area where the canisters are sitting is on the grounds of the nuclear power plant and is well guarded, by the way.

Unfortunately, most reactors are running out of space. So what about the storage place?

"Truth is," replies Dr. William J. Garland (Department of Engineering Physics, McMaster University), "we are not running out of space. In fact, from birth to grave, nuclear reactors use up the least amount of space per energy generated than any power system. Spent fuel storage is technically a non-issue. We can quantify it accurately, measure it with far more sensitivity than most other things (radiation is easy to monitor), and the radioactive material is densely contained in a very small volume. It is a proven, known technology. This is a big advantage of nuclear power, not a disadvantage. Coal-fired plants typically emit more radioactive material than a nuclear power plant, by the way. And good luck trying to confine coal-fired plant emissions to the extent that we can easily do with nuclear. It is just too diffuse, making it too expensive to achieve. For most technologies, our solution to pollution is dilution. We dump our garbage all over the place! Not so with nuclear." [9]

Dr. Robert.E. Jervis, (Nuclear Science and Engineering, University of Toronto) concurs: "Spent nuclear fuel requires less than one millionth the space for storage than the ash from coal-fired power plants. All of the spent fuel generated during the last half-century from nuclear power, if stacked on the field of a super stadium, would make a pile not much higher than the seating. That it is still stored and has not been permanently disposed of, such as in geological repositories, is because it still contains reuseable nuclear fuel, if reprocessed. An alternative is to utilize nuclear power reactors of a more fuel-efficient design that convert most of the nuclear fuel on a 'once-through' cycle, leaving spent fuel that is not economic to reprocess (such as the natural uranium-fuelled, heavy water Canadian CANDU reactors)." [10]

Dr. Patrick Bailey (President, Institute for New Energy [INE]), also concurs, somewhat: "The problem of storage space for spent fuel rods in commercial nuclear plants is not a problem. The problem is in the politics of energy production, and government. A nuclear fuel rod becomes very radioactive and very hot when it has been used inside a nuclear reactor. At a certain point, the rod must be removed, as it has reached its 'maximum burn-up point', or point of usefulness. Currently, these spent fuel rods are stored in spent fuel pools, in water, usually located on the property of the nuclear plant. The U.S. federal government had promised the nuclear utilities since the inception of the 'nuclear power plant age' that a central storage facility would be created for the storage of both 'low-level' and 'high-level' nuclear waste. Unfortunately, the discussions and decisions in these areas has been all about greed, profit, and political positioning, and not about getting a waste facility planned, dedicated, and constructed. One of the biggest nuclear waste issues still today is what to do with the low-level radioactive waste that comes from hospitals! The problem is not about how to store the waste—the problem is in the willingness to get the waste stored! While short-term profit can be made—with total disregard to the responsibility of the consequences—this situation will continue.

Nuclear utilities have their own storage pools and have been able to store their own spent fuel rods. In the past, they had looked to the Federal Government to create a better storage facility to store this high-level nuclear waste. Now that the U.S. utilities have become deregulated—i.e., acting as their own cost centers—they now have the capability to provide for more storage space, if approved by the NRC, at your expense—just by raising the price for the electricity generated at that plant.

The bottom line is, like in so many other areas today: If you do not get involved and assist in the solution to a problem, you will have to pay for someone else to handle the problem for you. Unfortunately, if a means is found for money to be made by those overseeing that problem, then you can be assured that the problem will never be addressed or solved." [11]

Yucca Mountain Storage Site

A long time ago, the government said it would build a permanent place to put the radioactive waste from nuclear power plants. This has not happened yet. There is a place called Yucca Mountain in the Nevada desert where the U.S. is trying to set up the place to bury the waste. [8]

It has been a roiling controversy for years: where to store the nation's thousands of tons of nuclear waste, which remains radioactive for centuries. The issue has become even more pressing in the wake of the September 11 attacks and those frightening, but unconfirmed reports that terrorists are trawling for radioactive ingredients to make so-called dirty bombs, not to mention, hitting the nation's nuclear facilities. In 2002, the Senate approved a controversial plan to bury as much as 70,000 mettle tons of spent nuclear fuel by 2036 in Yucca Mountain, about 90 miles northwest of Las Vegas; it is now stored at 131 aboveground sites in 39 states. None of these facilities was designed to safely store that waste on a permanent basis, and leaving spent fuel in temporary storage around the nation poses both a security threat and an environmental hazard. The Bush administration plan calls for nuclear waste to be shipped to Yucca beginning in 2010. The White House indicates that the $58 billion project is "scientifically sound" and would better safeguard the nation's nuclear waste by placing it in a single, secure underground location. [8]

Hold on! The government needs to rethink their decision about terrorism and the possible uses for and attacks on these nuclear canisters as they are traveling across our country. [8]

In any event, based on these facts, a lot of people have been arguing over whether or not this is a good idea, and it still hasn't been completed yet. The argument centers around if the place (which is deep inside a mountain) can really hold the waste where it is for thousands and thousands of years safely. [8]

Nuclear Waste at Hanford

A lot could be said about Hanford—in fact many books have been written about it, some good, some bad. Here's the basic deal. When the U.S. was trying to make nuclear bombs for the first time, there wasn't a handy "how to make a nuclear weapon" textbook. So a lot of experiments were done to try to understand all the chemical stuff about uranium and this new (at the time) element plutonium. In the process of doing all these experiments, the U.S. made a mess. As you probably know, there are a lot of big waste pits with thousands of gallons of liquid mess that is both chemically dangerous and radioactive. There are a lot of people trying to figure out what to do and the answer so far is: it's going to take a long time to figure out what to do. [8]

NOTE

Short aside here—there's a lot of politics mixed up with Hanford. A lot of people are upset because of what was done there. It really doesn't do any good to get angry about what was done and why. That is in the past and the answer to why these things were done should be left to a history class on the Cold War. Now this country has a mess, and scientists have to say, "What next?" and "How do we clean this up now?"

Each pit has its own unique chemical mix and different radioactive elements in them. Each one will be an operation of its own. Basically what will have to happen is this: The contents of a particular pit will have to be analyzed to determine exactly what is in it. From there, the waste will have to be taken out in little bits at a time to be processed. The nasty chemical part of the waste will have to be removed or treated in some way to remove that part of the hazard. [8]

NOTE

Some of the waste might be benzene, which is poisonous and may also cause cancer, so that has to be broken down first.

For example, here is a chemical destruction method that works very well. There is a method of chemical hazard removal called UV-peroxidation. This is a two-part process to remove the chemical hazard from liquid waste that has both chemicals and radioactive materials in it. First, hydrogen peroxide is mixed into a tank with the waste in it and is stirred around. Hydrogen peroxide is a strange form of water that has an extra atom of oxygen attached to it. When the hydrogen peroxide mixes with the chemical, the extra oxygen will get taken away to react with the chemicals and break them down into things that either aren't dangerous or are much less dangerous, and the hydrogen peroxide then becomes ordinary water. [8]

"UV-peroxidation is a promising technology," claims Dr. Chang. "However, we need more efficient UV lump technology, since efficiency will only be 15 to 35% in the near future. This type of indirect plasma water treatment generates UV from plasma in low to moderate pressure lumps, and then treats water via a glass tube lump. It is not efficient and currently direct plasma water treatment generates paraoxide and UV directly in water. It was supposed to replace this indirect method." [2]

TIP

The water is mixed in with all this other stuff including the radioactive material, but at least it's only water.

In the second part, the mix is pumped past very strong lights that give off lots of ultra-violet (UV) light. Imagine getting a sun tan in less than a minute. The UV can also break the chemicals down into other less harmful things. So this tag team of peroxide and UV can leave you with a liquid that is mostly water, simple chemicals, and the radioactive stuff.

"The UV-peroxidation process is marvelous at breaking down the organic compounds that are very harmful in mixed waste," explains Scott Kniffin. "The question then really becomes what to do with what is left. Heavy metals (e.g., arsenic) that are not necessarily radioactive, and any long-lived or specific hazard radioisotopes (e.g., Strontium-90) must be chemically precipitated out each in turn; this is much easier to do after the UV-p process. Those things that have been precipitated out may have use in other industries, even the radioactive materials. Several of the radioactive isotopes found in the slurry at Hanford have medical applications. If there is a need for a toxic metal (non-radioactive), why dig up and refine more when you can recycle what has been used there? After the materials that you can't safely release are removed, the rest can be safely diluted and flushed in accordance with EPA and NRC laws. The thing to remember is that this is going to be a slow and lengthy process because it is complicated, messy, and dangerous. No one wants to have the waste, but this was and is a cost of winning the Cold War." [8]

The problem at Hanford is that some of the chemicals might not react well with peroxide, or they might form flammable or explosive gases, or the gases might even be radioactive. So other methods will have to be developed for the chemical and radioactive goop in those tanks. It will take a long time, but the U.S. has to clean it up.

Now, back to the previous example. So now you have a radioactive liquid that's mostly water. What next? There are two possibilities here. The most unlikely is that the radioactive part is very small compared to the total amount of liquid and the thing that is radioactive is not dangerous on its own (for example, the element arsenic is a poison, and that fact is more important than that it might be radioactive). If it's okay, then the liquid can be diluted and poured down the drain. [8]

A more realistic solution would go something like this. Now, you add a chemical to the liquid to make the radioactive material have a chemical reaction and turn from a liquid into a solid. Then you separate the solid radioactive material from the liquid (there are lots of ways to do this). [8]

If there are several different radioactive materials in the liquid, you may have to repeat this several times. Eventually you are left with some radioactive material in a solid form and mostly water. The mostly water can be further cleaned to meet the laws regarding how clean it has to be, and then you can dump it down the drain. The radioactive material is another matter. [8]

In some cases, the material might have another use, so it can go that way. For the others, again this has to be buried somewhere for a long time. One way is to mix the radioactive material into special glass and make it into little marbles. These marbles would then be put into a different type of concrete and steel canister and buried just like the canisters from the nuclear power plants. There are other ways to seal up the waste, and several new methods are being researched. There is always plenty of room for improvement.

"Confined radioactive waste inside unleachable glass materials for long-term storage is now almost commercilized," explains Dr. Chang. "The next step that is needed is the development of neutron transmutation techniques to convert radioactive materials to non- or low-level radioactive materials before being disposed." [2]

Finally, other countries (especially those that didn't build nuclear weapons or started building them much later) have different ways of dealing with their radioactive waste.

The 10,000 Year Curse

Sure, right now Nevada's Yucca Mountain (which the Bush administration in 2002 designated America's official dump for the deadliest nuclear waste) sits in a desert sparsely populated by English-speaking homo sapiens. But, care to make a wager about the year 12,002? Then, Yucca might rise from the suburbs of some post-human society straight from SciFi Channel's *Stargate SG-1* TV series. So how can scientists warn Yucca's future neighbors against poking around this nuclear tomb, which will still be radioactive 10,000 years hence?

"As far as worrying about the long term storage issue," explains Scott Kniffin, "it is foolish to assume that we will never have a use for the radioactive material in the spent fuel. In the case of American power reactors, the uranium is typically enriched to ~4-6% U-235, depending on the individual reactor. Even after 4-5 years in the core, the concentration of U-235 in the fuel has changed very little. What has happened is that those atoms of U-235 that were split have an annoying tendency to become elements that absorb neutrons and disrupt the neutron cycle in the reactor and make it shut down. Eventually, we will run out of uranium that we can mine from the earth and refine. Why would we not then go back to the reactor waste and extract this perfectly good uranium and use it again? This is far too valuable a resource to assume that we would be foolish enough to simply bury and forget it. President Carter's decision NOT to reprocess spent nuclear fuel from civilian power reactors was a political one, not one based on science (France and Germany have multi-billion Euro industries based on reprocessing reactor fuel). This will likely be a problem that will solve itself out of necessity. A further advantage is that the volume reduction of the waste is on the order of 99%, and the uranium itself is the longest lived isotope in spent fuel." [8]

On the other hand, according to Dr. Bailey, "There is no guarantee that the waste disposal sites of today will not pose a threat to people in the future. It is a matter of degree. On one hand, we would like to have a society that does not harm the environment, and then we promote and buy all of the products that pollute it. Our society must have a sense of responsibility! What we create, we must be responsible for! We have huge dumps for other waste: off-shore sewage dumps in the oceans, land fill dumps for trash, weapons dumps for guns and bombs, nuclear storage dumps for atomic bombs, and chemical weapons dumps for poison gasses. Are these all safe? What is the definition of the safety of these dump sites for the people of the future? And what about the anthrax burial sites where farmers bury diseased cattle? These can be and have been easily found by others. If we agree to make waste, then we must be responsible for the waste. The technology for handling, transporting, and storing high-level nuclear waste has been known since the 1980s. The technology exists. The waste is there. Where would you like to place it?

"We must keep a balanced viewpoint. It is not fair to know only one side of an argument, or to become so emotionally involved that one cannot or will not listen to the views of the other side. I have seen people spend hours discussing the dangers of nuclear power in formal debates, while chain smoking. I have seen people purify their drinking water, and then chug diet sodas from cans. We need to look at all of our decisions, keep a balanced viewpoint, and come to rational and scientific solutions—not emotional, short-term, and irresponsible money making schemes." [11]

"Ah," says Dr. Garland, "why are people worried about this? I suspect they are not. Rather, it is a scare factoid thrown out by the anti-nukes. We are doing all sorts of things that leave a legacy for the future and we have no way of knowing whether it will be good or bad. Let's see, over-fishing of the oceans, paving the best farm land in Canada, agriculture on a massive scale worldwide, and on and on and on and on. That legacy is far, far more damning than a pile of spent fuel stuck in a rock. This is what the anti-nukes and others are really about. We are leaving a really big footprint on the world, and they view that as wrong. I agree with them, but their revulsion is misplaced. Nuclear is a convenient whipping boy for all of society's ills, especially the marginization of the individual. Individuals need to feel empowered. Most are not. In the end, I have to believe that we will not solve our problems, but we can't expect taking the childish approach of just stopping anything that has any negative aspect to it, to make the problem go away. It has been shown again and again that by 'powering' a society, it can support the infrastructure needed to have a democracy, a stable government, a medical system, and the like. Stopping nuclear would mean a few thousand excess deaths a year in Ontario due to air pollution. Sure we can do better than we are doing right now, but we need the luxury of a reasonable standard of living before we can even contemplate improvements. That is hard to do when your quality of life is low. Look at Iraq right now. Give the working stiff a choice of a job or democracy and guess what the answer will be? We are perhaps the first generation to have the luxury of such debate, grandstanding, and grousing on such a large scale. My dad worried about feeding 6 kids, not about the negative impact of technology. He had no choice. His legacy to me is the luxury of choice." [9]

Not a single spent nuclear-fuel rod will be shipped to the mountain until construction of the repository is finished—perhaps by 2010—and the Nuclear Regulatory Commission and Environmental Protection Agency approve a plan for storing the waste. Environmentalists and Nevada officials are vowing to fight the project in court, but the Department of Energy is already starting to ponder how to comply with a federal requirement to mark the site for the next 100 centuries. It has plenty of ideas to choose from.

In 2002, the University of Nevada–Las Vegas hosted an exhibit of potential Yucca warning concepts—some sarcastic, some whimsical, such as seeding the mountain with genetically modified, blue-colored yucca shrubs or transforming it into a simulated volcano. But, more likely, they'll take a look at what other scientists are doing at WIPP.

That would be the Waste Isolation Pilot Plant (WIPP) near Carlsbad, NM, which since 1999 has been storing waste from nuclear weapons production in an old salt mine. WIPP consulted panels of academics (including archaeologists, astrobiologists, and materials experts) about the warning-marker conundrum.

Speaking of Volcanoes

According to Dr. Garland, "Aerial surveys will look for evidence of hidden volcanoes around the southern Nevada site tapped to be the nation's nuclear waste repository. The field studies, expected in late 2004 around Yucca Mountain, will use aircraft equipped with magnetic-sensing instruments to find where workers should drill in search of evidence of volcanic activity. Opponents of the project say they fear dormant volcanoes could become active and spawn earthquakes that could rupture containers buried in the repository, possibly releasing lethal radioactivity. The question is, 'Are there indeed buried volcanoes?'

"I couldn't say. The more important question is: 'Can we proceed with the project even though there is a small chance of volcanic activity?' This opens up the whole topic of risk and how it is perceived." [9]

"Nuclear waste," explains Dr. Jervis, "once converted to a ceramic form and efficiently sealed and cladded for deep geological disposal, should pose no appreciable risk to surface dwellers in the future. Neither volcanic actions (which regularly spew natural radioactivity into the terrestrial environment) nor actions of underground waters, should bring appreciable radioactivity back to the surface. This has been demonstrated through realistic modeling of such processes over periods greater than 10,000 years and which have used directly measured parameters for fuel corrosion, diffusion, etc. If society can set apart, and keep undisturbed for centuries, historical sites and sacred burial grounds, it is surely not beyond the wit of man to maintain records in perpetuity of a few nuclear waste depositories." [10]

The studies were approved by the Energy Department even though previous scientific work found that extinct volcanoes and cinder cones near the mountain posed no credible threat to the government's plans for burying spent nuclear fuel and highly radioactive waste at Yucca Mountain. New studies show that scientific work that was used to support recommending the site in 2001 was not finished.

Scare Tactics

Scientists noted that although it might seem sensible to fashion markers from a durable material like titanium, ancient Egypt teaches a different lesson. The fine limestone that originally cloaked the Great Pyramid of Cheops, for instance, was pried off and reused. And scare tactics like the curses on Egyptian tombs can backfire. So, it shouldn't say, "Touch this rock and die." Inevitably, someone would touch a rock, and not die, undermining the warning.

Of course, writing may not get the message across in the distant future. So the scientific teams also considered ways to embed a warning in the marker structure itself. One proposal: "menacing earthworks" resembling the jagged lightning-bolt insignia of Hitler's SS. That one would scare the bejesus out of you. But, costly, elaborate structures can draw the wrong kind of attention. The more grandiose you make it, the more likely people will wonder what you're hiding.

WIPP eventually chose a plan to surround the site with a plain, 33-foot-high, 100-foot-wide berm of rock, soft, and salt. Inside the berm, to be built sometime after the site closes in 2035, will be 16 granite monuments (shades of Stonehenge) and many buried markers. Some will carry warnings in the six official languages of the United Nations, as well as Navajo; others will feature Edvard Munch-esque distorted faces to represent horror, and changing star positions to illustrate when the waste was buried.

At Yucca, where the buried radioactivity will be fiercer, project managers are leaning toward edgier concepts, according to those familiar with their thinking. Two favorites are the menacing earthworks and a field of giant concrete thorns bursting from the ground near the mountain.

Finally, the major disadvantages of nuclear fusion are the vast amounts of time and money that will be required before any electricity is generated by fusion. Fusion produces no greenhouse gases, but it will not be able to contribute to reducing carbon dioxide emissions until close to the middle of the next century. If the world does nothing and waits for fusion as the solution to the global warming problem, it may well be too late. Similarly, every dollar spent on nuclear fusion would have a much greater impact on reducing global warming if it were spent on reducing the demand for electricity. There are dozens of ways to reduce electricity use through efficiency improvements and conservation efforts, which are much less expensive than producing more electricity through nuclear fusion. Development of other electricity supply technologies, such as photovoltaic cells that convert sunlight directly into electricity, could also eliminate the need for fusion before it is operational.

Conclusion

The U.S. alone has spent $60.1 billion in the last 45 years on nuclear fusion research and it has been estimated that another $180 billion will be needed in the next 20 years just to get to the stage of a demonstration power plant. Although fusion has a bright long-term potential, it will not be a major factor as a supplier of electricity in the next few decades. During this time, major changes in the world's patterns of energy supply and use may take place due to shortages of fossil fuels such as oil and concerns over the environmental issues of global warming and acid rain. Finally, whether fusion will ever become an important source of electricity will depend on the results of future fusion research, the costs of electricity generated by fusion, and the demand for electricity in the mid 21st century.

Finally, explains Dr. Jackson, "The scientific feasibility of fusion energy has been demonstrated in the current generation of large tokamaks. Nevertheless, a great deal of engineering development lasting for at least 50 years will be needed for an economic and reliable fusion power system. Therefore, we should not expect to see fusion producing any substantial percentage of the world's energy before the end of this century." [3]

References

[1] Dr. Harold Aspden, Energy Science Ltd., c/o PO Box 35 Southampton, SO16 7RB, England.

[2] Dr. Jen-Shih Chang, Department of Engineering Physics, McMaster University, Office: NRB 118, 1280 Main Street West, Hamilton, Ontario, Canada L8S 4L7.

[3] Dr. Dave Jackson, Department of Engineering Physics, McMaster University, Office: JHE/A321, 1280 Main Street West, Hamilton, Ontario, Canada L8S 4L7.

[4] "Nuclear Chemistry: Like Regular Chemistry, Only Different," [*www .jozie.net/JF/chemclass/nuclearchemistry/fission_and_fusion.htm*], Copyright © March, 2001 Jozie's Web. All rights reserved.

[5] Dr. Bob Hieronimus, Dr. Eugene Mallove, Professor John O'M. Bockris, and Dr. Hal Puthoff, "Cold Fusion and Zero Point Energy," Transcripts from 21st Century Radio's Hieronimus & Co. [*www.planetarymysteries.com/hieronimus/zeropoint.html*], June 23, 1996.

[6] Murray, K.A., "The Gevaltig: An Inertial Fusion Powered Manned Spacecraft Design for Outer Solar System Missions," Report Number(s) NASA-CR-185163;ETEC—89-7, Rockwell International Corp., Canoga Park, CA (USA). Energy Technology Engineering Center, 1989 Oct 01.

[7] "Astronomy Picture of the Day," [http://apod.gsfc.nasa.gov/apod/archivepix.html], NASA Headquarters, 300 E Street SW, Washing-

ton DC 20024; Jet Propulsion Laboratory, 4800 Oak Grove Drive, Pasadena, CA 91109; NASA/Goddard Space Flight Center, Greenbelt, MD 20771, 2003.

[8] Scott Kniffin, "How Can You Store Nuclear Waste?" Radiation Effects Engineer, Radiation Effects and Analysis Group, Component Technologies and Radiation Effects Branch, Office of System Safety and Mission Assurance, NASA Goddard Space Flight Center, Code 562, 2000.

[9] Dr. William J. Garland, Department of Engineering Physics, McMaster University, Office: NRB 117, 1280 Main Street West, Hamilton, Ontario, Canada L8S 4L7.

[10] Dr. Robert E. Jervis, P. Eng., Professor Emeritus, Nuclear Science and Engineering, Dept. of Chem. & Environ., Eng. and Applied Chem., University of Toronto, Toronto, ON, Canada.

[11] Dr. Patrick Bailey, President, Institute for New Energy (INE), Los Altos, CA.

Index

C

D

E

F

G

H

I

J

K

L

M

N

O

P

Q

R

S

T

U

V

W

X

Y

Z

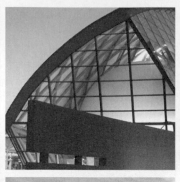

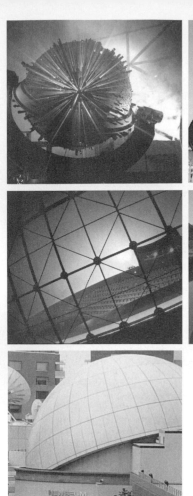

Laser Control of Molecular Dynamics and Chemical Processes

Marcos Dantus

Marcos Dantus
Professor and Adjunct Professor in Physics
(b. 1962)
B.A. & M.A., 1985, Brandeis University;
Ph.D., 1991, California Institute of Technology;
Postdoctoral Research Fellow, 1991-1993, California Institute of Technology

517-355-9715, Ext. 314

The scientific goal of our group is to understand and control laser-molecule interactions. This understanding is used to control molecular reactivity, and to develop a number of novel applications for lasers. A dedicated state-of-the-art femtosecond laser laboratory, with two non-collinear optical parametric amplifiers and pulse shaping capabilities, is devoted to this work.

We are presently working on the following exciting projects:

- Controlling chemical reaction dynamics and intramolecular vibrational energy redistribution
- Controlling matrix-assisted laser desorption and ionization mass spectrometry for the development of an automated pulse sequencer
- Exploring novel biomedical and analytical applications of nonlinear photonic control, for example selective two-photon microscopy
- Quantum computation and quantum information processing with molecular eigenstates

Ultrafast Lasers. The time scale for the formation and cleavage of chemical bonds is 10^{-12} to 10^{-14} s. With pulses as short as 10 fs (one fs = 10^{-15} s), we are able to follow the motion of atoms inside molecules and influence it as it takes place. Because of the uncertainty principle, these ultrashort pulses have a broad spectral bandwidth. Using a pulse shaper, we are able to tailor the phase of the individual wavelength components. These shaped pulses can be used to control the quantum-mechanical aspects of laser-molecule interactions. For example, we can control the efficiency or two-photon excitation, an aspect we have used to achieve selective two-photon microscopy.

Nonlinear optical methods are used in our group to encode and manipulate information onto superpositions of quantum states. Time-dependent quantum-mechanical simulations of our experiments are carried out to refine our understanding and to help us push the limits of the system. This work is carried out in collaboration with the Institute of Quantum Science at MSU.

Proteins Sequencing. Protein sequencing is part of a major technological challenge under the umbrella of Proteomics. Advances in this field are essential to a variety of disciplines including chemistry, biochemistry, pharmacology and medicine. Our projects are aimed at improving and automating the elucidation of protein sequence using our shaped laser pulses in combination with the matrix-assisted laser desorption and ionization mass spectrometry.

We collaborate with groups in biochemistry, electrical and computer engineering, and physics. Students in our group have interests in analytical, biochemical, and physical chemistry as well as the physics of atomic, molecular and optical systems. Our group is highly energetic and diverse in nature. We have collaborations with other groups at MSU and around the world.

Representative Publications

Multiphoton intrapulse interference I; Control of nonlinear optical processes in condensed phase, K. A. Walowicz, I. Pastirk, V. V. Lozovoy, and M. Dantus, *J. Phys. Chem. A* 106, 9369 (2002).

Multiphoton intrapulse interference II; Control of two- and three-photon laser induced fluorescence with shaped pulses, V. V. Lozovoy, I. Pastirk, K. A. Walowicz, and M. Dantus, *J. Chem. Phys.*,118, 3187 (2003).

Ultrafast four-wave mixing in the gas phase, M. Dantus, *Annu. Rev. Phys. Chem.*, 52, 639 (2001).

Femtosecond ground state dynamics of gas phase N_2O_4 and NO_2, I. Pastirk, M. Comstock, and M. Dantus, *Chem. Phys.*, Letters, 349, 71 (2001).

Photon echo pulse sequences with femtosecond shaped laser pulses as a vehicle for molecule-based quantum computation, V. V. Lozovoy, and M. Dantus, *Chem. Phys*. Letters, 351, 213 (2001).

Selective two-photon microscopy with shaped femtosecond pulses, I. Pastirk, J.M. Dela Cruz, K.A. Walowicz, V.V. Lozovoy, and M. Dantus, *Optics Express* 11, 1695 (2003).

www.chemistry.msu.edu/dantus

dantus@msu.edu